MW00445921

HR

Adelmo Brottenani

Oct. 26. 1995

C. W. P.

HR

A BIOGRAPHY

of H.R. MacMILLAN

KEN DRUSHKA

Harbour Publishing
Madeira Park, BC

Harbour Publishing
P.O. Box 219
Madeira Park, BC Canada V0N 2H0

Published with the assistance of the Canada Council and
the Government of British Columbia, Tourism and Ministry Responsible for
Culture, Cultural Services Branch.

Edited by Daniel Francis
Cover, page design and composition by Roger Handling, Terra Firma Design
Cover photograph by Karsh
Printed and bound in Canada

Canadian Cataloguing in Publication Data

Drushka, Ken
HR

ISBN 1-55017-129-1

MacMillan, H. R. (Harvey Reginald), 1885-1976. 2.
Foresters—British Columbia—Biography. 3. Businessmen—
British Columbia—Biography. I. Title.
SD129.M32D78 1995   634.9'092   C95-910418-6

Photo Credits
Pp. 21, 22, 23, 26, 31, 33, 41, 43,47, 50, 52, 54, 61, 72, 75, 106, 108, 110, 113,
141, 167, 196, 218, 224, 242, 254, 261, 266, 268, 274, 284, 295, 303, 308, 315,
324, 340, 343, 350, 357, 363, 367, 375, 379: Courtesy H.R. MacMillan papers,
University of British Columbia Library, Special Collections (UBC). P. 199:
Photo by Jacobys, courtesy Charles Dee. All other photos are from the
MacMillan family's private collection.

to the memory of
Harvey S. Southam

contents

Introduction 9

chapter 1
Ontario Boyhood 16

chapter 2
Birth of a Forester 29

chapter 3
Graduate School 45

chapter 4
Between Life and Death 56

chapter 5
Chief Forester 68

chapter 6
Around the World 83

chapter 7
In the Private Sector 98

chapter 8
Founding the Export Company 114

chapter 9
Uneasy Partnership 127

chapter 10
Taking On the Depression 146

chapter 11
Lumber Wars 160

chapter 12
Life Becomes Art 175

chapter 13
Dollar-a-Year Man 193

contents

chapter 14
Power Struggles in Ottawa 208

chapter 15
Canada's "Buzz Saw" 221

chapter 16
Debating Forest Policy 237

chapter 17
The Good Life 253

chapter 18
The Elder Statesman 272

chapter 19
Cold Warrior 288

chapter 20
The Second Sloan Commission 307

chapter 21
Succession 328

chapter 22
The Philanthropist 344

chapter 23
Losing Control 362

chapter 24
Final Years 378

Epilogue 386

Notes 396

Sources 409

Index 410

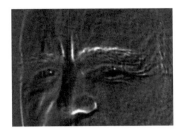

For much of this century, H.R. MacMillan has had the most widely recognized name in British Columbia. Not everyone knows him for the same achievements, which were numerous and varied over the long course of his life. Ironically, the accomplishment for which he is best known, the building of MacMillan Bloedel Ltd., one of the world's major forest companies, is one from which he threatened to dissociate himself near the end of his life.

MacMillan attached his name to the first business venture he set up, partly because he believed in the British lumber tradition of using one's name to show customers and the public that one is proud of the way one conducts business. For MacMillan, this edict became a rule that he applied to all of his entrepreneurial and philanthropic activities.

Born and raised in obscure poverty on a small farm north of Toronto, he first distinguished himself at an Ontario agricultural college and then at Yale, one of the continent's preeminent universities. As British Columbia's first chief forester, MacMillan earned a reputation as a dedicated, professional public servant who embodied the most elevated principles of scientific methodology and impartial honesty. He had attained this status—an achievement most of his peers would have been proud to realize after a long and satisfying career—before he reached the age of thirty.

MacMillan was instrumental in salvaging British Columbia's devastated economy during World War One; and for an encore, he oversaw the organization of BC's primary contribution to the war effort, the use of Sitka spruce in building Allied fighting planes.

In 1919, placing all his frugally-acquired financial assets on the line, he established a lumber export company and became one of a handful of local entrepreneurs to wrest control of the province's business firms from the hands of distant financial interests. He then took control of BC's largest fish packing company, which was on the verge of bankruptcy, and turned it into a resounding financial success. Through the long depression of the 1930s, MacMillan was the voice of business in BC and, increasingly, beyond. In World War Two, as an important economic leader of the country, he stood out for his single-minded dedication to the Allied cause.

MacMillan came home after the war convinced of the need for social reform, and equally convinced that he must firmly oppose totalitarian political movements. His speeches and arguments exerted an enormous influence on the province's business and political leaders. A few years later, he broke ranks with other established commercial and political interests, and articulated a far-sighted approach to the management of BC's natural resources. During the last fifteen to twenty years of his life, MacMillan's primary public presence was that of the philanthropist, as he devoted his energies to redistributing some of the wealth he had generated.

MacMillan himself never appeared particularly impressed with his accomplishments: throughout his life, he shrugged them off as incidental. According to him, he was made chief forester because he happened to be the only Canadian forester available; he went into business because he needed a job; he expanded this business merely as a defensive measure. The fact that he was enormously talented and worked hard to achieve the results he obtained was not the point, as far as he was concerned. Playing the game well was more significant than the results. Or so he liked to say. It is equally true that he kept as close an eye on the bottom line as the most parsimonious accountant in his employ.

MacMillan's public accomplishments reflect only certain facets of his character. There were more private, personal aspects of his life that were as important to him as those for which he was well known. At heart, he was a dedicated conservationist, cast from the same mould as Gifford Pinchot,

Teddy Roosevelt, Aldo Leopold and other great conservationists of the late nineteenth and early twentieth centuries. His conservationism found expression in hunting and fishing, which led to his lobbying for the creation of numerous parks, his opposition to building hydroelectric dams on BC rivers, and his major efforts to develop and influence forest management practices and policy.

MacMillan was one of that rare breed of successful businessmen with genuine intellectual capabilities. He had a passion for books and an intense interest in ideas that was more than a hobby: he read voraciously throughout his life. Moreover, he had the resources to act on his ideas. He was a strong supporter of pioneering anthropological and archaeological work in BC, and his fascination with marine biology led to his funding the University of British Columbia's fisheries department and the Vancouver public aquarium.

He read and understood history, and could see well beyond the demands of daily business to ponder the larger context of whatever he undertook. He was therefore able to predict the consequences of current events—even, some people believed, to foresee the future. In the world in which he functioned, this was a rare ability and was likely the key to his success.

Magazine and newspaper profiles that appeared at various times throughout MacMillan's life attributed his strength of character to the austere moral precepts of his antecedents. It was often suggested that the values of hard work, diligence and moral rectitude he learned as a young boy growing up in a predominantly Quaker community were what provided him with the strength and fortitude to accomplish what he did. This is an exaggeration, if not an outright distortion, of his growth as a human being. At an early age he appears to have lost most of his interest in institutionalized religion, even in the form offered by the Quakers. The impetus for MacMillan's actions in the world owed more to the fertility of his own inquiring mind than from any external moral force. He paid no serious attention to any church during his life, rarely attended services and, when he generously supported them with money, did so more for social reasons than spiritual ones. His spiritual centre was a very private and personal affair, revealing itself in the world through the filter of his intellect.

It should not be inferred that MacMillan was in any sense a saint. He was not. He was an exceptional human being, but he was a man with most of the ordinary human strengths and weaknesses. He had an active and quick sense of humour, offset by an explosive and occasionally foul temper, both of which he inflicted freely on his family, friends, colleagues and employees. His most ignoble acts tended to be visited on those closest to him. MacMillan was a big, burly man, well over six feet tall and weighing more than 200 pounds. His intense blue eyes were shaded by huge, shaggy eyebrows. In good humour, laughing and joking, he was delightful, warm and gregarious. In bad temper, he was frightening, intimidating and occasionally ruthless.

MacMillan usually knew what he wanted. He wanted things done his way—now. If he didn't get his way, if someone obstructed or opposed him, he would plough them under. Amazingly, he made few enemies, even among those unfortunate enough to get in his way.

Beneath his intellectual reserve, imposing physical presence and domineering manner, MacMillan could be an intensely emotional person. This was revealed in his visceral attachment to British Columbia, particularly Vancouver Island, which he dominated for almost three decades in almost baronial fashion. At one stage, he owned the biggest forest operation, most of the island's fish processing plants and the largest farm, with the most visible public park named after him. From his Qualicum summer home, he prowled the back roads and highways of the island at the wheel of a Bentley, a hunting rifle and fishing rod in the back seat and, with any luck, a dressed deer in the trunk. This was his element, as much as his Vancouver office. Here, the natural shyness that led him to avoid cocktail parties, funerals, weddings and other social events, disappeared. He was a gregarious, familiar figure, known to hundreds of people from all walks of life. Because he firmly believed that the best way to learn how something was done was to go and ask someone who did the job, he was a constant and unannounced visitor to his sawmill and logging operations, where he was just as apt to talk to the lowest man on the ladder as to the divisional manager.

He was a man capable of forming strong emotional attachments to people, places and animals. In certain, intense situations he was easily moved to tears, such as when he found himself forced to fire a favoured employee.

His emotional attachment to his family, particularly his two daughters, was intense and enduring. But for him, for much of his life, the emotional centre of his being focussed on someone outside his family.

MacMillan had an enormous personal impact on people. He attracted others' attention; people reacted to him far more strongly than he did to them.

He was like a loaded freight train, thundering through the quiet villages of their lives. People who met him once or twice came away from these encounters convinced they knew him well.

Is there a consistent thread in MacMillan's life? If one peels away the superficial actions and accomplishments, is there an inner being, an essential H.R. MacMillan against which everything he did can be measured and understood? Perhaps, but it is not easy to find. He was a complex human being, with many facets which were not always consistent with each other. In the end, perhaps the best way to understand or appreciate him may be to forgo the search for a unifying essence. In the words of Walt Whitman:

> Do I contradict myself?
> Very well then I contradict myself,
> (I am large, I contain multitudes.)

A book such as this is rarely the work of one person. As the author, I have drawn on the time and energies of many others, without whose assistance I would have accomplished little.

The person who has made this book possible in so many ways is Peter Pearse. If it were not for his efforts and abilities, the work would never have been started. He made it possible for me to work on it uninterrupted for more than two years, and his advice and criticisms throughout the project have been of immense value to me.

Jean Southam, H.R. MacMillan's youngest daughter, has also provided enormous assistance. She opened doors that would otherwise have remained closed, dredged up documents and photographs, and gave unstintingly of her time. Her one wish, constantly reinforced, has been to assist in the publication of a faithful biography of H.R. MacMillan, the man.

Several others who once knew MacMillan also provided much-needed help. My good friend Ian Mahood, whose courageous memoirs I helped edit a few years ago, and who as a young forester spent many days tramping through Vancouver Island forests with MacMillan, has provided steady support and a substantial fund of ideas and information. Bert Hoffmeister spent long hours patiently answering questions and rummaging through his memories. Ralph Shaw and Norman Hyland went far out of their way to assist in unearthing details of MacMillan's corporate life. Dewar Cooke provided accuracy, insights and constructive advice. I was also helped by Gordon Southam, Charles Dee, Peter Larkin, Marion Hawley, John Lecky, Duncan MacFayden, Gary Bowell, Bill Manson, and Ross Tolmie.

Any failings found herein are my own, in spite of the best efforts of all these people.

Much of the information in this book, as well as many insights into MacMillan's character, came from material collected and a manuscript written by MacMillan's grandson, the late Harvey Southam. Beginning in 1980 and until his untimely death in 1991, he spent several years working on a biography of his grandfather. Of particular value were numerous audiotapes of interviews he conducted with several of MacMillan's friends and colleagues, most of whom died before I began this project. In many ways this book is a completion of the work begun by Harvey Southam.

There are two collections of H.R. MacMillan's papers at the University of British Columbia Library, Special Collections, although the division between them is rather arbitrary. One consists of his personal papers, and the other makes up a portion of the much larger MacMillan Bloedel collection. These are extensive holdings, consisting of about 250 boxes of material, meticulously sorted and filed by MacMillan's assistant, Dorothy Dee, in anticipation of MacMillan's biographer. Unfortunately, it appears, Miss Dee considered it part of her task to cull the files of any material she felt might cast MacMillan in a bad light. Consequently, there is a dearth of documentation in some subject areas; for example, any observations or comments MacMillan might have made on the infamous Robert Sommers affair. Nevertheless, these are truly valuable and enormously useful collections, and I was greatly assisted in my use of them by George Brandak and the staff at Special Collections.

Alan Mattes undertook research for me at the University of Guelph and in the National Archives of Canada. Paul Marsden at the National Archives of Canada and Brian Young at the BC Archives and Records Service provided invaluable assistance. John Lecky generously loaned me his collection of correspondence with his grandfather.

I am grateful for financial assistance from the MacMillan Family Fund, the Stewart Fund, the McLean Foundation and the Pacific Reference Foundation, as well as Ian Mahood, Dewar Cooke, Olaf Fedje, Jack McKercher and Peter Pearse, without which the writing of this book would not have been possible.

And, finally, I would like to thank my publisher, Howard White, whose idea this book was in the first place, and Dan Francis, for being the kind of editor every writer needs.

O ntario boyhood

I n 1885, the year of Harvey Reginald MacMillan's birth, the immediate world into which he was born was about to be transformed. The great movement west to settle the Canadian Prairie was just beginning, and the industrialization of Canada was under way. The rural farming communities surrounding Aurora and Newmarket on the northern outskirts of Toronto, where MacMillan spent his childhood, were among the first to experience the effects of these changes.

During the previous century, the rolling forest land between Lake Ontario and Lake Simcoe had been settled. It was cleared and cultivated by backwoods pioneers, refugees from Yankee republicanism to the south and immigrants from Britain, MacMillan's ancestors among them. Its most heavily used trail was an important route into the interior of North America. It was along this trail that the lieutenant governor of Upper Canada, John Graves Simcoe, travelled in 1793. As part of his scheme to settle the area, he ordered the construction of a road running north from his administrative capital at York. Yonge Street was cleared, and a rough road was constructed to Holland Landing, primarily as a military measure in case of an American invasion.

Although a military incursion from the south failed to materialize, the new road did open the way for invaders of a different sort—settlers. The first significant wave of settlers were American Quakers. The Quaker religion, more formally known as the Society of Friends, had begun in England in the mid-seventeenth century. About thirty years later, William Penn brought a colony of Quakers to Pennsylvania; because their future was uncertain under the post-revolution United States government, many Quakers emigrated north to British territory. In 1793 a group of somewhat tardy United Empire Loyalists moved from Vermont and founded a settlement centred around a meeting house on Yonge Street just west of the present site of Newmarket. A second colony was established in Whitchurch Township, four miles southeast of Newmarket.

The character and capabilities of these Quaker settlers made them the dominant social influence in the Newmarket area for more than a century. Theirs was a plain, austere religion whose followers believed each person must find his or her own religious truths. As a consequence, there were no priests or ministers. Religious services, or meetings, were conducted in silence, interrupted only by members who felt a divine inspiration. In later years MacMillan recalled "being at a Meeting one Sunday and sitting alone on a back seat when my Uncle got up and began to speak a sermon. His first words were 'the iron has entered my soul.'"[1]

Quakers were also pacifists, adhering to the Christian principle of turning the other cheek. It was a hard-earned principle, born out of the persecution they had suffered in England in the past. But it was not a pacifism that discouraged their participation in the social and political conflicts of their time and place. The American Quakers had been heavily involved in the Revolution, and their descendants who settled around Newmarket became embroiled in the political upheaval that eventually engulfed that corner of the world. Another central feature of Quaker culture was a profound respect for education. In an era of almost universal illiteracy, most of the settlers who trudged through the mud up Yonge Street to build a new life were able to read and write. Soon after their arrival, Quakers in both settlements had opened schools.

Unlike most other settlers, the Quakers were descendants of several generations of pioneers; they knew what needed doing to settle a new land, and had the skills to realize their dreams. As other immigrants arrived from England, Scotland and Ireland, many chose to join the various Quaker sects whose spiritual stability seemed so at home in this new land. One who followed this path was David Willson, a New Yorker of Irish descent who in 1801 obtained a land grant about two miles north of Newmarket, at Sharon. He joined the Quaker Church shortly after his arrival and began to take an active part in its meetings. Within a few years, Willson's religious views diverged from those of the other members, including his unorthodox belief in the use of music in religious ceremonies. He was expelled from the church, along with half a dozen supporters, and in 1814 he started a separate sect of the Quaker faith called the Children of Peace. As the sect grew, it built a series of meeting halls at Sharon, one of which housed Canada's first pipe organ. The Children of Peace also organized the country's first brass band and brought in a renowned choirmaster from Boston. By the time Willson died in 1866, the Children of Peace numbered more than 300 and had built the Temple of Peace, an unusual building of elegant simplicity that still stands at Sharon.

Various members of Willson's family also settled in the area, including a separate branch of his father's family which included H.R. MacMillan's maternal ancestors. By the mid-nineteenth century, a huge extended family of Willsons, related by blood and marriage, lived in the farming country north of Toronto.

During the years of early settlement of York County, Upper Canada was ruled by a conservative aristocracy established by Lieutenant Governor Simcoe. Known to its detractors as the Family Compact, this governing class was anti-American, Tory and dominated by the clerics and professional men of Toronto and other larger centres. In Lower Canada, opposition to a similar oligarchy grew under Louis-Joseph Papineau and eventually clashed with troops loyal to the colonial government. When troops garrisoned in Upper Canada went east to help quell the rebellion in Lower Canada, radical farmers led by William Lyon Mackenzie declared independence. On December 5, 1837, several hundred rebels from the area north of Toronto marched down Yonge Street, intent on capturing the capital. They were dispersed by the militia and two of their members were executed. Several others were jailed, including two sons of David Willson. Although the rebellion was defeated, the independent spirit from which it grew remained strong in the northern, rural reaches of York County. Fifty years later, when MacMillan was a child, memories of this rebellion were still vivid in his family and he was familiar with mementoes of it — cannonballs used for door stops and treasured boxes carved by prisoners from the area while they languished in the Toronto jail.

Until the 1850s, growth and development in the Newmarket-Aurora area was slow but steady. Yonge Street remained a major transportation route to the north and west, with two to three hundred teams of horse-drawn vehicles a day passing through. All this changed after May 16, 1853, when the first train came up the newly laid tracks of the Ontario, Simcoe and Huron Railway, the first railway of any significance built in Canada. This railway marked the beginning of the industrialization of southern Ontario, including York County and the hamlets of Aurora and Newmarket. Within a few years a series of foundries and factories opened in York County, some of them employing as many as three or four hundred workers. Urban populations increased rapidly as hotels and retail establishments were opened. In 1854 Robert Simpson opened the first of his stores in Newmarket, which had become the principal trading and market centre of the region.

For many years this urban-based industrialization had little impact on the established rural residents. Most of them were self-sufficient farmers who had little involvement with the cash economy developing in the towns. To some degree, they looked upon themselves as a sort of landed gentry, and were considered as such by the town people. The fact that they had little cash was not important. They were devout, hardworking and educated. In his later years, MacMillan always acknowledged the lack of money in his family during his childhood, but he did not grow up with any sense of inferiority because of it.

It is almost impossible, a century and a half later, to imagine the work involved in clearing the trees from a hundred acres of York County land. A highly skilled and exceptionally energetic axeman—this was before the advent of crosscut saws—was able to fall one acre of trees in a week. Then they had to be cut up, piled and, in large part, burned. Stumps had to be removed, the land broken with a team and plough, and the rocks picked and hauled away. Only then could a crop be planted. Few of the settlers had the resources to hire help or to undertake these tasks in a leisurely fashion. If they were not to starve, food had to be grown before it could be put on the table. A century later, MacMillan occasionally marvelled at this monumental undertaking, torn between admiration for the accomplishment and wonder at the laborious liquidation of a forest which, properly harvested, would have provided continued economic returns on a par with agriculture.

In the 1850s there was still land available for ambitious settlers who were willing to endure the hard work of gaining title to it, and who had no interest in taking employment in the industrialized urban centres. One immigrant who arrived at this time was John McMillan, son of a Highland Scottish farmer who had served in the Black Watch and fought at Waterloo. He was the paternal grandfather of MacMillan, who as a college student would change the "Mc" in his family name to "Mac." John McMillan and three of his brothers settled in the Newmarket area about 1850. They followed the usual practice of working for prosperous farmers until they had saved enough to make the down payment on farms of their own, the rest of the money being borrowed from local banks. Decades later, after serving for several years as a director of the Canadian Imperial Bank of Commerce, MacMillan commented on the banking practices of the period:

> These bankers, who were really "blood suckers" would have been wiped out if the farmers had not been so law abiding. In my opinion, these money lenders should have died a miserable death.[2]

Shortly after his arrival, John McMillan joined the Hicksite branch of the Quaker church and married the daughter of another Quaker family, Edith Willson.[3] John and Edith McMillan acquired a farm on Lot 32, Concession 4, southeast of Newmarket, near Pine Orchard. Their first son, John Alfred, was born in 1858, and two more sons and a daughter followed.

In the mid-1870s, Edith died of tuberculosis. This usually fatal disease was rampant in North America in the late nineteenth and early twentieth centuries: in 1910 in Quebec, 200 deaths per 100,000 population were attributed to it. Initially it was thought to be a hereditary disease, because once a family member contracted it, it usually spread to others in the household. Eventually doctors learned TB was contagious, and that it could lie dormant in its victims for many years. The closed-in and often crowded housing conditions in rural Canada during the late nineteenth century, combined with the bitter Ontario and Quebec winters, provided ideal conditions for the spread of tuberculosis. Entire families were devastated over several generations until the causes were discovered and treatments developed in the mid-1900s. Edith, as it was put in a common phrase of the time, which MacMillan repeated throughout his life, "brought TB into the family." John remarried and had four more children. Of the eight children, only two lived beyond their twenties, the rest succumbing to the disease.

In 1883 the eldest son, John Alfred, married Joanna Willson. She was the daughter of J. Wellington Willson and a great granddaughter of Hugh L. Willson, a step-brother to David Willson, founder of the Children of Peace. Some years before this wedding, Wellington and his family had moved a few miles north into Simcoe County, where there were no Quaker churches. In those days there was usually only one church within travelling distance from a family farm, so the Wellington Willson family joined the local Presbyterian church. John Alfred, when he married Joanna, was a member of the Pine Orchard Quaker church, which promptly expelled him for marrying outside the faith. Apparently this was not a serious transgression, as the McMillan family connections with the Quaker community and church continued. MacMillan later recalled attending both the Presbyterian Sunday school and Quaker meetings, where he was "trained to sit quietly on a bench up against

John Alfred McMillan, H.R. MacMillan's father, c. 1883.

Joanna Willson (centre, at rear), H.R. MacMillan's mother, with her sisters, (left to right) Almana, Martha and Emily, c. 1883.

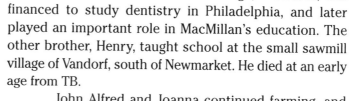

the outside wall of the Meetinghouse."[4]

    After his marriage, John Alfred took over the farm settled by his father, who moved on to an adjacent property. He and his brothers had no intention of spending the rest of their days toiling at the backbreaking and mind-numbing occupation that farming was at this time. They turned to one of the few means available to earn cash—cutting firewood for Newmarket residents. On top of their farming chores it was exhausting work, but it provided enough money for them to get the education they needed to escape from the farm.

    One of the brothers from their father's first marriage, Ellsworth, was financed to study dentistry in Philadelphia, and later played an important role in MacMillan's education. The other brother, Henry, taught school at the small sawmill village of Vandorf, south of Newmarket. He died at an early age from TB.

    John Alfred and Joanna continued farming, and he kept on cutting firewood to finance his study of medicine. On September 9, 1885, their son Harvey Reginald was born at home. The only surviving comment on this event is a remark by a maternal uncle who said his new nephew was "as healthy as a setter pup."[5] From birth, MacMillan was known as "Reggie" among his large, extended family. Not long after Reggie's birth, his father developed pulmonary tuberculosis, also known as consumption. He died six weeks after his infant son's second birthday.

H.R. MacMillan as an infant, 1886.

    The death of his father created a situation which influenced the course of MacMillan's life. Joanna McMillan was destitute. Settlement of the estate left her with no money, and she was forced to rely on the charity of the McMillan and Willson families. MacMillan would never forget what it was like to be poor.

> I know something about it because my father died when I was two years old, and I was brought up by my grandparents and my mother who lived in what today would be called "extreme poverty." The chief source of revenue was milk, which was produced by hand and shipped about 30 miles to Toronto, for which the producer got 8 cents a quart. . . I can remember my mother and her sisters worked

H.R. MacMillan, c. 1898.

very hard milking cows . . . [and] my mother going out to milk the cows early in the morning while I was still in bed.[6]

For three years, as Joanna struggled to support her son, they lived on a succession of farms with various members of her extended family. In the fall of 1891, she went to work as the housekeeper for the Austin Doane family at Sharon, and Reggie was left with his paternal grandparents so he could attend the nearby Bogarttown school. Almost seventy years later, in response to an inquiry from a Newmarket schoolboy working on a class assignment, MacMillan wrote:

> The only thing I can remember about Bogarttown school now is that I got into a fight in the school yard with a boy—whose name I now forget—and it seems that I beat him, which probably is something I did not expect to do.[7]

For some reason, the arrangement with the McMillans did not work out. A few months later, MacMillan moved to the home of Joanna's parents three miles outside Aurora, where he lived for the rest of his childhood.

The nature of MacMillan's life during the decade he lived with his grandparents on the farm near Aurora is a matter of some speculation. Almost forty years later it became the subject of a novel, *Growth of a Man*, written by his cousin, Mazo de la Roche. The book told the story of young Shaw

Manifold, from the time of his mother's departure to work on a distant farm, until his appointment as British Columbia's chief forester twenty years later. In some respects the book is true to MacMillan's life, but in many others it is pure fiction. For example, Manifold lives with his maternal grandparents, who severely mistreat him. There is no evidence MacMillan was ever abused by any of his grandparents, particularly Joanna's parents. The most MacMillan himself ever recorded on the subject was much later when the Willson farm came up for sale and a cousin asked if he wanted to buy it.

> I am getting pretty old to buy a farm in Ontario; also my childhood there was not as happy as yours . . . my grandparents were about sixty or more years old when I went to live with them, so I was pretty well isolated as to companionship.[8]

When MacMillan went to live with the Willsons, several of his aunts and uncles were still living on the farm. The daily life of a York County farm family at that time can probably be captured in one word: drudgery. MacMillan recalled endless hours, merging into days and weeks, of hoeing the vegetable garden, feeding livestock and caring for chickens. He hated chickens for the rest of his life, failed poultry classes in university and avoided eating them whenever possible.

In their intellectual activity, the Willsons were probably typical of farm families of that time and place.

> I can well remember my grandfather Willson reading "The Weekly Globe" regularly. However, I cannot remember ever seeing him read the bible or any other book. No farmer in those days read books as I remember it. As a matter of fact, I doubt if anyone today would read books after following a farmer around at his work and chores every day of the year.[9]

Joanna lived less than four miles away, but in pre-motorized Ontario, this was a considerable distance. "My neighbourhood," MacMillan once wrote, "extended for a radius of how far one could drive on a gravel road with a horse and buggy in a day, and get home in time to do the chores."[10] He stayed with his mother and the Doane family, of whom he became quite fond, during summer and Easter vacations. Joanna spent Christmas at the Aurora farm, and tried to arrange a weekend visit every two months. For a time, when MacMillan was older, they spent Sunday afternoons walking to meet halfway between the two farms. On rare occasions they took vacation trips by train to Toronto, and once they went to Niagara Falls and the Buffalo World's Fair. They wrote short, chatty letters to each other every two weeks or so, in which

Joanna admonished her son to work hard at school and be a good boy. The dozen letters from her that MacMillan kept all his life convey the concerns of a loving, caring mother, reporting inconsequential news of her day-to-day life and reminding him of things he needed to do, such as remembering to visit his McMillan grandparents. At the end of June 1895, she wrote:

> My dear boy. I suppose you will be looking for a letter from me before this. I suppose you think you are free now for a good long time. You must try and have all the fun you can during the holidays as they will soon fly around and then for a good long spell of work at school. The way to enjoy yourself is to be a good boy and do all you can to make other people enjoy themselves and you will be sure to have a good time.[11]

This advice, to find pleasure in providing pleasure to others, was a lesson MacMillan took to heart; as he grew older, it became one of the central traits of his character.

During the course of that summer in Sharon, Reggie convinced Joanna they should take a trip to Toronto to attend the Canadian National Exhibition, the biggest event of the year in rural Ontario. She wrote to him in early September.

> I suppose you are counting on going to the exhibition. I don't know what we will do if we can't get some one to go with us. I don't know very much about the place and we might get lost you know and have to ask a policeman to look for us.[12]

Apparently Joanna overcame her fears, for a few days later they took the train to Toronto, where they stayed with her aunt, Louisa Lundy, on Dunne Avenue near the Exhibition grounds. Louisa's husband Daniel was the warden of Toronto's Central Prison. Also living at this Parkdale house were the Lundys' daughter and her family, the Roches. Here, MacMillan first met his cousin Mazo, along with her lifelong companion, generally acknowledged to be her lover, Caroline Clement, also a cousin of Joanna's. This was the beginning of a lifelong relationship between MacMillan and these two eccentric women, the only members of his family with whom he maintained an ongoing connection.

From the outset, MacMillan's feelings about Mazo and Caroline were ambivalent.

> They lent me the first novel I ever read. I was then passing into high school and was reading nothing but serious books. Mazo was a very clever girl and won

prizes for her stories and also the Governor General's prize for poetry, more than once. I was very fond of her. She was two or three years older than I.[13]

But on another occasion, he described Mazo as "snooty" and said that her addition of "de la" to her family name of Roche was an affectation.

I remember once when Mazo and Caroline brought me to Toronto to visit them, a country boy off a farm—their standard of living was above me and my clothes were not good enough so they dressed me up in a suit of Walter Lundy's.[14]

In her first letter after they returned to their respective homes, Joanna asked Reggie to have his grandmother take his measurements so she could buy him a new suit of his own, and a pair of boots. As an afterthought she added, "you had better give up going barefooted for this summer."[15]

As MacMillan reached an age when he became aware of the world around him, events were occurring beyond the confines of York County that were to have a profound effect on his life. In the year of his birth, the last spike on the first trans-continental railway was driven in a remote mountain pass of British Columbia. In 1896 Clifford Sifton became minister of the interior in Laurier's government and launched an immigration program to colonize western Canada. The constricted future MacMillan faced in the settled rural Ontario community of his birth was offset by the expanding horizons in the

H.R. MacMillan as a boy, c. 1895.

west. Hundreds of thousands of European immigrants arrived, which pro-
foundly affected central and eastern Canadians, many of whom were swept up
in the westward migration.

With Joanna's encouragement and in keeping with his own inclina-
tions, MacMillan realized at an early age that his only hope of escaping the
drudgery of farm labour was through education. That he was aided and abet-
ted by his grandparents was a reflection of the value the Quaker community
placed on learning. MacMillan was not an exceptionally brilliant student, at
least as far as his marks were concerned. Charles Mulloy, principal of the
Aurora High School, remarked later that MacMillan was "one of the most brilliant students who had ever passed through the school,"[16] but this comment
needs to be taken with a grain of salt: when he made the observation, Mulloy
had become MacMillan's father-in-law. At age twelve, MacMillan recorded in
his diary that Mulloy "hit me on the head with a stick." A few days later, he
noted, his father-in-law-to-be threw a Latin book at him and hit him on the
head. The Aurora High School records reveal that when he passed his first
year of high school in 1898, MacMillan was not one of the class's six honour
students.

Enrolled in the same high school class were Mulloy's children, Allan
and Edna. A romance between Edna and Mac, as MacMillan was now called by
his friends, began in the first year or two of high school. He would pick a rose-
bud on his long walk to school each morning and leave it on her desk. Allan
became a close friend, and they eventually went to college together. Another
close friend of many years, Clare Hartman, lived across the road from the
Willson farm and later married Edna's sister, Florence. During his teenage
years the Mulloy house was a second home to MacMillan.

As high school graduation approached, MacMillan and his mother
had to face the questions of where he would go for further education and,
more important, how it would be paid for. Joanna's meagre earnings of $4 a
week could provide him with some assistance, but it would not cover tuition.
One day during the summer of 1900, when MacMillan was hoeing potatoes in
his grandparents' garden, a man driving past in a horse-drawn buggy stopped
to talk. He was an extension education teacher from the Ontario Agricultural
College at Guelph, and he informed MacMillan that OAC had scholarships
available. As it turned out, MacMillan already had a connection with the college through a cousin who was married to Charles Zavitz, a member of the

teaching staff. Equally encouraging, the curriculum required students to work on the college farms, for which they were paid a small wage. That solved the problem of living costs, but it did not cover all of the $16 annual tuition fee. In the end, MacMillan's uncle, Ellsworth McMillan, who had become a dentist and was practising in Brazil, offered to provide some financial aid.[17] That settled the matter. MacMillan applied and was accepted as a first-year student in agriculture in the fall of 1902.

During his last year of high school, the Willson grandparents left the farm and moved into Aurora. MacMillan moved with them, and when he finished school that year, he obtained a job selling stereopticon slides for an entertainment device popular at the time. After acquiring a bicycle, he travelled from farm to farm and on to the various small towns and villages of York County and beyond. It must have been a great summer for the young MacMillan. He was free of the farm work he hated so much; he had graduated from high school and was bound for college; for the first time in his life he was gainfully employed and could help bear the financial burden his mother had carried for so many years. And he was in love. The romance with Edna had bloomed, and at the end of the summer, before he left for Guelph, they became engaged.

Birth of a forester

Freed from his ties to farm life, MacMillan was impatient to get on with his life. During the summer of 1902, he learned of a job at the Ontario Agricultural College tending experimental seed crops for his distant relative, Charles Zavitz. He arrived at Guelph and began work in mid-August. So did Leonard Klinck, a fourth-year agriculture student who took an interest in MacMillan after learning that they were from the same rural area near Aurora. Their lives would be intertwined for many years, and as MacMillan often remarked, Klinck exercised a great influence on him.

When classes began in mid-September, MacMillan threw himself into college life with all his energy. Judging from college publications and his own letters and diaries, he was a gregarious and popular young man. He was active in sports and played hockey and football on a number of class and college teams. He had the build of an athlete; at age seventeen he was six feet tall and weighed 155 pounds. Before the first year ended, he was elected secretary-treasurer of the OAC athletic association. It was the only time in his life he participated in athletics.

At the time of MacMillan's arrival, the college had been in operation for more than twenty-five years and earned a solid reputation as one of Canada's most respected educational institutions. Its stated purposes were to train and broaden the outlook of young men who intended to become farmers, to provide young women with instruction in the domestic sciences, and to train teachers. The faculty also conducted most of the agricultural and horticultural research in Ontario at the time.

The campus, on the edge of the small city of Guelph, consisted of about seventy acres, several stately instructional and administrative buildings, barns, workshops and field plots. The main building contained a men's residence for 180 students, where MacMillan lived for most of his time at the college. Room and board cost $3 a month, including an unimaginable luxury for a farm boy who had grown up bathing once a week in a tub in front of the kitchen stove—unlimited access to showers and bathtubs with hot and cold running water.

Total enrollment when MacMillan began classes was 212 students. Women were first admitted to the college during MacMillan's term there, and Macdonald Hall, the women's residence, was built in 1904. By today's standards, relations between male and female students were rigidly Victorian; contact between the sexes was only permitted during formal events. At the age of eighteen, MacMillan appears to have had little interest in the opposite sex, apart from his ongoing relationship with Edna. They exchanged one or more letters a week until she finally suggested they limit themselves to one every two weeks. He and Joanna continued their well-established habit of exchanging regular letters, and she maintained a trickle of money to cover his incidental expenses.

His financial situation, particularly during the first year or two, was precarious. Joanna sent money for fees and board, while Uncle Ellsworth provided an occasional five-dollar bill. MacMillan kept meticulous monthly cash accounts, recording every bit of income and expense. Apart from his tuition and board, these rarely came to more than $3 or $4 a month. The state of his finances is indicated in an entry dated February 13, 1903: "Lost: 1 cent." On rare occasions, he spent a penny on some gum or, when he was feeling extravagant, 5 cents on ice cream. He was probably not alone in his poverty, and it did not cause him much concern. In fact, he seems to have been having the time of his life.

For the first year or so, MacMillan was apparently unclear about his reasons for attending OAC. He had no intention of becoming a farmer, even though most of the courses were designed for students with an agricultural future. In his first couple of years, he was not a devoted student. He was regularly called on the carpet for skipping classes, and once he was fined for being absent. With the exception of one subject, he was no more than an average student. At the end of first year he stood sixth in overall standings, well behind the head of the class, John Bracken, the future premier of Manitoba. He did very poorly in most subjects related to farming, particularly poultry, and was clearly more interested in the academic arts and sciences courses.

MacMillan during his first year at Ontario Agricultural College, c. 1903.

Where MacMillan really shone was in English. He beat Bracken out for the first-year English scholarship, which earned him $20 toward his second-year fees. During his first year, he became an active member of the literary society, and college friends like R.J. (Bob) Deachman predicted he would embark on some writing-related occupation.

> He had a facile pen. He had a bent toward poetry. To tell the truth I thought HR would make a name for himself in literature, I didn't expect him to become wealthy. If he had gone in the direction I anticipated he might have been an artist in words.[1]

A great deal of MacMillan's social life at OAC centred on the Zavitz household. Most weeks he spent one or two evenings visiting, often staying for dinner. Although Zavitz was in poor health, he and his wife regularly hosted students and, probably because of their family relationship, MacMillan saw them often, particularly when members of the widespread Willson clan were visiting.

During his first year, MacMillan went through a religious conversion of sorts, probably a result of Edna's influence. Raised as a Quaker, he was confirmed in the winter of 1903 as a member of the Presbyterian Church. He attended church services at Knox Presbyterian, as well as attending Sunday classes. Later in the year, he held Sunday school classes in his room for other OAC students. MacMillan's religious interests probably focussed more on the moral teachings of the church than on spiritual matters. He was never a devout man, but he maintained a lifelong concern with how he and others should behave. Still, his new-found interest in Christian morality did not mean he was a prig. He was quite active in, and often the instigator of, student pranks and other activities frowned upon by college authorities. His diary is dotted with gleeful entries recording the tossing of student beds, the diversion of rain gutters into residents' rooms, and various other undergraduate hijinks. The mischievous sense of humour for which he became well known in later years was probably developed at this time (it would not have had much room to grow while he lived with his grandparents).

MacMillan spent his first Christmas vacation in Aurora, engaged in a round of teas, card games and other social events. Much of his time was occupied visiting the Mulloys, and his diary contains numerous rows of discreet x's, presumably when he and Edna found the opportunity to exchange kisses. In February he wrote to the federal Forestry Branch in Ottawa to enquire about a job for the following summer, but received a negative answer. He immediately asked Zavitz for another job and a few days later was hired for the summer, with the additional duty of shooting birds that raided the experimental seed plots. After working a few weeks at 9 cents an hour, he negotiated a raise to 11 cents. His monthly earnings over the summer averaged about $25.

He spent most of that summer living and working at OAC. Joanna and Edna wrote numerous letters suggesting he visit Aurora, but he kept putting it off. After a brief visit from Joanna, he flipped a coin and determined he would go to Aurora on the weekend of July 1st. He purchased a class pin and caught the morning train to Aurora via Toronto. He spent most of his first evening sitting on the Mulloys' front porch making up with Edna, who by this time was upset at his absence. The nature of their relationship was already set: Edna, the homebody, and MacMillan, full of ideas and energy, roaming the world. Before he returned to Guelph, they exchanged class pins.

MacMillan spent the rest of the summer in a state bordering on euphoria. His diary is full of entries describing "glorious" and "lovely" mornings, full days and busy evenings enjoying the social life of OAC and Guelph. Each day he rose at 4:00 a.m. to ambush the crows making dawn raids on Zavitz's plots. (After he was reprimanded for using too many shells, his marksmanship improved; it was a skill he would retain for most of his life.) The rest of his days were spent on tasks ranging from hard physical labour on the handle of a shovel, to the tedious work of sorting and counting seeds. He also arranged to serve as the college agent for a Toronto correspondence school offering business courses by mail. He badgered OAC students, faculty and college staff to sign up for courses, putting a lot of miles on his "wheel," as he called his bicycle.

A great source of irritation to OAC students was the food provided to residents. There was not enough of it, and they did not like what there was. Late that summer, MacMillan and three other students working at the college decided to take matters into their own hands. They pooled their money and bought a tent, which they set up on vacant land near campus. They built a rough cook shack fitted out with a stove and stocked with food. The shack leaked like a sieve and their cooking failed to meet the most primitive culinary

The summer camp at Ontario Agricultural College, c. 1903. Left to right: R.K. Monkman, H.W. Scott, A.J. Logsdail, A.P. Smith, C.F. Bailey, MacMillan, F.C. Scott.

standards, but they were happy with the arrangement. MacMillan and his camping companions spent most of their evenings until classes resumed foraging for food. From nearby farmers and the college fields they stole barrels of apples—six in one night's raid—which they hid for evening snacks during the winter, as well as vegetables, chickens, eggs and, on one occasion, a lamb they feasted on at their camp. For a month they lived their semi-vagabond existence. When classes began in September, MacMillan moved back into residence, sharing a room with Allan Mulloy, who had registered in first year.

During his second year on campus, MacMillan was a well-known and popular figure. He was selected to organize the freshman initiation, and noted in his diary: "Got them into gymnasium. I went in basement with sulphur. Burnt sulphur. Drove them out, soaked them, hustled them around campus. Beat them. Wrote notes of it for paper." Even though two of the freshmen were "pretty badly hurt" the event was judged a success.[2]

He was active in sports again, and in January was elected assistant editor of *The Review*. In the second issue following his appointment, he changed the spelling of his name from "Mc" to "Mac." He kept this spelling for the rest of his life, although he did not bother to register the change until shortly before his death. When someone asked why he altered the spelling, he replied, "I liked it better." By this time his childhood nickname, "Reggie," had given way to "Mac."

Midway through his second year at OAC, MacMillan began to develop an interest in forestry. Only a few years before, the very idea of forestry—as a scientific approach to forest management and as a profession—was unknown in North America. Forests were considered an obstacle to settlement and the development of agriculture; they were something to be liquidated at great effort and expense. The lumber industry was considered for the most part a transient enterprise that flourished while the forests were being cleared and moved west ahead of the settlers.

The idea of managing forests to provide timber for a permanent forest products industry—as well as for other benefits such as water supplies, flood control, windbreaks, influences on climate and wildlife habitat—was brought to North America by Bernhard Fernow, a German forester who took up residence in the United States in 1876. Although initially he had difficulty finding work as a forester, Fernow arrived at an opportune time. In the United States during the late nineteenth century, alarm was growing over the destruction of the country's vast forests. Increasing concern was expressed about shortages of timber for railway construction, mining props, and residential and commercial construction. The only American professionals with an interest in the situation were lumbermen, botanists, nurserymen and arborists, all of whom lacked a coherent, comprehensive understanding of forests. What Fernow had to offer was a full-blown, scientifically based system of understanding and analyzing forests, and a theory of how to manage them. He was a visionary conservationist, arguing for the scientific management of forests for perpetual and multiple uses. His vision of forestry had an enormous appeal to many young men at the time. By the turn of the century, the first generation of Fernow disciples were well established in academic and government positions. It was through these people that MacMillan became attracted to Fernow's ideas.

Fernow's arrival coincided with the birth of a conservation movement that swept North America between 1880 and World War One. He rose to prominence rapidly, becoming one of the founders of the American Forest Congress in 1881 and participating in the second congress held in Montreal the following year. In 1886 he was appointed first chief of the Division of Forestry in the US Department of Agriculture. In this position he established the first US forest reserves, initiated the first major reforestation project in Nebraska, established the first US forest products laboratory in Wisconsin and opened the first forestry school at Cornell University in New York. When Fernow resigned his position in 1898, he was succeeded by Gifford Pinchot, the son of a wealthy American lumberman who, on Fernow's advice, had trained as a forester in Europe. In 1900 the Pinchot family founded the Yale Forestry School at New Haven, Connecticut.

Fernow, after leaving his government position, took the job of dean at the Cornell forestry school. One of the first professors he hired was Judson F. Clark, a graduate of OAC who had taught at Guelph before doing graduate work at Cornell. One of Fernow's earliest acts as dean was to acquire a 30,000-acre demonstration forest in the Adirondack Forest Preserve. This preserve had been assembled by the State of New York after intense lobbying on the part of local conservationists, and it was covered with second-growth timber that had replaced the natural stands logged early in the nineteenth century. The forestry school began to implement a plan of sustained yield forestry devised by Fernow. It involved clear-cutting substantial areas of the forest and replanting them with other species. There was an immediate public outcry over clear-cutting in a forest preserve. Conservationists throughout the state lobbied for cancellation of Cornell's use of the forest and, in the end, the governor not only granted that request but also cut off state funding of the forest school in 1903, forcing its closure.

One student registered in the Cornell school in the fall of 1903 was from Ontario. He was Edmund J. Zavitz, a nephew of Charles Zavitz and a graduate of McMaster University, and he became the first Canadian to attend the Yale Forestry School. He later became Ontario's chief forester and launched a massive reforestation program on the province's devastated forest lands. In the spring of 1904 he visited OAC, where he met MacMillan. It was an event that set the course of MacMillan's life.

The younger Zavitz came to OAC after learning from three of the college's graduates who had studied under Fernow—Judson Clark, Roland Craig and Norman Ross—that the Guelph institution was developing an interest in forestry. The previous summer, Ross had established a forest nursery at Indian Head, Saskatchewan, and Zavitz had been offered a summer job working there. Instead, he accepted an offer from OAC's new president, George Creelman, to establish a nursery at OAC. MacMillan recalled the incident in a letter he wrote to Zavitz sixty-four years later, just before Zavitz died.

> When you dropped the idea [of going to Indian Head], Creelman mentioned to me that it might be a good thing for me to do. I would get my fare paid out and back; I would be given $25 a month and board (I think) for the summer, in addition to which it would give me a chance to see something of the West and get acquainted with the Forest Branch of the Department of the Interior and perhaps get a start with them. This turned out to be the turning point in my life.[3]

The evolution of professional forestry in Canada lagged a decade or so behind the United States. As was the case in the US, initial measures were attempts to limit the outright destruction of forests or their conversion to agricultural uses and settlement. The first undertakings in Canada were the creation of Rocky Mountain and Glacier national parks in the late 1880s. The next step was to establish forest reserves, some of which were designated as parks, for the protection of water and timber supplies.

In the present era, when the timber industry is fighting to justify its existence, it is difficult to imagine a time when some of the most enlightened voices for conservation and preservation of forests came from the ranks of the timber processing industry. In fact, the forestry movement that appeared in North America in the late nineteenth and early twentieth centuries was just as widespread, as fervent and as evangelical as the environmental movement of today. Forestry was more than a mere occupation, it was a calling or a cause, attracting some of the best and brightest young men of the time to its ranks. Foresters saw themselves as saviours of the forests, rational scientists rescuing forests from the depredations of ignorant settlers and the greed of politicians. Forestry was seen as an enlightened form of public service, guided by a higher, conservationist ethic. This ethic, with its scientific base and devotion to the public good, captured MacMillan's imagination. Here was a real alternative to farming.

During the 1890s, federal timber reserves were set aside at Turtle Mountain, Riding Mountain, Lake Manitoba and Spruce Woods in Manitoba, where by 1902 the entire northwestern quadrant of the province had been reserved from settlement. Before Alberta and Saskatchewan entered Confederation in 1905, Ottawa established reserves at Waterton Lake, Bow River, Moose Mountain and Beaver Hills. Ontario had reserved Algonquin Park, Rondeau Park and the 1.5-million-acre Temagami Reserve, among others. Lands acquired by Canada from British Columbia for railway construction were also set aside, and Yoho Park was created in 1901. Similar actions were taken in Quebec and, somewhat later, in New Brunswick.

The beginnings of a forest bureaucracy were established in 1899 when Elihu Stewart, a former mayor of Collingwood, Ontario, was appointed chief inspector of lumber and forestry for Canada. In 1901, when the Forestry Branch of the Department of the Interior was established, Stewart became superintendent. A year earlier he, along with a group of lumbermen, scientists and civil servants, organized the Canadian Forestry Association to draw public attention to timber losses from fire and other causes, and to agitate for increased planting of trees.

The most tireless advocate of the forest movement in both Canada and the US was Fernow. In an article published in the *OAC Review* during MacMillan's first year at Guelph, he provided a compelling argument for the need for forestry.

> Why shall we not let Nature continue to produce again, what she can do unaided? . . . There can only be one answer, that Nature alone, after all, is not able to do what man requires; that for some reason man must aid. And the reason, the direction in

which Nature fails, is that she has no idea of economy in time or space, being most wasteful of either; while man, with the constant increase of population, is forced to use both time and space more and more economically, and for the strict purposes of his needs. Forestry is the child of necessity.[4]

Citing per capita timber consumption figures, he estimated that the US would run out of timber within forty years and, until it was able to introduce widespread forest management practices, would have to rely on Canada for supplies. Thus it was important to begin managing Canada's remaining forests.

Fernow ended his article with an emotionally charged argument based on an assumption that has confounded those who make forest policy in Canada ever since.

The practice of forestry means present expenditure for the sake of future returns, and that a distant future. Who but the State can afford such an investment, which cannot bring returns much short of a century? If for no other reason than the time involved, forestry is a business which principally interests the State, the representative of the community at large, the guardian of the future!

Fernow's notions concerning forest ownership flew in the face of the entire European experience to that time, as well as his own personal background. He had been trained as a forester to take control of his family's long-established, privately owned forests. In university, Fernow had become something of a socialist, and brought his half-formed social and political ideas with him to North America. Here, particularly in Canada, they were picked up by a generation of young foresters, all of whom worked for government. Starting in Ontario and Quebec, the idea of public ownership as a virtue born out of necessity was used to support a policy with quite different objectives. These two provinces, and later others, adopted policies of retaining public ownership of forest land in order to increase revenues, by first selling the timber to lumbermen, and then selling the denuded land to farmers, miners or other industrial users. Later it would be argued this was a rational decision, made to protect public interest in the forests. The implications of the policy eventually would embroil MacMillan in the greatest public debate of his career.[5]

One of the first undertakings of the fledgling federal forest bureaucracy was the politically astute task of assisting settlers: advertising the benefits of planting trees and providing seedlings to interested farmers. A series of nurseries was established, and MacMillan went to one of these—at Indian

Head, fifty miles east of Regina—for the summer of 1904.

Although he later talked and wrote about most of his experiences as a youth, MacMillan never referred to the events of that summer. Mazo de la Roche also failed to shed any light on his first real job, sending her fictional version of MacMillan to work on a Great Lakes freighter for the summer. The experience could not have been a failure, since MacMillan continued to work in the Forestry Branch in subsequent years. Given the enthusiasm with which he undertook most things at this time, it is likely that he thoroughly enjoyed the long train ride through the forests of northern Ontario and across the Prairies to Indian Head. While the work itself was probably not very exciting, MacMillan would have experienced it in its larger and more meaningful context as part of the new world of forestry in which he was becoming a player. Whatever occurred, he was edging toward a career that would engage him, in one form or another, for the rest of his life.

He returned to Guelph in September and, in addition to his academic responsibilities, took up his duties as editor of *The Review*. This job captured most of his attention during his third year at OAC. Apart from writing his regular letters to Edna and Joanna and playing on a college hockey team, he apparently did not take part in any other extracurricular activities. Instead, he devoted himself to transforming the publication. Bob Deachman, who had been editor during the previous term, was now MacMillan's managing editor. Together, they turned the publication into an editorially substantial and financially successful enterprise. At the end of his term, the magazine published an unusually effusive tribute to MacMillan's work:

> This year our loss is MacMillan, one of the most enthusiastic supporters the REVIEW has ever had. . . . During the first year, in his capacity of Associate Editor, he was no small factor in the evolution of the REVIEW from the college sheet to the magazine as it stands today. Later, when he fell heir to the editorship, the problems which confronted him were many, varied, and vexatious and unquestionably the most difficult ever faced by any of our editors. With characteristic unselfishness and energy, however, he sacrificed himself as well as his course to a large extent in the endeavour . . . and has amply demonstrated the executive and constructive ability which is certain to carry him high in whatever field of activity he may choose to enter.[6]

Although none of the issues MacMillan edited contained articles that ran under his byline, the unsigned editorials bear the mark of his personality.

His first issue rejected a proposal to levy a duty on imported publications. It also carried a warm tribute to Mrs. Sara Craig, matron of the residence where MacMillan lived, who was moving to Ottawa with her son. Roland Craig was the former OAC graduate who had studied forestry under Fernow at Cornell and, after a year of working for the newly formed US Service, was hired by the Canadian Forest Branch as its first inspector of tree planting. A subsequent edition of the magazine noted with approval the appointment of another OAC graduate, Judson Clark, as Ontario's first provincial forester. All of Fernow's Canadian disciples were back, working in Canada to further the cause of scientific forest management. All would play significant roles in MacMillan's future.

The added work load did not adversely affect MacMillan's marks; on average, he earned higher grades than in previous years. Ironically, one of his best showings on the final exams was in a course on cold storage. He had not attended a single lecture all year in this subject, and on the night before the exam, some fellow students briefed him on the course content in his room. When the marks were posted, MacMillan stood in first place at 84 percent.

He spent the summer of 1905 back with the Forestry Branch, this time as chief of a field party collecting data on the forest reserves at Turtle and Moose mountains, along the Manitoba–US border. It was the first survey of a Canadian forest reserve. In contrast to the previous year, this year's activities were well documented. The first issue of *The Review* that fall contained MacMillan's first published article, "Life on a Field Party." It was well illustrated with photographs he had taken himself.

> There was freedom and variety in the work. During the whole summer, both day and night were spent outside. From morning until night the work kept us out in the open, dinner at night we ate in the shade of trees, and the evenings were spent reading, writing and mending clothes around the camp fire. When we wished to retire, we rolled in our sleeping bags, not even in a tent, and slept the sleep of the just until morning.

Later, MacMillan traced his love of camping, hunting and fishing, which he enjoyed for another fifty years, to this field trip.

By the time he returned to OAC in September 1905 for his fourth and final year, MacMillan's decision to pursue forestry as a career had jelled. That year the college offered its first courses in forestry, and MacMillan took them. Edmund J. Zavitz, having completed graduate work at Yale and the University

The 1905 summer field party, which surveyed forest lands along the Manitoba–US border for the federal Forest Branch. MacMillan, in charge of the party, is at right.

Members of the 1905 summer field party. The young MacMillan is at left.

of Michigan, gave most of the lectures. In January, a special convention of the Canadian Forestry Association was held in Ottawa. This historic meeting marked the recognition of forestry by mainstream Canadian politicians and its acceptance into the power structure. The gathering had been called by Prime Minister Wilfrid Laurier and was opened by Governor General Earl Grey. It was attended by almost everyone in the country who had an interest or a role in forestry. Several prominent Americans attended as well, including Gifford Pinchot, who by now was the first US chief forester. Papers were delivered by a score of prominent foresters, including Zavitz, Ross and Clark.

Clark spoke on "Canadian Forest Policy," advocating an approach proposed in the US by Fernow and considered radical at the time. A program of forest protection should be organized, Clark argued, including slash disposal to prevent fires. Tax laws should be changed to encourage sound forest management. Public lands should be classified and forest land designated. Municipal forest reserves should be established. Regulations should be adopted to limit annual timber harvesting, which would ensure forest renewal. Clark described the core of his proposed policy as a "system of practical forest management, having for its aim the perpetuation and improvement of the forest by judicious lumbering."[7]

As usual, Fernow and his ideas were the centre of attention. Both Laurier and the leader of the opposition, Robert Borden, quoted him. Fernow argued in his address that governments should not only own forest lands, but should actively manage them with a variety of objectives, including the production of timber. He and Clark both called for the opening of forestry schools in Canada, a number of which were under consideration at the time.

MacMillan, whose two summers' work with the Forestry Branch had attracted the attention of his superiors, was invited to attend this illustrious convention. It was an unusual opportunity for a twenty-year-old student to rub shoulders with all the great names of North American forestry. MacMillan would later play down the significance of his presence there, saying "I was trying to earn a few dollars to help me through my college course at Ottawa, a friend got me a messenger job . . . that gave me an opportunity to attend all the meetings."[8] Referring to an invitation to visit Government House, which he received during the convention, he said:

> Pinchot, who was the chief guest, had undoubtedly asked the host to invite young, would-be-foresters such as I. I never lost my conviction that Pinchot was a very great civil servant who performed most important duties for the USA and, after sixty-odd years, I am still proud that I had the opportunity to benefit from his inspiration.[9]

The up and coming young student was becoming known elsewhere as an articulate spokesman for the cause of forestry. Bob Deachman, his friend from his days on *The Review*, had obtained a job editing *The Farmer's Advocate*, one of the country's leading agricultural publications, and in early 1906 it began running articles written by MacMillan. In one, "Why Not Plant Trees?," he produced a clear, well-reasoned argument that farmers should plant and maintain woodlots. Another, "Fire in the Prairie Forest Reserves," was a passionate plea for the protection of the remaining prairie forests. MacMillan had a natural ability to write clearly and forcefully, an invaluable skill when practically all communications were in written form.

Some of the material in these articles was extracted from MacMillan's fourth-year thesis, which he submitted in March to a committee that included E.J. Zavitz. It was a clear, concise account of "Forest Conditions in Turtle Mountains," describing the results of his previous summer's work. Following a lengthy scientific—but highly readable, even after ninety years—description of the forest, the author expressed a few of his own opinions. He was highly critical of the damage inflicted on the forest by fires and careless logging, and he condemned squatters who illegally settled in forests and who "inculcate in others a contempt for order, and thus constitute a source of moral danger in the community." Pointing the finger at his employer, he said the remaining forest was there in spite of, rather than because of, the efforts of the Department of the Interior. MacMillan included an interesting and, in the context of his later career, curious paragraph about the Turtle Mountains forest:

> The aesthetic value [of the forest] cannot be over estimated. By the determined action of the earliest settlers who drove past three hundred miles of arable prairie to reach the forested hills, by the ever increasing Summer camps, an appreciation of the presence of the forest is amply demonstrated. Even now there are from fifty to one hundred people who spend the Summer camping in Turtle Mountains. As the population increases, days of ease grow nearer, and the active, booming age of realism gives way to a more human sentiment of appreciation this

MacMillan's 1906 summer field party, at Riding Mountain, Manitoba.
MacMillan is second from right.

number will yearly grow larger. Here they send their children during the long Summer vacations, here they entertain their visitor, instead of remaining cooped up in the little sun-baked red-hot prairie towns, and here, on the timber reserve, is the one spot, which all citizens of such favoured localities value most highly.[10]

MacMillan graduated a few months later. Examinations for fourth-year OAC students were set by the University of Toronto, and it was from there that he received his degree. Much to Joanna's chagrin, he did not wait around for convocation. He had another summer job with the Forestry Branch and was anxious to accumulate as much money as possible to finance further studies.

Again, he was in charge of a survey party, taking an inventory of the Riding Mountain forest reserve in southwest Manitoba. By now there was no question in his mind about his future. He was going to be a forester. MacMillan never stated why he chose forestry, even during introspective moments later in his life. The opportunity presented itself, and he took it. Or, perhaps more accurately, the profession chose MacMillan. He was an intelligent, capable young man, attracted to the challenges offered by forestry, intrigued by its scientific aspects and attuned to the moralistic overtones of its conservationist ethic. MacMillan impressed established members of the profession, and they went out of their way to involve and court him.

The next, obvious step on the ladder was to further his education, and in this he did make a deliberate choice. He chose Yale.

Graduate school

At the end of summer of 1906, MacMillan was anxious to get to university, choosing not to stop in Aurora to see Joanna and Edna.

> I took the train from Dauphin direct to New Haven—having planned things so that I would work until the last day possible so as to get every dollar—which I did. I knew I would need it when I got to Yale, which turned out to be the case. . . . I went by train to Port Arthur and then by boat to an Ontario port on Georgian Bay, and then by train to New Haven. I could not afford to take a room in which to sleep, so slept on the cargo deck wrapped up in my new overcoat, which I had bought in Winnipeg. The whole season was a big adventure and it turned out very well.[1]

Yale, where MacMillan was the only Canadian registered, was quite a different experience than the Ontario Agricultural College. Admission to the two-year masters program required a degree from another university, and students were much more focussed on their studies. The forestry school had been established with a grant from the

Pinchot family, who also donated a 200-acre tract of forest and a country estate at Milford, Pennsylvania, where a field school was conducted. Although junior students were required to attend a summer term at Milford, MacMillan was exempted, probably because the courses taught were in subjects he was putting into practice at Riding Mountain.

The forestry school operated out of Marsh Hall, which contained well-equipped laboratories, a library, an herbarium and a large collection of domestic and exotic woods. MacMillan lived in a student residence at 74 Lake Place, the present site of the Yale Gymnasium, and roomed for part of his term with Willis "Pud" Millar, who later taught forestry at the University of Toronto. There was a heavy course load in the forestry school, and MacMillan, now sure of his plans for the future, reserved most of his energies for his studies. His professors included Henry Graves, James Toumey, Herman Chapman and Ralph Bryant; Gifford Pinchot, while serving as the US Chief Forester, taught forest policy during one term.

Forgoing the expense of returning to Aurora for the Christmas holidays, MacMillan found a job at a pulp wood camp near Moose Head Lake in Maine. He took a train from New Haven and walked the last few miles into camp across the frozen lake. Although most of the workers were Canadians, they were all French-speaking. MacMillan enjoyed the work and was grateful for the money, as he was by now determined to pay his own way without relying on assistance from Joanna and his Uncle Ellsworth. In February he attended a highly publicized forestry convention in Fredericton, where he delivered a paper on North American forestry. The wide coverage of his speech in the press showed that he was recognized in Canada as a promising member of the forestry community.

Instead of returning to work with the Forest Branch in the summer of 1907, MacMillan accepted another offer from a syndicate of Lindsay, Ontario businessmen formed to stake timber in British Columbia. The syndicate was taking advantage of a new form of timber lease, made available by the government of Richard McBride. The new system had precipitated a timber-staking frenzy, attracting speculators from all over North America. The previous year Judson Clark had left his job as the Ontario provincial forester to move to Vancouver as manager of the syndicate, called the Continental Timber Company. Roland Craig was hired to organize a party to spend the summer staking claims, and Craig offered MacMillan a job. The third member of the party was Aird Flavelle, a young man MacMillan's age who was a nephew of the syndicate's organizer. As soon as his classes ended at Yale, MacMillan boarded a train for Vancouver.

The city, then only twenty years old, was filled with the bustle of a frontier town. It was a memorable experience for the impressionable MacMillan, whose knowledge of cities had been shaped by visits to large eastern centres. Although a minor recession had slowed economic activity somewhat, Vancouver's business sector was busily engaged in promoting real estate schemes, mining ventures and the formation of numerous forest companies attempting to cash in on the McBride timber leases. Underneath it all, there was growing tension over the increasing numbers of Asian immigrants. The Asiatic Exclusion League held large public meetings which in the fall precipitated a mob attack on Chinese and Japanese residents.

MacMillan (at right) on a Sunday walk with classmates who became prominent North American foresters. Left to right: W.N. Millar, Thornton Munger and Nelson Brown. New Haven, February 1907.

By the time MacMillan arrived, the other members of the crew had acquired a boat they christened the *Ailsa Craig*. It was a twenty-foot gill-net fishing boat, with a 10-horsepower Easthope engine, a small cabin with wooden benches for bunks and a coal stove for cooking. Craig described their progress in a report to the syndicate:

> The crew consists of H.R. MacMillan at $75 per month and expenses, Aird Flavelle at $60 per month, and I presume travelling expenses also, and a cook at $45. We intend going up Jervis Inlet first to look at some land which is still vacant and I understand is timbered, then I think we will try Powell Lake which is separated from the sea by only 3/4 mile of river and up to the present is fairly free from locations.[2]

The party packed its gear aboard the *Ailsa Craig* and headed north along the mainland coast on April 9. They staked some claims on Powell Lake, where the marksmanship MacMillan had developed protecting Zavitz's seed plots proved useful. Years later, in a letter to the director of the Museum of American Indians in New York, he explained that "in the early days the mountain goats were quite numerous and in the spring were low down on the steep rocky mountains, which in so many places raise almost precipitously from the water . . . I shot one with a revolver from a canoe."[3]

From Powell River the party headed north to Jervis Inlet, but failed to find suitable timber stands. They continued north through the Yuculta Rapids to the area around Minstrel Island, which was then a major logging centre.

> The first were handloggers who lived in shacks, afloat or on shore, and worked on steep side hills, singly or in pairs. They knew exactly what they could do— they sought the largest straight trees, mostly Douglas fir, the rest cedar (hemlock and balsam weren't worth cutting). The trees selected were those so situated that they could be run, end on, down the steep side hill to the water. The loggers, equipped with falling axes, cross cut saws and Gilchrist jacks accomplished marvels in guiding and turning heavy logs. They sniped the down hill end of the log, peeled the bark off . . . and seldom failed to get the log to water. After the hand logger came the donkey crew—very few men, living in shacks on a float, with a small steam donkey ashore or on a float.[4]

Among the many timber cruising parties working in and around Minstrel Island that summer was one composed of Martin Allerdale Grainger and his future brother-in-law, Leonard Higgs. MacMillan never mentioned encountering Grainger on that occasion, but within a few years they became close associates. The experiences and observations of that summer were reconstructed by Grainger the following winter in his classic novel about logging on the BC coast, *Woodsmen of the West*. As MacMillan affirmed on many occasions, this book was a full and accurate description of the era.

MacMillan was intrigued by the frontier lifestyle of the BC coast. During their stay in the vicinity of Minstrel Island,

> Aird Flavelle and I made our calls for mail at the old saloon [at Port Harvey] which was open every hour in the year for handloggers patronage—served weekly by the SS Cassiar of the Union Line. It was at Port Harvey that I saw loggers who had sold the product of months' work for $5–$6–$8 per M—maybe a total of $200 or more— drink it up in one bout—sometimes going to the beach to drink sea water, so as to start at the bar again with an empty stomach. The only real cure seemed to be an empty pocket—tho' I have seen some try drinking Worcester Sauce.[5]

July found MacMillan's crew working up the Salmon River, where

> owing to axe cut in my left hand I stayed on fish boat 2 or 3 miles up Salmon River from Port Khusam Indian village while Aird Flavelle and Roland Craig went up White River to stake timber licenses behind existing pulp leases. While I waited

there the Hastings Co. moved in their first logging outfit on scows—to build a camp, railroad, whatever else needed. They began production by early fall. Before that not a tree had been sent to market from the Salmon River drainage. In August Craig, Flavelle and I staked the license in the nearby Adams River Valley now owned by MB&PR.[6]

When classes resumed in the fall, MacMillan returned to Yale. Although he had just spent a successful and invigorating four months in the forests where he would eventually make his name and fortune, he did not at this time appear to harbour thoughts of returning to British Columbia. Apart from the business of staking timber, in 1907 there was little opportunity for a professional forester in the province. The timber industry was operating on the "cut and get out" model perfected in North America over decades, and little or no thought was given to the future. Considering MacMillan's growing dedication to the cause of forestry, BC had little to offer him at this time.

Flush with his summer's earnings and energized by months of hard physical labour, he returned to his studies at Yale with a single-minded dedication. He spent the two-week Christmas vacation back in Ottawa, working at a job created for him by the Forestry Branch. By the end of the year he stood near the top of his class and was one of the most popular students in the forestry school, chosen by his classmates to give the class speech at graduation and elected class vice-president. In the Yale Forestry School's senior year, the spring term was spent on field work at a logging camp at Wesgufka, Alabama. MacMillan later described his experiences there in a letter to Mazo.

> We were given about ten days from the close of the winter term until work opened up in Alabama. These ten days were no good to me as I had nothing to do and no place to go. [Willis Millar] was in about the same financial position as I, and we decided to work our way to Alabama. Some one said we could not do it. I had about $25 left, and I made a bet of $25 that we could do it. We did it, but it involved sleeping two nights in Central Park in New York City in the month of March, and going down from New York City to Charleston in the steerage of a boat with a crowd of negroes, which was a cheap but long-remembered trip, and walking or beating our way on freight trains from Charleston to our final location. . . . We picked up a good dog, which I sold later for $25. . . . Incidentally, we walked 240 miles in six days, which I have always thought was pretty good, and carried enough in a pack to enable us to sleep and cook wherever night and meal time found us. This was a very interesting term, excepting that I got what I thought at that time was an attack of malaria, but which I have since decided was tuberculosis. I was quite sick for a week or two.[7]

Yale Forest School, class of 1908, at Wesgufka, Alabama. MacMillan is circled.

But at age twenty-three, MacMillan was in peak physical condition and recovered quickly. He graduated cum laude, receiving a Master of Forestry degree. At this time there were only a few dozen professional foresters in North America, and he had a choice of jobs.

It was no trouble to get a job when I left Yale. As a matter of fact, when I left Yale I had the opportunity of going to New Brunswick to inaugurate a forest school at the University of New Brunswick, or to Oregon to take charge of the Forest Department at Salem, or of going back to the Department of the Interior, in which I had worked in the summers and at odd times for several years. . . . I chose the Department of the Interior, although the salary was less than one-half what I would have got elsewhere, because the scope of the work appealed to me and the experience which I would gain in Western Canada, which was then developing at a great rate, would, in my judgement be of very much more value to me. This turned out, so far as I can see to have been correct. The job I went to was Assistant Inspector of Forest Reserves, at a salary of $100 per month. My head-quarters were in Ottawa, but I spent most of the year in the west.[8]

The year MacMillan graduated, 1908, was an auspicious one for Canadian forestry. The country's first forestry school went into full operation with twenty-eight students at the University of Toronto. The school had

opened the previous year with five students and Fernow as dean. Following
the closure of the Cornell school, Fernow had spent a year establishing a
forestry school in Pennsylvania, but was attracted to Toronto because of the
opportunities offered by the vast Canadian forests. During the first year
Edmund Zavitz taught some courses, splitting his time between Toronto and
Guelph. In 1908 the illustrious American forester Clifton Durand Howe, director of the private Biltmore forest school, joined the Toronto faculty. In addition, in the fall of 1908, a second Canadian forest school opened at the
University of New Brunswick.

On a visit to the new Toronto school in the early winter of 1908,
MacMillan was introduced by Fernow to a young electrical engineering student, Whitford Julien VanDusen. MacMillan painted such an exciting future for
Canadian forestry during their one-hour conversation that VanDusen
switched to the forestry school. Before the year was out, MacMillan had hired
him to work on a summer survey crew in Manitoba. Many years later,
VanDusen described his summer:

> When I got there the leader of the survey party was sick, and there was a letter
> awaiting me from HR ordering me to take charge of the party. Well, I got a raise of
> $5 and went up to $35 a month as leader of the party. A first-year forestry
> student![9]

In March 1908, most of Canada's professional foresters, who numbered less than a dozen, met in Montreal under Fernow's chairmanship to
establish a professional association, distinct from the Canadian Forestry
Association. Seeking to add a touch of glamour to the relatively unknown des-
ignation of "forester," they called the organization the Canadian Society of
Forest Engineers. During the next few months virtually every professional
forester in Canada was elected to membership, including MacMillan, who
joined in June. The primary purpose of the organization for its first dozen
years was social. The members constituted a close-knit fraternity devoted to
furthering scientific forest management throughout the Dominion.

Given the state of forestry in Canada, MacMillan's decision to work for
the federal government was a logical one. At this time the administration of
natural resources in the three prairie provinces was under federal control, so
the Department of the Interior established and administered forest reserves.
British Columbia, with by far the largest forest resources of any province, had
undertaken no forestry initiatives, and therefore had nothing to offer a man

MacMillan visiting Willis Millar at Priest Lake, Idaho, October 1908.

with MacMillan's interests and training. Most owners of industrial operations showed little or no interest in managing forests to provide themselves with a long-term timber supply.

During his first summer of full-time employment, MacMillan's work took him west again.

> The work in sight was so interesting it seemed like a holiday. The Department of the Interior at that time was full of optimism in keeping with the expansion in the West. Invasion [of forests by settlers] was taking place everywhere and one of the Government policies was to set aside as forest reserves those tracts of land which were unsuitable for settlement. My job was to take a general look at the country in different definite areas which had been withheld from settlement and to decide approximately the boundaries beyond which settlement should not be permitted.[10]

He spent most of the summer in Saskatchewan, surveying Prince Albert National Park, and in the Beaver Hills near Yorkton. As he had during his years at OAC, he maintained a steady correspondence with Edna, who was still in Aurora, and he apologized occasionally for writing too often. They had decided to get married after his graduation, and to live in the west. In mid-September, from Edmonton, he sent her a set of photographs of the Rocky Mountains, including one of the Banff Springs Hotel.

> I think we had better have a home built near this large hotel, if we don't like to live either at Calgary or any of the other towns out here—what say you— Edmonton is my choice if it turns out to be in the right place. This is the country I shall be working in for the rest of the fall.[11]

His final task for the season was to examine the area around Waterton Lake in southwestern Alberta to determine its suitability for a national park. He travelled by rail to Cardston, arrived in late September and set out on foot.

> I had no previous experience in the high mountains and was not aware of the sudden changing conditions that might take place. I had never been in that territory before. . . . I was alone and did not know anybody. I went out to one of the Foothill ranches in the valley of the Belly River, which flows from Montana into Alberta, thinking that at some cattle ranch I would find somebody who could supply me with help and horses, with which I could spend a few days or a week in the mountains looking the situation over. The fall roundup was in progress and amongst the scattered population of only one or two families every ten or fifteen miles I found no men to take on this temporary job. However, at one ranch after I had made several attempts elsewhere, they told me that if I would go along with a

young lad about seventeen with two saddle horses and two pack horses, they could spare me that much help, which I accepted. The boy was inexperienced and had never been up the mountains.

The country in that part of southern Alberta changes very quickly from low-lying prairie and foothills to steep mountains. There were old trails from the time of Indian trading up most streams and over passes through the mountains. It was possible to take horses up these trails to the 6,000- or 7,000-foot level, at tree line. MacMillan and his young helper followed a trail high into the mountains.

MacMillan (left) at Kaniksu National Forest, Idaho, 1908.

On the last day going up, the lad with me, who was looking after the pack horses and his own saddle horse, fell behind. As the day was warm, I had taken off my coat, or sweater, or whatever outer clothing I had with me, and put it on the pack horse and consequently had nothing on over my shirt. The country was very interesting and as the trail was easy to follow, I went on without paying much attention to the young fellow with the pack horses.

In the evening I was on the summit above the timber line. The permanent snow and ice were practically down to the trail, which ran in a narrow valley in which were shrubs, shale and boulders and through which the wind continuously blew quite a gale.

As soon as the sun dropped it got cold and I began to wonder about the boy with the camp equipment and food. After dark he showed up with his saddle horse having lost the pack horses. He was pretty badly frightened. He had not been out in this kind of country before. I was a fool for ever having gone with him. We had only two saddle blankets and nothing to eat.

We tethered the horses and found that we had no matches, which again was terribly foolish of me, after the amount of time I had spent in the woods. We cut some brush with our pocketknives, broke it off with our hands and worked out a place where we could sleep sheltered from the wind, huddled together in our saddle blankets, intending to start down first thing in the morning.

It snowed during the night and when daylight came there was two feet of snow on the ground, and it was still snowing. We were quite alive but pretty cold.

The snow had obliterated the sharp rocks which made it impossible to ride the horses, and, while a person could not be lost because the water ran down hill, it took us two days to get out to where we could get anything to eat, by which time the horses hooves were split up to the quick and we were soaked with perspiration and wet snow.

Following his harrowing ordeal in the mountains, MacMillan went to the Crow's Nest Pass and spent several weeks examining forest conditions there. He travelled into BC to examine the aftermath of an enormous fire which had destroyed a huge swath of forest around Fernie. His work completed, he made a brief trip into the US to visit Willis Millar, who was working as a forest supervisor at Priest Lake. In late November he returned to Ottawa, suffering from a cold that had plagued him since his unfortunate experience in the mountains.

Ottawa was a beehive of activity. The Laurier government had been re-elected and, as MacMillan described it, "the stage was set for a great expansion in that portion of the civil service having to do with conservation and care of natural resources."[12] With nothing to distract him and seeing a chance to get ahead in the department, he worked almost every night, weekend and holiday. But the cold he had when he returned from Alberta was getting worse. He saw several doctors, who gave him medicine which did not help. Eventually he found one who made a startling diagnosis: he had tuberculosis in a very progressed state. MacMillan, whose future had seemed assured, was facing the greatest crisis of his life.

Between life and death

When MacMillan was diagnosed as having tuberculosis in 1908, there was no medicinal cure for the disease. However, there was treatment that was sometimes successful, although rarely in cases as advanced as his. Twenty-five years earlier, a young New York doctor named E.L. Trudeau had been told he was incurably ill with TB. Trudeau decided he would rather die in congenial rural surroundings, so he moved for the summer to Saranac Lake in the Adirondack Mountains. By fall, when he should have been dead, he found he was gaining strength, and he decided to stay there for the winter. In the end he cured himself, and a sanatorium was built at Saranac which he operated. It became the model for institutions built throughout Canada and the US. Within days of being diagnosed a consumptive, MacMillan went to Saranac.

On arrival he was examined by Dr. Hugh Kinghorn, the son of a wealthy Montreal family and one of the first patients to be cured at Saranac. MacMillan later wrote to Mazo about this interview.

> Kinghorn was a dark, slim, lame man with a caustic tongue and a very dry sense of humour. I thought I was pretty well until I met him and in about five minutes he told me that I had one chance in ten of surviving and then only if I had a good constitution, good lungs and observed his instructions literally and continuously, and that instead of the cure taking two or three months, it would probably take two or three years; that there were many people who had been in the sanatorium for five to fifteen years. I felt quite different after I had heard all this.[1]

MacMillan moved into a boarding house attached to the sanatorium and began his treatment.

During his convalescence, he wrote a detailed account of his experiences as a TB patient, which was published anonymously in the *Canadian Magazine* under the title "Fighting the White Plague."[2] The treatment was tediously simple: eat lots of nourishing food, avoid any and all physical activity and breathe pure, fresh air, even in the coldest winter weather.

> From other patients I learned how to be comfortable, how to avoid chills, how to keep warm in cold weather. The open porches are all partially glassed in the winter to afford protection from wet and drifting snow—but they are always sufficiently open to allow the entrance of fresh air, and are consequently the same temperature as the outside air. This makes the problem of keeping warm a difficult one. I kept comfortable on the coldest days, wearing a hockey cap, woollen lined leather mittens, felt boots, and over my ordinary clothing a fur coat and two heavy woollen blankets. If the weather was extra cold I had a hot soapstone placed at my feet. At night, sleeping in a bedroom as cold as outdoors, I wore a woollen cap, woollen socks and used the soapstone.

It took MacMillan some time to realize and accept the seriousness of his circumstances and the task that lay ahead of him.

> I expected to be able to walk around a little, play cards whenever I felt like it, write multitudes of letters, read, study and enjoy myself. I soon found from the doctor and other patients at my boarding-house that I was far wrong, that I could do none of these things, must not, for instance, climb stairs oftener than absolutely necessary, must take as few steps as possible and take those few slowly. I also learned in the first day or two that I must stay quiet in my chair eight or ten hours a day and spend the rest of the time in bed, except the hour or two devoted to meals. And such meals! My appetite, which had been poor, increased to such an extent that I found myself eating more than ever I had on hard exploration trips.

After six weeks diligently observing the prescribed cure, MacMillan went to see Dr. Kinghorn, convinced he was probably half cured already. On being told there was no difference in the condition of his lungs, he realized he was in for a long haul. He could not afford the $1,000 a year Saranac charged, but the doctor told him about a new sanatorium which had opened at Ste-Agathe-des-Monts, a small village in the Laurentians north of Montreal. MacMillan could afford the $600 annual cost because the Department of the Interior allowed him to keep his position in the Forestry Branch and continued to pay his salary.

The regimen at Ste-Agathe was similar to that at Saranac.

> I continued to gain in weight and condition, gained thirty pounds in sixty days. I still "took the cure" as faithfully as ever, keeping perfectly quiet all the time, sleeping ten hours at night in an open room, two hours in the afternoon, lying in a reclining chair on an open porch all the rest of the time, administering to a healthy appetite five times a day, with three regular meals of plain, nourishing, easily digested food, and two light lunches of milk, cocoa or tea and a biscuit morning and afternoon. . . . Just when I was beginning to make improvement I was exposed to a case of influenza. I took it easily and plentifully, it kept me in an unhealthy condition and for months I made no progress. Then in the fall, when I was again getting in good condition, a cold which I contracted made matters worse. My temperature started to rise and I suffered a relapse which left me sicker than ever before. . . . I remained in bed nearly three months, until my temperature was quite normal. During this time I did nothing but read novels, eat and sleep. I was always alone.[3]

MacMillan had been at Ste-Agathe for five months when his mother quit her housekeeping job and came to live in the village. They found a small flat to rent and she devoted herself to looking after him. This attention, he said later, "undoubtedly saved my life."[4] But even with the comforting presence of Joanna, it was some time before he began to gain on the disease.

> I got worse for about a year, at the end of which time it was apparently very doubtful which way I was going to travel. The outstanding event of that length of time from my standpoint was one grey day in the winter: I was lying in my room, which was a narrow one, upstairs, facing the northwest, from which the windows had been removed and through which I could see nothing but the grey sky, and, generally speaking, I was feeling very sorry for myself. The doctor came to see me. He had a talk with me and judging by my symptoms, of which he had been keeping a record, the time had come when I had to give the job all I had or it would be too late. I remember feeling very low. I had taken the cure with all the

concentration and care of which I was capable. I remember that for months I kept a chart of what hour of the day I first coughed and how many times a day I coughed and every day I tried to delay the hour of the next cough and to reduce the number of coughs—I very nearly suffocated myself in the process. Suddenly I began to recover very rapidly and at the end of fourteen months I was allowed five minutes exercise once a day, which consisted of walking ten or fifteen steps and back again, and for the next six months the exercise was increased until it reached an hour twice a day, which was practically all the freedom in the world.[5]

His doctor at Ste-Agathe was Rodderick Byers, another early survivor of Saranac, who later became one of Canada's foremost experts on tuberculosis. Of Byers, MacMillan wrote many years later:

He is a great friend of mine, and saved me more by his philosophy than by any treatments. He gave me some pieces of advice which were invaluable. One was that nearly all people who fail, do so by attempting the cure in a 80% or 90% fashion instead of in a 100% fashion—the last 10% to 20% being what counts.[6]

After eighteen months at Ste-Agathe, MacMillan was clearly on the mend. Byers advised him in the fall of 1910 to spend another year at the sanatorium. Although his physical activities were still drastically restricted, he was now able to work, writing and editing articles for a variety of trade and professional publications, and for the Forestry Branch.

While MacMillan was engaged in the protracted struggle for his life in the Laurentians, North America's professional forestry establishment continued to grow, in step with the broader based conservation movement sweeping the continent. Public enthusiasm for careful use of natural resources played a significant part in the re-election of Laurier's government in Canada in 1911 and the Republicans under Taft in the US the following year. The forestry schools of both countries were attempting to define what constituted forestry and, as usual, Fernow was in the forefront of discussions. In his view, the core forestry curriculum should consist of courses on soils, forest pathology (insects and diseases) and forest zoology (fish and wildlife). The main tasks were taking forest inventories (determining the nature and extent of the forests), protecting forests (primarily from fires) and reforesting, either naturally or by planting. In Fernow's view, and in the minds of those around him, including MacMillan, engineering had little or nothing to do with the theory and practice of forestry. The work of the forester was to study, to understand, to protect, to restore and to manage the forest for a variety of purposes and uses. The various tasks associated with harvesting timber from the forests—road building, and the cutting, transporting and processing of timber—were peripheral to forestry, and were not considered important subjects in forestry schools. MacMillan would never abandon this point of view, even when engineering came to dominate the study of forestry.

The growing belief in the virtues of scientific resource management required an increased knowledge of the resource base, including the quantification of various resources. Conservation commissions were established in both countries, the Canadian one in 1909 under the chairmanship of Clifford Sifton, and including Bernhard Fernow as a member. The commissions were to take inventory of resources and disseminate the information required to make informed decisions on resource management. Just before he was released from Ste-Agathe, MacMillan sought the job of chief forester with the commission, but he was unsuccessful. The position went to a prominent US forester, Clyde Leavitt.

Also in 1909, the British Columbia government began to express an interest in developing a coherent set of forest policies. After corresponding with Fernow, the province established a commission of inquiry under the commissioner of lands, Fred Fulton. Judson Clark, by this time living in Vancouver, was active in establishing and operating the commission, for which Martin Grainger served as secretary. Clark wrote to Fernow that "I would not be at all surprised if British Columbia should within a year or two have more advanced forest laws than any other Province."[7] For the next three years, Fernow was the leading advisor to the BC government on forestry matters. One of his recommendations was to establish a provincial forest service.[8] In part the government's interest was due to the enormous revenues it had been collecting through the sale of the McBride timber leases. Under a long-standing policy, acquisition of Crown timber previously had been restricted to those who owned a mill or were willing to build one. After 1905, anyone was able to obtain timber in BC and hold it for later use or resale. The fees for these leases provided the government with $13 million over a six-year period, and the politicians were anxious to sustain this source of revenue.

Meanwhile, MacMillan still faced another year in the Ste-Agathe sanatorium. Heartened by the knowledge that he was on the road to recovery, he became increasingly active in the world of forestry. In June 1910, he asked for and received permission to represent the Forestry Branch at a meeting in Belgium on forest research, but did not attend: Byers told him he should spend another year recovering. So MacMillan devoted himself to writing. He produced a lengthy series of magazine articles on Canadian trees, a series of statistical pamphlets on Canada's forests and forest industries for the Forest Branch, as well as several articles on tuberculosis. At this stage in the development of forest policy, one of the major objectives of professionals and government departments was to prevent forest fires. MacMillan wrote several articles on the subject, enhancing his status as an ardent and eloquent spokesman for the cause of forest conservation.

MacMillan, the young forester, c. 1909.

> The amount of timber yearly destroyed in Canada by fires, would, if known, stagger Canadians. A much greater amount has been destroyed than has ever been cut. Senator Edwards, an experienced Ottawa Valley lumberman, has stated that in the Ottawa Valley, where over large areas stumpage has been for years the most valuable of any in Canada, twenty feet of white pine have been destroyed by fire for every foot cut.[9]

MacMillan also published an article which, in view of his subsequent career and the importance of the subject to the Laurier government, is both a historical curiosity and a measure of his independent attitude. At the time it was published, the Liberals were fighting for their political lives on the issue of free trade, or reciprocity, with the US. MacMillan's article opposed the government position of free trade in timber products.

> Taking into consideration that the forests of Canada are not unlimited, and that their maintenance is absolutely necessary for successful homebuilding throughout the country, it is just as well that a tariff prevents their rapid exploitation to supply the voracious United States market.

Going further, he called for a prohibition on exporting pulpwood to the US.

> If the export of raw pulpwood were prohibited and no annual crop were found to be a satisfactory substitute, these mills or even a greater proportion of the mills in the United States would in a few years be forced through a lack of raw materials to close their doors. The market they supplied would then be supplied by Canadian mills cutting from the lands which, if scientifically managed, will support a huge permanent pulp industry.[10]

The winter of 1910–11 was one of the hardest for MacMillan. He was recovering in body, but emotionally he was at low ebb. On Christmas Eve he wrote to Edna, whose letters and support had helped sustain him for almost two years.

> Little one mine. I cannot help writing to you again, even at the risk of making you weary of me. Oh but I have had the blues—yesterday and last night—so much so that last night I could hardly sleep. I am ashamed to tell you—and did not intend to—I am so afraid you will think it cowardice of me, foolishness or some such thing—to feel so downhearted. But I tried not to. I think probably it was chiefly loneliness and homesickness for home, homesickness to see you. I am terribly tired of this, tired to death of lying around when everyone else works and has a grand time, tired of the neverending uncertainty of it all.
>
> If you had asked me in that letter I got yesterday I would have gone to Aurora for Xmas. I am so anxious to see you, to get away—just as well you didn't, you will be having a grand time, and if I had been there I would have broken in upon it. Oh, I am better off this way, I know it myself, but it makes me sick to think it.
>
> But it will be different another year. I feel sure that if I am careful we can get married this coming year and so long as I think that you may be sure I'll be careful. Otherwise I think sometimes I would just let things go—this neverending strain, this constant stretch of uncertainty and monotony are too much.
>
> I feel I must see you before long. I do hope I can. I always get so much comfort from your company and so much encouragement from your love. Oh dearie, you do love me, don't you? If only I could see you—only I could hear you tell me so—if only I could feel the warmth of your caress, the pleasant intoxication of your presence—I would be so much better. A merry Xmas dearie. Yours, Reginald.[11]

Finally, in mid-1911, MacMillan, now twenty-five years old, left Ste-Agathe. During his last session with Dr. Byers,

> he told me I could work or I could play, but I could not work and play, and, inasmuch as I had to work, I went to bed at dinner time every night for years and

rested Saturday afternoons and Sundays. . . . When I left to go back to work he told me not to have any general ideas about living a long time, but to fix my mind on five years and not to do anything that in my judgement would interfere with that objective and then to come back in five years and if he found me then as well as I was, he would say I was cured.[12]

From Ste-Agathe, MacMillan went straight back to his job as assistant director of the Forestry Branch in Ottawa, while Joanna returned to the Newmarket district.

I lived in a boarding house kept by an old Irish lady whose husband had a tremendous library. I think he must have been a lawyer or some sort of student. He had passed away and left her with nothing to do but look after someone like myself.[13]

MacMillan never looked back with regret on his thirty months at Saranac and Ste-Agathe. He did not view the lengthy interruption of his career as a waste; he used his time in the sanatoriums well, reading voraciously and, when he was able, writing. In this way he maintained his professional standing. His convalescence does not appear to have affected his career adversely, and when he was released, he simply carried on from where he had left off.

A few weeks after returning to Ottawa, on August 2, MacMillan and Edna were married in Aurora. The fifty invited guests were entertained at a reception following the ceremony, at a cost to Edna's father of $45. The newlyweds took up residence in Ottawa, and later that month MacMillan headed west, to northern Saskatchewan and western Alberta. During his illness the Forest Branch had grown rapidly, and a great deal of organization and supervision of new men was required.

He returned to Ottawa and his new bride in mid-fall, adhering faithfully to his doctor's instructions, going to bed in the early evening and resting on weekends. It was an exciting and promising time for a young forester.

At that time, 1911–12, everything was booming in Canada. Expansion was the order of the day. Trans-continental railroads were being built and municipalities and cities were growing rapidly, money was pouring into the country and organizations were rising like magic. My college friends were reaching important positions in all directions. British Columbia was one of the provinces with great resources, and its government was organizing to look after the important functions of government. An imaginative Minister of Lands, W.R. Ross, had just been appointed by Premier Sir Richard McBride.[14]

Ross was one of BC's most colourful and respected cabinet ministers.

Over six feet tall, he was 240 pounds of solid muscle, topped with a crown of curly black hair. He was the grandson of a famous Hudson's Bay trader and his Cree wife, and, although well educated, had spent most of his early manhood working in BC logging and mining camps. He was an early convert to the causes of conservation and forestry, and before his love of fine liquor killed him at an early age, he introduced sweeping changes to forest legislation.

The Fulton commission had tendered its report in 1910, calling for the creation of a provincial forest service and government fire protection measures. Ross presented his legislation in January 1912, in a dramatic, lengthy speech to the legislature. After speaking in support of continued public ownership of natural resources as a means of providing government revenues, he went on to announce the formation of the Forest Service and the creation of a joint industry–government fire protection fund. The enthusiastic response from members of all parties, as well as the press and public, indicated that Ross had struck a chord, and he proceeded with a clear mandate to establish a strong government arm that would care for the forests.

Ross also announced that the government had engaged a consulting forester, Overton Price, assistant forester in the US forest service and the person primarily responsible for its organization. Price came to Victoria in early 1912 and, backed by the long-distance advice of Gifford Pinchot and Fernow, began organizing the BC Forest Service.

One of the first people Price contacted was MacMillan. He looked him up on a trip to Ottawa in early February and caught him at an opportune time. MacMillan had spent the winter working at his office in Ottawa, with occasional trips to Toronto to confer with Fernow and give lectures on fire protection to forestry students. The outlook for forestry in Ottawa that spring was bleak. The Laurier government had been defeated the previous September over the reciprocity issue, and Robert Borden's Conservative government was dragging its heels on the expansion plans of the Forest Branch, where a mood of discontent had set in.

Price wrote to R.H. Campbell, MacMillan's boss, to apologize for his offer to hire away a valuable employee.

> Until I met MacMillan, he was not prominent in my mind as a possibility, since I had never met him before, and had made no inquiries about him. Having seen him and having made the inquiries, I feel that he is fully up to the high standard of efficiency, initiative, and training, as well as to the high personal standard which in

my judgement a man must possess to be really useful in pioneer forest work in British Columbia. . . . I have reached the conviction and I have not reached it easily, that forest work in Canada will be more benefited should MacMillan take up work where urgent work cannot even begin for the lack of professional foresters of his calibre, than if MacMillan should continue where he is greatly needed, where he is doing remarkably well, but where an organization will not be disrupted if he goes.[15]

On February 14, Price sent MacMillan two letters. The first contained a formal offer of employment at $200 per month, a position on a new Forest Board and responsibility for the fire protection division. The second, more personal letter expressed Price's expectation that MacMillan would become head of a planned Forest Branch. The letters confused MacMillan. He was unclear about Price's offer and was not prepared to make a hasty decision. On February 16 he wrote to Grainger, expressing his reservations, and adding:

It would help me to make up my mind if I knew who was going to take charge of the work on the Coast. If they are going to get a first class man as Provincial Forester I would be inclined to go.[16]

The next day he sent a memo to Campbell, pointing out that the BC offer was for substantially higher pay than he was getting in Ottawa. Three days later he sent another memo, asking for a raise. Campbell responded a few days later, offering a small raise in pay and the possibility of a civil service appointment. MacMillan then wrote to Price, telling him he could not take the job in BC. This letter crossed in the mail with an answer from Grainger, who made a strong pitch for him to accept Price's offer.

There is no one humdrum routine to be followed out, the field is open, and the force attacking the forestry problem is free to do its level best to find a decent solution. A man can, therefore, throw himself into the work with some enthusiasm—he isn't up against a brick wall. As for serving the public—well, you couldn't find a country where prosperity depends more upon the proper handling of its forests than it does in British Columbia. There is satisfaction in helping on work that is worth while. I would be awfully glad if you fix it up with Price to join the B.C. service.

Grainger went on to discuss MacMillan's concerns about the nature of the work and who would be in charge:

You don't quite catch on, I think. The new Forest Act provides for a Forest Board. Price, I take it, is offering you a job on the Board—and as a member of that Board you would take charge, as I interpret your letter, of his "section of operation." They'll get a man to run the timber sales branch; and he will also be on the Board. From among the members of the Board there will in due time be appointed the man you call "Provincial Forester," who will take charge generally. Why shouldn't this man prove to be yourself? Leave your mouldy East, and come and do something really important in God's own country. We will take our coats off and go to it.[17]

On February 29, MacMillan wrote back to Grainger:

I suppose you know before this that I turned down Price's offer because it was no advance on what I can get here. I did not know at the time that there was to be no provincial forester or I might have regarded the matter a little different. I was under the impression that it was some kind of secondary job.[18]

The next day he received another letter from Price, asking him to reconsider his rejection of the BC offer, suggesting that any reasonable salary request would be accepted. MacMillan sent another memo to Campbell, stating that even with the proposed raise, he was still underpaid and that if he accepted a civil service appointment it would hinder his promotion in the Forest Branch.

The issue was still unresolved near the end of March, when MacMillan went on an inspection tour of the prairie forest reserves. However, the political situation in the Forestry Branch and the relationship between the director and his assistant were deteriorating. On April 20, MacMillan wrote to Campbell to say he was encountering difficulties in the exercise of his duties. Four days later Campbell wrote back, his annoyance showing:

I do not know why you should have any feeling that you have not the confidence of the Department, because you certainly would not have been sent out to do the work that you have been doing unless the Department had confidence in you. Of course there are matters in connection with which I may feel that I have not the confidence of the Department myself at the present time, and it makes the situation rather a difficult one. When the question arises in regard to anything in connection with which I am not sure of having the full jurisdiction entirely in my own hands I cannot always give you full power to go ahead, however much I might like to do so.[19]

From the field, MacMillan continued his negotiations with Price, a process made difficult by the fact that he was constantly on the move. Grainger maintained a stream of telegrams, attempting to persuade MacMillan to accept the job in BC. His request for a salary of $3,000 a year was accepted, and the only remaining question was when he could begin work.

MacMillan's uncertainty and confusion in this situation were not typical of him. In fact, one of his distinguishing traits was the ability to think and act decisively. It may be that he was genuinely confused, especially when letters began crossing in the mail. He may have been trying to play Price off against Campbell in hopes of increasing his salary, although that was not his style. He may have been seeking a stronger commitment to his appointment as chief forester. But it is also possible he was not his usual self because he was about to become a father.

Edna was pregnant and due to deliver their first baby in June. MacMillan was reluctant to take on the new job until she was able to accompany him west, while Ross was insisting that work begin soon on organizing the Forest Service, and that summer work get under way before the 1912 season was lost. Finally MacMillan and Ross agreed on a starting date, and the minister wired MacMillan an improved offer.

> Willing to appoint you acting forester in charge forest branch provided you can assume duties within thirty days. Salary three thousand a year provided your work is satisfactory your designation will be changed to chief forester expiration six months must have your answer by wire at once.[20]

The same day, Grainger wired MacMillan in Ottawa: "Minister has wired you here is your final chance don't muff it."[21] He followed up with another telegram the next day, urging MacMillan to "cut period of delay down to absolute essential minimum regarding your wife. Don't prejudice matter by extending period by a single day for office consideration."[22] MacMillan wired his acceptance on May 26, and the next day he submitted his written resignation to Campbell. That same day, he got yet another telegram from Grainger, which ended, "Bully for you welcome from all the boys."[23]

He was on his way west.

Chief forester

W hen H.R. MacMillan became the province's first chief forester in 1912, British Columbia was in the midst of a decade-long economic boom. The province's population of 400,000 had more than doubled in ten years, and one-quarter of BC residents lived in Vancouver. The expanding economy was built almost exclusively on natural resource exploitation. The long-dominant mining sector shared economic primacy with fishing and lumbering, and all three industries developed an enormous appetite for workers as markets in Canada and the rest of the world continued to grow.

Overheated real estate markets in Victoria, the Lower Mainland and the Okanagan fuelled a construction boom that could not keep up with the demand for commercial and residential buildings, as

MacMillan discovered when he inquired about accommodation. Grainger wrote back:

> Re your arrival here, perhaps Mrs. MacMillan would like to write to my wife regarding temporary quarters—it isn't altogether an easy matter in Victoria in summer owing to the heavy tourist travel and the boom. . . . The existence of a baby makes things very hard as far as boarding houses go.[1]

Since 1910, BC had been the only province in Canada to show a budgetary surplus, largely as a result of revenues from the new timber leases. The Conservative government of Richard McBride, riding a wave of popularity since its first electoral victory in 1903, claimed full credit for the province's financial buoyancy, and was swept back into power with a huge majority in 1912.

BC forests, in the first decade of the twentieth century, were among the largest and most valuable unexploited forests in the world. Although a forest industry had existed in the province since the 1860s, it was still small and undeveloped. BC was located a great distance from the large overseas export markets, and most of the accessible US markets in California and the southwest were serviced from Washington and Oregon. But the standing timber reserves were enormous, making up about half of the total Canadian timber supply. As MacMillan wrote in his first annual report as chief forester:

> British Columbia contains one of the few great bodies of commercial timber left in the world which are not yet materially reduced by destructive lumbering. With the possible exception of Siberia, Brazil and the North-western United States, the timber wealth of British Columbia is unparalleled in any other country.[2]

The job of overseeing this vast forest wealth was undoubtedly the most prestigious that existed for a professional forester when MacMillan accepted it. But he had little time to savour his good fortune. The daunting task of bringing this huge, diverse area under the type of scientific management envisioned by Fernow and his protégés required immediate attention. A ready-made model for the organization of the BC Forest Service existed in the US Forest Service; the task at hand, even before MacMillan left Ottawa, was to adapt that model to British Columbia. The first hurdle was to assemble a staff, and the major difficulty in this regard was to find professional foresters to head all of the eleven forest districts he planned to establish. By the time MacMillan came on board, several other staff members had been appointed, most of them from the federal lands branch. Grainger was placed in charge of records and John Lafon, an American, was appointed chief of management.

Early in June, when Edna was in the final stages of pregnancy, MacMillan was engaged in an extensive and far-flung correspondence in an effort to find staff. On June 3 he wrote to Fernow, telling him of his decision to accept the job and asking about foresters. Fernow responded quickly with congratulations, but little in the way of useful information. "All our men that graduated this year have found at last permanent employment," he wrote. "I therefore do not know of any foresters that are 'lying around loose.'"[3] MacMillan wrote back a week later to say he was

> extremely anxious to find 8 or 10 Canadian trained foresters for permanent reconaissance and administrative work at about $1400 or $1500 to start. I am unfortunate in appearing so late in the field and will have difficulty in getting men. Competition for this Forestry Branch and the C.P.R. will be strong. I shall certainly be early in the field for your men next year.[4]

Two days later he got a letter from Roland Craig in Vancouver, with some cautions:

> As might be expected there is considerable soreness about the appointment of so many Yankees and they would be wise to keep quiet and saw wood for a while. I believe Lafon advocates employing only Oregon loggers and cutting down wages and has been expressing his opinions of B.C. lumbermen rather freely. He will no doubt get wise soon however. There is good work to be done and I think you will have good government support if you do not hurt any of the political supporters of the govt.[5]

MacMillan engaged in almost daily correspondence with Grainger in Victoria and Overton Price, who shuttled between Victoria, Washington and Ottawa, in a frantic attempt to assemble enough staff to begin field work before the season was over. Grainger's advice was usually blunt and to the point:

> I hope to goodness the reconaissance men will arrive on the very earliest date possible, otherwise we shall have the present field season a wreck, and then will come the winter, and all these men on our hands, giving a not very favourable impression of effective activity.[6]

Price visited MacMillan in Ottawa and informed him he had authority to hire immediately seventeen foresters at $2000 a year and fifteen at $1500. He could find few, he informed Grainger.

There are three organizations seeking foresters, B.C., the Dominion and the C.P.R. Every man in Canada now has a job at $1200 with expenses or better. There will only be 3 good men graduate in Canada next year. With our fairly big programme of work it is I believe absolutely essential that we secure several men at once. Developments are very rapid now. For instance, R.H. Campbell hired today at $1600 J.D. Gilmour of the C.P.R, whom I figured on getting at $1500.[7]

One by one MacMillan tracked them down. He applied the persuasive powers for which he was already well known and the financial resources of the provincial treasury to persuade more than half of the country's qualified foresters to throw over their present jobs and come to BC. He enticed Gilmour away from the federal government for $1,900 and hired him as district forester in Cranbrook, beginning a lifelong association with one of the country's most brilliant foresters. Among the others he lured to the West Coast in his first wave of appointments was P.Z. Caverhill, a University of New Brunswick graduate working for Ottawa, who became Kamloops district forester and BC chief forester eight years later. F.W. Beard was hired away from the federal forest branch at Dauphin and placed in charge of cruising timber sales. W.J. VanDusen, whom MacMillan had persuaded to switch to forestry at the University of Toronto, was enticed from his federal job in Pincher Creek.

HR called me up in the fall of 1912 and wanted to know if I would go over and see him. He wanted me to go to Nelson. I said no, but I would go to Victoria. I agreed and resigned from the Dominion service . . . when I got to Victoria, I never was District Forester. HR appointed me assistant to the Chief of Management, John Lafon.[8]

In the midst of this frantic activity, on June 16, Edna gave birth to their first daughter, Marion, in Ottawa. Three days later MacMillan informed Grainger he was leaving for Washington to see Price for a few days, and that he would arrive in Victoria in mid-July.

Leaving Edna and their new daughter in Ottawa, with plans to join him in the late summer, he went by train to the West Coast. On July 14, MacMillan, who at age twenty-six was one of the BC Forest Service's youngest employees, assumed the office of chief forester. The local newspapers published lengthy articles about him when he arrived in Victoria, and he told a *Daily Colonist* reporter that "wisely and properly handled, British Columbia can grow timber and ship the products of the forest in perpetuity."[9]

The Forest Service had been occupying office space in an old wooden building behind Victoria's legislative buildings, and when MacMillan arrived the place was a scene of agitated activity. Grainger, a more than slightly eccentric English mathematician, was ten years older than his new boss and his precise opposite in character—talkative, sociable, slightly disorganized and, when the occasion demanded, frenetically active. He had an irreverent sense of humour and dedicated his novel about coastal loggers "To my creditors, affectionately." Tall and thin, Grainger looked positively skinny beside the more heavily built MacMillan. He had chronic foot problems and always wore beaded moccasins, even with a tuxedo. But in spite of these differences—or perhaps because of them—the two men formed a close personal and working relationship that lasted until Grainger died.

At first they shared a cramped office with two stenographers and a fat young red-headed Scotsman who delivered messages. MacMillan, who hated the sound of pounding typewriters, found a tiny cubbyhole down the hall. "He was very serious, very businesslike," one of the stenographers later recalled.

Chief Forester MacMillan near Kamloops, BC, August 1912.

"His eye-brows! I used to sit and look at him, and I thought 'Well, I've never seen a man like you before.'"[10]

The immediate task facing the new chief forester was to get his staff into the field and produce some tangible results during the summer and fall of 1912. In a letter from Toronto, Fernow cautioned against "driving too hard and undertaking too much at once." MacMillan dismissed this go-slow approach.

> Just at present the Government is inclined to push the forest policy here strongly, to agree to the appointment of a large number of new men, to give them a free hand in administrative work, to give them great responsibility, and, so far as I can see, a salary commensurate with that responsibility. We are the only Forest Branch in Canada having actual control of all timber business of one province, or jurisdiction. For this reason we shall require a large number of foresters of administrative ability, and I think will be able to pay a large number of men more highly than any other organization. . . . Each of our present districts is as large as the whole forest reserve area of the Dominion Government. . . . The Minister has agreed that all ranger appointments shall be taken out of politics, and that only those men shall be appointed as rangers who have passed a civil service examination set by the Forest Board.[11]

Top priority was given to the reorganization of the old fire protection force that had operated under the Lands Department. As the district foresters came on board, MacMillan began to plan for a province-wide survey of the forests to determine more accurately the size and nature of the resource. Over the summer his growing staff assumed the other duties of the old forest branch—issuing and administering timber licences and leases, collecting royalties and rents, scaling timber, examining preemption applications and thousands of other details.

By fall the operation was up and running. Edna had arrived with baby Marion and MacMillan's mother, Joanna, who at her son's insistence had agreed to accompany the family to Victoria. They moved into a rented house on Moss Street, near Beacon Hill Park, within walking distance of the office and in the neighbourhood where most other senior Forest Service staff lived.

MacMillan devoted the winter months to building and organizing the administrative apparatus of the Forest Service—the bureaucracy. In his first annual report, submitted at the end of the year, he announced that the Forest Service was "now organized and at work." In the first few paragraphs, he stated in precise terms the direction in which he intended to proceed.

> British Columbia contains not less than one hundred million acres of forest land,

and the total stand of commercial timber is certainly not less than three hundred billion feet, and probably much more. At present the lumber output of British Columbia is about a billion and a quarter feet a year. At the present rate of cutting, making no allowance at all for annual growth, it would take the lumber industry of British Columbia nearly 250 years to use up merely the mature timber now standing. The annual growth of the forests of British Columbia is even now, before they are either adequately protected from fire or from waste, certainly not less than five times the present annual lumber cut. . . .

What is not cut is wasted in the end. It is not merely advisable to encourage the growth of our lumber industry until it equals the production of our forests—it is our clear duty to do so, in order that timber which otherwise will soon rot on the ground may furnish the basis for industry, for reasonable profits to operator and Government, for home-building, and, in the last analysis, for the growth of British Columbia.[12]

This statement of intent, coming as it did from the chief forester and not the minister or the government, had a profound effect on the evolution of the Forest Service for the next six or seven decades. It provided a clear and concise rationale for the development of a provincial forest administrative agency, devoted to facilitating the growth of a forest industry that would be the driving force behind the provincial economy for the rest of the twentieth century.

MacMillan's primary objective was to bring BC's forests under the kind of management systems advocated by Fernow, Pinchot and their conservationist-minded professional colleagues. The means to this end, he believed, was the provincial Forest Service, which could acquire the authority and funds it needed only if the forest industry stayed at the centre of the provincial economy. This interpretation of the role of the Forest Service was a departure from what MacMillan had learned at Yale, and it reflected an attempt to resolve the contradiction between conservationist and industrialist ethics in a developing economy—a contradiction he would live with for the rest of his life. MacMillan's position distanced him from those of his colleagues—including Aldo Leopold, his former Yale classmate—who would formulate a preservationist argument against unrestricted industrial use of forests. Seventy-five years later, the BC Forest Service would come under heavy criticism for maintaining too close and supportive a relationship with the timber industry, but 1912 was a different era, in which such a role was seen as enlightened and proper.

Early in 1913 MacMillan called a conference of his district foresters

BC Forest Branch Personnel, April 1, 1913. Front row, left to right: Percy Le Mare (District Forester Lillooet), R.E. Benedict (HQ), A.M.O. Gold (Cruiser), J.R. Martin (DF, Nelson), Walter L. Loveland (VI Ranger), M.A. Grainger (Assistant Chief Forester), S.W. Barclay (Royalty Inspector), P.Z. Caverhill (DF, Kamloops), J.G. Sutherland (Cruiser), R.L. Campbell (Librarian & Publicity), Thomas Mouat (Acting Superintendent of Scalers), H.R. MacMillan, J.D. Gilmour (DF, Cranbrook), H.R. Christie (HQ), A.K. Shives (Forestry Assistant), H.G. Marvin (DF, Ft. George). Back row, left to right: W.C. Gladwin (HQ, Records), H.K. Robinson (DF, Vancouver Island and Surveys), W.J. VanDusen (Assistant Chief, Management), Jack Stilwell (VI Patrols), L.R. Andrews (DF, Vernon), P.S. Bonney (Forestry Assistant), H.G. Kinghorn (Forestry Assistant), C. MacFayden (DF, Tête Jaune Cache), H.S. (Buck) Irwin (Acting DF, Prince Rupert), H.B. Murray (Forestry Assistant, Kamloops), G.H. Prince (Forestry Assistant).

and their senior staff in Victoria, where he elaborated on the timber-use theme. By this time he was able to announce a few key decisions. One was that a survey of BC's forests would be conducted by the federal Conservation Commission, under the direction of the famous American botanist H.N. Whitford and the ubiquitous Roland Craig. MacMillan also introduced a new task for the Forest Service: promoting and developing a timber export market. This latest decision revealed a characteristic of MacMillan which many believe to have been the key to his subsequent success—his ability to foresee events far in advance of those around him. Although it took him another three years to develop a full understanding of global timber markets and the potential for the British Columbia lumber industry, he had enough of a grasp of the situation in early 1913 to devote some of the scarce resources of the fledgling Forest Service to their development.

He was also occupied on other fronts. An active member and director of the Canadian Society of Forest Engineers, he was the prime mover in the formation of the British Columbia Forest Club, which elected Craig as its first president. The April issue of the *Yale Forest School News* carried a brief description of the new BC Forest Service, written by MacMillan, which ended with a revealing sentence. "In order that the rangers and foresters may be secured it is the intention of the Government to establish a forestry school in connection with the new provincial university."[13] At this point the University of British Columbia existed only on paper and its first president, Frank Wesbrook, had been appointed only two months earlier. He and MacMillan became acquainted through Leonard Klinck, who came from Guelph to UBC as dean of agriculture. MacMillan, not the type to wait for matters to take their own course, had convinced the government—or, at least, Lands Minister Ross—that the university should have a forestry school. Fernow's evangelical forestry had gained one of its strongest adherents in MacMillan, and he was doing his bit for the cause in British Columbia, laying the foundations for the profession to be expanded.

MacMillan also had a personal interest in the proposed forestry school. From the outset, Wesbrook prevailed upon him to leave the Forest Service and become dean of forestry when the school opened. Throughout his tenure as chief forester, MacMillan actively considered the offer and discussed it extensively with Wesbrook.

The chief forester spent much of 1913 in the field, taking a firsthand look at the province's forests and making inspection tours of district offices. Wherever he went he took his camera, compiling a photographic legacy of BC's forests, its logging operations and sawmills. He was still careful with his health, retiring early, eating well and getting lots of fresh air. A habit recalled by many who travelled with him along the coast on Forest Service launches was that MacMillan never slept below in the cramped, stuffy cabins, but laid out his sleeping bag on deck or in the wheelhouse. On one of these trips he dealt with a request, passed on by the premier's office, from the director of Kew Gardens in London, for a Douglas fir flagpole.

> I left the Forestry Branch boat in Jervis Inlet, climbed the hill, crossed the ridge, walked down to old John O'Brien's camp (Brooks, Scanlon & O'Brien) on Gordon Pasha Lake—not a tree had been cut all the way—very little logging had been done anywhere in this region. . . .

I saw Mr. John O'Brien Sr., who was then running that operation. I asked him to get out the tallest, most perfect flag pole that he could produce. By the time the flag pole was delivered in Vancouver, World War I had broken out and the pole lay in Vancouver for many months before a ship arrived with space long enough to take it—even that ship was too short and several feet had to be cut off the pole. Mr. O'Brien charged me $3,000 for it, and I was severely reprimanded for having paid so much money for one tree.[14]

By the end of 1913 MacMillan was fully convinced that the future of forestry and the Forest Service depended on the growth of the forest industry. Although public revenues from the forests had once again broken all records, and the forest industry had become by far the largest economic sector of the provincial economy, he realized that still more growth was needed for the government to maintain its support of forestry and allow the chief forester to execute his plans. In his second report to the Legislature, he elaborated on the idea he had raised at his first staff meeting, and assigned a further task to the Forest Service: "The study of ways in which the sale of our timber products may be pushed more actively in foreign markets is therefore a prime duty of the Provincial Forest Service."[15]

Ironically, most BC sawmills were not interested in exporting lumber in 1913. They had grown fat and complacent serving the domestic market that flourished with the settlement of the prairie provinces during the first decade of the century and had all but turned their backs on overseas markets. The majority of BC's sawmills had been founded during the late 1800s on lumber exporting, particularly to nations in the British Empire. In 1900, for instance, more than 44 percent of the lumber shipped from the Pacific Northwest to British markets was from BC; in 1913 BC's share had dropped to 10 percent. What MacMillan foresaw, and what few mill owners realized, was that little further growth was possible in the domestic market; if the industry was going to thrive, it would have to sell its products abroad.

When VanDusen came on staff in January 1913, it became his responsibility to develop the export market. He spent most of the year visiting the mills and getting to know the lumber business.

All the mills were serving the prairie market and the Canadian market, as well as the local market here. The people who were servicing the foreign market at that time included Hastings sawmills, Chemainus, a little bit from Genoa Bay. And that was about it. I remember one of my assignments, shortly after I got to Victoria, was to come over to Vancouver and stimulate some interest in developing the markets to Australia and other Commonwealth areas. I didn't get much interest, as a matter of fact, at that time.[16]

Understandably, the established mill owners, veterans of the continental timber business, were not impressed by the arguments of a youthful chief forester and his twenty-four-year-old subordinate.

The first months of 1914 were uneasy ones for British Columbia. Although the opening of the Panama Canal in June offered the promise of economic expansion, other factors offset the potential benefits. The growing rift between Britain and Germany affected the money markets during that year, and the British investment which had underwritten so much of BC's past economic growth began to dry up. On the home front, the McBride government, less than two years into its strongest mandate ever, plummeted in popularity. Public confidence in the government was shaken further with the collapse of the Dominion Trust Company and the discovery that its depositors were not protected by the government. Issues of race and immigration policy headed the list of public discontents, culminating in the July expulsion of a shipload of 376 Sikhs who intended to immigrate to Canada. Underlying these events was a growing labour unrest that had been building over the past three years. Large parts of the provincial economy were strangled by strikes, and the city of Nanaimo was under martial law. Then, in August, war broke out in Europe and everything changed in British Columbia.

The initial reaction of many West Coast residents was to sign up for service overseas. Within days of the declaration of war, 3,000 BC militia members headed east by train, bound for Europe. A fifth of the students registered for the first year of classes at the University of British Columbia went off to war instead. And a growing number of MacMillan's hard-found staff were drawn into the fray, depleting the ranks of the Forest Service and putting most of his plans on hold. To make matters worse, the lumber industry came to a virtual standstill as domestic markets sagged and what was left of the export market collapsed. US ship owners, who dominated the cargo trade, feared their vessels would be sunk by German warships cruising the Pacific if they carried British cargo, so they began withdrawing service to BC or raising rates to unaffordable levels.

In the face of these setbacks, MacMillan persevered with the job of building the Forest Service. Over the summer, Klinck and Wesbrook accompanied him on a forestry boat up the inner coast to Stuart Island, and in

October the three of them went on a lengthy inspection of the university's endowment lands in the Cariboo. They worked out of the Gang Ranch and travelled for several days by horseback, camping out at nights. For MacMillan, this trip was a return to the kind of outdoor activity he loved but had not been able to enjoy since he had contracted tuberculosis six years earlier. In the athletic and intellectually inclined Wesbrook, he found a kindred spirit and a close friend. Clearly he had recovered from his illness, as indicated by an entry in Wesbrook's diary: "MacMillan and I went for deer for 2 hours after camping. Rode and walked 20 miles and walked after camping about 6 or 8 more."[17]

That winter MacMillan sent Wesbrook a budget estimate for the proposed forestry school. He calculated it would cost less than $45,000 to operate the school for its first two years.[18] A few weeks later he delivered a speech to the fifth annual meeting of the Conservation Commission in Ottawa. Although his address was titled "Essential Features of a Successful Fire Protection Organization," he used the occasion to express his views on the state of forestry in Canada. He confronted head-on those who were critical of forestry and opposed forest management.

> There are an astonishing number of people in this country who misinterpret the term forestry, and oppose any extension of forest administration because of that misinterpretation. Such people believe that forestry is a conceit of sentimental persons who desire to protect woodlands, to prohibit the cutting of trees. . . . Or they may believe sincerely that forestry involves the expenditure of a great deal of Government money in the planting or cultivation of forests, expenditure which could never be repaid either as to capital or interest. Still another misconception of forestry exists in the minds of many persons connected with the lumber industry, who believe that a forest department would import from Europe or elsewhere absurd regulations requiring the planting of a tree to replace every one cut, or of hedging logging operations about with killing restrictions. Such conceptions are of course founded on misinformation. They are nevertheless prevalent, and are responsible for the fact that Canada is now, of all the countries dependent largely upon forest industries, doing the least for the protection of the timber-lands.

He went on at greater length to explain what was needed.

> The future of this country depends upon our making every acre productive. Broadly speaking, the earth's surface can be made productive in two ways only, by producing agricultural or timber crops. . . . Optimistic as we have been in this country we seem to have been unable to see any value worth caring for in our

non-agricultural lands. . . . We must realize that our present stands of merchantable timber cannot support our growing industries indefinitely. . . . The future forest industries . . . must be supported by the timber grown on the logged-over and burned-over non-agricultural lands. Looking at these lands we should see, not wastes, holding no promise for the future, but productive lands, needing only protection from fire to enable them to support logging camps, pulp mills, rural and industrial communities of a type which has done much for Canada.[19]

For perhaps the first time, MacMillan had articulated a mutually beneficial relationship between conservationists and the lumber industry. His speech was the high point of the meeting and did much to enhance his reputation outside the profession. The first on his feet to respond was Senator William Edwards, a prominent Ontario lumberman, politician and bank director. "I want to give him a word of advice," said Edwards. "Let him not go into politics, because if he does he will become Premier of Canada [sic]."

On March 3, 1915, Lands Minister Ross stood up in the Legislature to speak to proposed amendments to the land act. During the course of his speech, he reviewed the dismal state of the lumber industry, using essentially the same arguments MacMillan had advanced earlier. He then announced that the government intended to give some practical assistance to the marketing of timber abroad. To this end, Chief Forester MacMillan had been appointed a special commissioner by the federal minister of trade and commerce, and would soon be leaving on a world tour of potential markets. A March 17 meeting of the federal treasury board instructed MacMillan to "look into the requirements and possibilities of markets for Canadian lumber in China, India, Australasia, South America and certain European countries." He was to be paid $5,220 a year, his current salary as chief forester, and all expenses for the anticipated six- or seven-month trip.

Ross's announcement generated enormous excitement in BC. All the provincial dailies carried large front-page stories. Interest increased even more the following day when the San Francisco shipping companies controlling the transport of West Coast lumber announced a 50-percent increase in shipping rates between BC and Australia.

Until MacMillan left Vancouver four weeks later, the newspaper headlines expressed the hopes of the BC timber industry: "MacMillan Ready to Leave For Australia," "Chief Forester to Circle the Globe," "MacMillan Goes to the War," "Trade Commissioner to Leave Shortly." Plans changed repeatedly as MacMillan rushed back and forth between Victoria, Vancouver and Ottawa.

Initially, he was booked on a Japanese ship with plans to visit China, Japan and Australia before travelling on to India, Africa, Europe and South America. The built-up hopes, even before he left, were enormous. Apparently he was expected to single-handedly solve the timber marketing problem and restore prosperity to the province.

It is difficult, eighty years after the fact, to comprehend why politicians of all parties, seasoned lumbermen, the media and much of the general public were so excited about the appointment of a relatively inexperienced civil servant, not yet thirty years old, to such a nebulous mission. Partly it had to do with circumstances. The coastal lumber industry had demonstrated since 1905 that it could provide enormous economic benefits to the province, in the form of both jobs and government revenues. Combined with sawmills in the interior of BC, the industry could produce far more lumber than prairie markets could consume. At the same time, partially through neglect and to a great extent because lumber exports were controlled by US brokers, BC's share of the Pacific Northwestern export trade had declined substantially. The advent of war exacerbated these conditions, and by 1915 the enthusiasm and optimism surrounding the province's decade-long industrial expansion had withered. The boom had bust. People were desperate for a solution.

It is perhaps a vestige of Canada's colonial history that when something is amiss in the country, when a major problem arises, many people, including some of the most dedicated advocates of free-market economics, turn to some form of authority for help. With the Crown reduced to a symbolic entity and the Colonial Office defunct, Canadians habitually turn to their governments, or to someone sanctified by government—in this case, the chief forester—to make things right. They sometimes display a pathetic eagerness to believe in the power and efficacy of state-sanctioned authority, rather than in their own and their fellow citizens' capabilities. MacMillan had, in less than three years as the province's chief forester, established his reputation as a competent, personable, dedicated public servant. His predictions about the export market and his suggestion that the government, through the Forest Service, become involved in timber marketing, had fallen on deaf ears two years earlier. Now, his insights were endorsed with enthusiasm and his appointment greeted with jubilation. Just how realistic these expectations were remained to be seen.

At the last moment, when the war needs of the United Kingdom and its allies became more pressing, it was decided MacMillan should travel to Britain first. On April 7, with the entire Forest Service staff and scores of reporters and politicians standing by to see them off, he and Edna boarded a ship in Victoria's inner harbour, bound for Seattle and a trans-continental train east.

chapter 6

Around the world

The MacMillans began their 1915 trip by travelling to Chicago, then north for a brief visit to Aurora. Following meetings in Ottawa and Montreal, they went to New York to await the departure of their ship, the *Lusitania*. After two days of sightseeing, MacMillan succeeded in obtaining passage on another ship, the *Orduna*, which was scheduled to depart earlier. They arrived in Liverpool a few days before a German submarine torpedoed the *Lusitania*, killing 1,200 people.

In London, the MacMillans connected with Premier McBride, who was there on a visit. While MacMillan attended meetings, Edna, who was once again pregnant, rested. Between business engagements they were tourists, visiting museums, art galleries, theatres and other attractions. Life in London during World War One was little changed from peace time. Apart from one mention in her diary of a Zeppelin bombing raid that killed four people, Edna recorded little evidence of war. The cultural and social life was as active as in any city out of the combat zone. As MacMillan's circle of acquaintances widened, they were invited to dinners, lunches and other social events.

Within two weeks of his arrival, MacMillan closed his first deal: the British Admiralty ordered 10 million board feet of BC lumber. The Vancouver and Victoria papers were jubilant. This single order was equivalent to half the total shipments to British Empire markets during the previous year. But even more significant than the size of the order was the way it was made: this was the first time the British government had placed a direct order for timber with the BC industry. Previously, orders had come through agents in the US who, because they controlled shipping, were able to direct most of the business to affiliated American sawmills. In this case, MacMillan and McBride convinced the Admiralty that British vessels delivering coal to British ports on the Pacific could take lumber from BC as return cargo.[1]

For two months the MacMillans maintained their busy social and business schedule, including brief visits to Scotland and Ireland. In mid-July Edna decided not to continue the trip with her husband and, plagued by headaches, returned by ship to Montreal. En route, she learned the *Orduna* had been torpedoed.

In London, MacMillan continued his rounds of meetings for another month. Occasionally he encountered members of the Forest Service who had joined the army and were on their way to the front, but he was lonely without Edna, as his letters to her conveyed.

> The buses run as ever, and it seems the same city, only my girlie has gone and left me alone in it. It is quite different coming back at night to the hotel and an empty room, to what it was when you were here and always had something interesting when I came back at night.[2]

He buried himself in his work, and by the end of July he had secured orders for an additional 20 million feet of BC lumber, much of it for railway ties and sleepers. Already he had vastly exceeded the wildest expectations of his mission. In BC, these large orders created a novel problem for the industry. VanDusen had been assigned the task of handling them on the home front.

> There was no way, no facility, for buying, no brokers here. Nobody could do it. So the companies went to the Forest Service, the government, and asked if we could

do it. We worked it around and said buyers didn't want to come from England to go out and talk to a whole bunch of small sawmills all over the country. And that was one of the first things that came to our department. We had to try to work out some kind of an organization where we could take a requisition from the timber people in England and place it with all the small mills. We couldn't go into the business because we were a government department. I forget just how we did it, but we did it anyway.[3]

At the end of July, MacMillan travelled to neutral Holland for a week, writing to Edna from the S.S. *Prinses Juliana*: "There appear to be very few English speaking on the boat and it is crowded with Dutch and with Germans who are being deported to their dear Fatherland."[4]

A few days later he wrote again, telling her of his visit to a popular swimming beach in Holland.

The water was full of great fat German women and men, all in tights, very edifying to view—disporting themselves like a lot of savages. I was wondering how mother would have enjoyed herself if she had been there. This hotel is also full of Germans, very sour and fat, who glare at me and Gott strafe me I suppose mentally when I speak English. They certainly are an amiable crowd. I saw the Queen this morning reviewing a portion of the Dutch army—she is fat and jolly.[5]

Two weeks later he wrote from Paris. His departure for South Africa had been delayed for a week when an opportunity arose to go to France. The war was not going well for the French, with the northern and eastern portions of the country occupied by the Germans. MacMillan spent a few days in Paris, which he found depressing. The sight of wounded and crippled soldiers and the large number of military hospitals impressed upon him the reality of the war. He managed to get out of Paris to inspect some forests on the Swiss border, and in spite of the unfavourable circumstances of his visit, he was enormously impressed by how the French managed their forests. He saw that many of their methods could be applied in Canada.

The people by using a forest wisely have acquired a forest imagination. The reverse is the case in Canada, where, especially in Eastern Canada, the population is concentrated in deforested areas. It seems hopeless to expect people so situated, without any living demonstrations of the profits of wise forest administration before their eyes, to insist upon the proper care and protection for vast public forest areas they have never seen. . . . Forest management in France supports as many families per square mile as agriculture in many parts of Canada. . . . A short visit to a few French forests convinces me that there are now certain regions in Canada where economic conditions are quite as favourable to intensive forest management as in many of the profitable French forests. The real obstacle in Canada is to be found in the public, not in the forests.[6]

Back in London, he felt much closer to the war. At one point he encountered a member of the Forest Service on his way to the front and the two men dined together. MacMillan then attempted to sign up for military service, but was rejected because of his bout with tuberculosis. His dinner guest went on to die at the front.

At the end of August, MacMillan sailed on a freighter, the *Norman*, bound for Cape Town. The three-week voyage was a long, tedious trip, broken by a week of rough weather during which he observed his thirtieth birthday. MacMillan spent most of his time reading, a habit he had acquired in the sanatorium, and one which was of great comfort for the rest of his life. What depressed him most was the knowledge that because of wartime conditions, his letters took six to eight weeks to reach Edna.

After arriving in South Africa, MacMillan threw himself into an extensive tour of the country. His primary purpose was to develop timber markets, but he also wanted to inspect forestry operations. This required much travel on close, dusty, uncomfortable trains. As the due date for their second child approached—"Alphonse" was the name MacMillan used in his letters and diary for the as-yet unborn infant—he became increasingly anxious, and frustrated that he could not communicate directly with Edna. With many hours to himself, riding on trains and sitting in hotel rooms, he sent a blizzard of letters to Victoria, some of them descriptions of his travels, others simply straightforward expressions of his love. In a letter written in late October, he mentioned meeting Generals Smuts and Botha—"won't speak to ordinary people when I get back," he added. He also told her of a job offer he had received by cable from London. "I cabled today I would [accept] if they would give me 5 months to do India, China, Australia on my way home. If they do I'll be in London before you get this and of course you'll know it."[7]

The following day, while eating dinner at the Grand Hotel in Pretoria, he was handed a telegram stating simply: "daughter well." In Victoria, Edna wrote in her diary for October 26: "Baby Jean arrives 2:45. Fine and fat, weighs 8 lbs. Has got black hair."[8] MacMillan, in Johannesburg, was fretting because he had not received a letter from home in almost four weeks. But with word of Jean's birth and the knowledge he was soon to leave the country, his spirits improved. The job in London did not materialize and, about to depart for Bombay, he felt he was on his way home.

One thing MacMillan's gruelling itinerary proved was that he had fully recovered from tuberculosis. He put in long days and sat up at night in hotel rooms, writing reports to Ottawa and articles on his travels for various trade journals. In a letter to Joanna from Johannesburg, he reassured her: "I feel very well indeed—never better—but I have to keep working to keep from becoming lonesome—it is a deuce of a job—new people every day."

Eventually his perseverance paid off. While waiting for his boat to India, he managed to conclude a major deal for sleepers with one of the South African railway companies, relieving him of his fears he had been wasting his time there. He was in high spirits two weeks later in Ladysmith.

> I think I am just about through with South Africa—24 hours more, that is all. I hope to receive some letters in Durban. The last I have had are about 10 weeks old so there must be some more somewhere. Of course I can't hope to hear anything about yourself and the new MacMillan for some weeks yet. . . . I feel extraordinarily well tho, and am as brown as a walnut—am getting fat too. Best love, Hubby.[9]

But writing from Durban the next day, on the eve of his scheduled departure, he was once again distressed at the war news.

> I am beginning to think I must go to this war. It drags on so we'll all have to help. I have thought about it very often. I often wonder now who else has gone of our friends. You see, my latest news from home is nearly 3 months old. Surely nearly all the unmarried have gone by this time—they must. It will be hard on the Forest Branch, tho'—it can't be helped however. It will be lucky if there is no more trouble here in South Africa. The Boers in some parts are still very rebellious. If everything goes well in Europe they will keep quiet tho.[10]

For more than three weeks his ship crawled up the east coast of Africa, stopping at several ports where MacMillan went ashore. He sent Edna long, detailed descriptions of the markets and the city streets. The strangeness and vitality of Africa, particularly those parts not dominated by colonial Europeans, he found exciting and stimulating. On board ship, the heat was almost unbearable; there was nothing to do but read and sleep. There was no lounge or sitting room, and the cabins were small with two to three passengers in each. Four hundred Indians crowded the deck. MacMillan passed the time boxing with two Australians.

He reached Bombay December 20 and, discovering that someone he wanted to meet was about to leave for England, made a forty-two-hour train trip 1,300 miles across India to Calcutta. He arrived Christmas Eve, and the following day he wrote Edna a long letter. "The first day or so I was in India I would have been glad under any pretext to get away but now I think, after getting used to the crowds, the dirt and the smells, I shall like it."[11]

That was an understatement; he loved India. The crowded streets and markets intrigued him, as did the masses of people bathing, washing, eating and living along the Ganges. His diary and his letters were full of enthusiastic accounts and vivid descriptions. He scoured the street markets, buying rugs, furniture and fabrics, which he shipped home.

MacMillan was enormously impressed with the Indian Forest Service and its accomplishments. Early in the visit, after a field trip to Assam with a forest economist, he wrote a letter to Wesbrook from Calcutta, describing a trip by dugout canoe and steamer, and commenting on Indian forestry.

> The forest work the British have done in India is wonderful in the courage of its conceptions, the patience and attention it has required through long years of development, and the magnificent results, an enriched country, an increased revenue, are due only to a trained permanent staff working without political interference and the active presence of a research institute (also a forest college) which for 40 years has conducted many valuable investigations. If India with its many other resources needs such a forest service, what does B.C. need. I only hope that soon we can get the government to realize that the forest revenue, which this year is one-half the total revenue of the province, can only be kept anywhere near its present level by expending more of it in the maintenance and protection of the forests. Let us not be citizens of B.C. for today only.[12]

The same day he wrote an equally enthusiastic letter to Edna, assuring her that "I feel as well as can be. Am down to 180 in weight but never was better in my life. Getting lots of air and exercise, 'tho I can't say the food is a constant temptation to over eat."[13]

While MacMillan was developing a dislike of curry and sweltering in the Indian heat, Edna was enduring the Great Snow of 1916 in Victoria. It was the worst winter storm in the history of southern Vancouver Island, snowing almost continuously for forty-seven days. City streets were clogged with deep drifts, and people sat huddled around small fires as fuel supplies dwindled. Edna struggled through the ordeal with the help of Forest Service staff living nearby.

MacMillan spent weeks at a time travelling through rural India by various kinds of boat, on the back of an elephant, on foot and by bicycle. Over the course of the three months he was there, he covered about 15,000 miles. As he had in South Africa, and would later do in Australia, he made speeches to several business organizations, propounding the virtues of Douglas fir and the benefits of purchasing lumber from British Columbia. As was the case at home, his talks received widespread coverage in these nations' newspapers.

In early February, having met with no success in selling BC timber, he went on a bear hunt in Orissa State with the British manager of a local paper mill. He shot two bears on this trip, one of them the "biggest bear shot here in recent years."[14]

After he was in India for a few weeks, MacMillan's mail began to catch up with him, and he had access to papers and other news from home. McBride had been forced to resign in December 1915, and the ineffectual William Bowser was now premier. Letters from Wesbrook indicated the university's future was far from certain. Although Ross had said "one day" he would agree to MacMillan leaving the Forest Service to be dean at UBC, there was no money allocated to establish a forestry school. Other letters tempted MacMillan. One informed him that E.J. Palmer, manager of Victoria Lumber & Manufacturing Co. in Chemainus, BC, was interested in hiring him. Another suggested that Fernow's position as dean at the University of Toronto forestry school was available to him. And, after a brief trip to Burma, he was offered a job there; he did not accept, but he kept the option open. His commitment to British Columbia was tenuous. Clearly, conditions had changed in BC. With McBride no longer serving as premier, even though Ross was still lands minister, the position of chief forester had lost some of its appeal.

MacMillan's relations with McBride had never been close, but there was a mutual respect between them which occasionally was severely tested. When MacMillan began his job he had a brief meeting with McBride, who told him his responsibility was to enforce the Forest Act. Some months later, MacMillan hauled into court a logger who had failed to pay the necessary timber royalties to the government. McBride pointed out to his new chief forester that the logger was a good friend and a supporter of the government. MacMillan reminded the premier of their first meeting. That was the end of the matter, and henceforth McBride found other means to reward his friends.

Although MacMillan's inspection tour of India's forests was filling his head with ideas for BC, he had little success in marketing lumber. His diary reflected his concern: "Perhaps I ask too many questions, but I seem to find scanty welcome in India from most companies, particularly with older men."[15] He wrote again to Wesbrook while travelling back to India from a brief trip to Burma.

> Increasingly I find myself burning to get back home to put at least a few of the new ideas into practice. Not the least of my impressions is strengthening of my conviction that B.C. is very much a forest province, that it is now being run very close to the margin of destruction, especially in the southern Interior, and that a most imperative investment is a forest school.[16]

He spent the last week of March preparing reports and writing an article for Fernow's *Forest Quarterly*, titled "Forestry in India from a Canadian Point of View."

> One should be permitted to dream a moment what would be the situation in North America today if we had possessed only a little of the Indian's characteristic of pausing to make each acre fertile before passing on to denude another. We should have been still somewhere East of the Appalachians and the beaver would not yet have been driven out of Canadian rivers to take refuge in the folds of the flag. . . .
>
> Indian governments differ in two respects from Canadian governments in their attitude towards forest finance, they do not push their yearly financial demands upon the forest beyond the true forest yield, and they devote practically half of each year's revenue to the improvement of the forest.[17]

MacMillan's comments on government expenditures on forestry may have been prompted by reading a draft of the 1915 BC Forest Service annual report, likely prepared by Grainger. The report indicated that because of the war, public revenues from forests had dropped by one-third, or $1 million. Only one-quarter of this revenue was spent on forestry, most of it for administration and protection.

With some sadness, he left India for Ceylon, where a huge pile of mail from Canada awaited him. He immediately wrote a long letter to Edna, lamenting the lengthy reports that were confining him to his hotel. He was not encouraged by news of the Forest Service.

> I don't like to think of what it will be like in Canada now. The Forest Branch is a pretty good indication. It will be a stiff job putting it on its feet again, so many men have gone to the war and the country will be broke.

He was also hungry for news of his children, and in answer to Edna's suggestion he might be disappointed in not having a son, he responded:

> I haven't worried at all about the femaleness of the youngest baby. That's the least of my worries. It can't be helped either before or after and she may turn out a perfect wonder yet. So what is the use. The name sounds very pretty—I like the Jean to call her.[18]

On April 2 he sailed for Australia and, after a restful voyage, arrived in Fremantle ten days later. He disliked the city so moved inland to Perth, where he found mail waiting, including a letter from Lands Minister Ross saying the promised salary increases had been cancelled. "Practically shows me nothing further for me in BC," he noted.[19] After a week of travel, observing Australian forests, logging and sawmilling operations, he concluded his daily diary entry: "Thinking this evening while sitting alone in boarding house of going into lumber business in B.C. if I get chance."[20]

Soon after he arrived in the country, he dispatched another report to the *Yale Forest School News* in which he gave full vent to his feelings about the state of Canadian forestry.

> The attitude of Governments in every British Dominion is more statesmanlike towards Forest administration than it is in Canada, chiefly because timber is high in price, and the people have not been educated in the destruction of it. A forester who has to work in a country where the people were born in log houses in fields surrounded by stump fences, and where the population in the early days always had a "back lot" to clear, and the youngsters had to swing an axe in winter time and pull roots out behind the ploughs in summer time, has a difficult proposition before him.[21]

MacMillan was in Australia for almost four months, with intentions to go to China and Japan before returning to Victoria. For the most part his impressions of the country were not favourable. With a few exceptions, he disliked Australians and was appalled at the way some of them behaved.

> Took logging train in 17 miles . . . very wretched trip. Train crowded with loggers—got off every station bought drinks nearly all drunk before noon—almost impossible to get anything to eat all day—no facilities except at counters where tea and sandwiches sold. This is State railroad—stations sell booze—men take it into trains—ride in 1st class carriages on 2nd class tickets—give guards drink—are generally a hard lot with blind trust in unionism to keep wages up, retain them their jobs.[22]

He found the Australian mills inefficient and poorly run. The country's forests, he noted, were being destroyed. "Forest Branch, according to lumber journals, has made big hit with lumbermen but very little forestry I fear."[23]

The news from home that continued to trickle in was mostly bad. In BC, the Bowser government was in trouble, and the feeling was that even its scant concern with forestry would not be matched by a new government. The Forest Service was continuing to lose men to the war in Europe.

MacMillan dutifully carried on, meeting political and industrial leaders in spite of their indifference to BC timber. He was, however, making valuable contacts and acquiring a detailed knowledge of the Australian timber business. For example, he established a close relationship with Lane Poole, western Australia's chief forester, who remained a friend for life. In late May he received a cable from Ottawa, instructing him to finish up in Australia quickly and return home, forgoing the planned trip to China and Japan.

In mid-June a cold he had been unable to shake for a couple of weeks worsened and he came down with a fever, which forced him to spend ten days in bed resting, eating and reading. After he recovered he attempted to find passage home but was unsuccessful for several weeks, during which he continued to meet with government officials and business leaders. In July he received a long, rambling letter from Wesbrook, full of news about the university, and with several references to MacMillan's future. At this time Wesbrook considered it almost a certainty that MacMillan would join him at UBC.

> I had not known how long you would be away and in view of the fact that inadvertently our estimates were cut, we have not included anything in the budget for forestry this year. I had supposed that you would want a number of months after your return to straighten matters out with the Dominion Government and also with the Provincial Government and I think you will find that most of this college year will be used up in that way . . . .
>
> In case you could be available for University purposes earlier, I am sure that some how or other we could make the adjustment and find the money. . . . I do hope that you will finally decide that you can do more for this Province and country with us than with the Forest Branch. . . . I do not wish to hurry you to any decision. I would not want you with us if you felt that you had made a wrong decision. I know that you will be helpful to the Province and to us wherever you are. We should all like the inspiration of your close presence but probably your wisest plan would be to make your decision after you come back and take stock. We want you to keep us and our future constantly in mind and I am sure that you will do that. We have felt a sort of proprietary interest in you wherever you are and at all times.[24]

Finally, in mid-July, MacMillan obtained passage on the *Makura* and arrived in Vancouver July 28. The following day he granted a lengthy interview to the *Daily Colonist*. In his view, the termination of the war would create new markets for lumber and, with the right approach, BC firms could gain a substantial share of that market. But he was quite blunt in assessing the measures that had to be taken to compete in the global marketplace.

> I believe that unless the mills here can get together and by co-operating in the supplying of lumber, by close study of market conditions—the lumber business is a speculative one to a great extent—and the securing of cargo space, we will not be able to compete with the exporters to the south.

There was a desire in British Empire markets to keep business within the Empire, he noted, but only if prices were competitive, consumer specifications were met and the quality was high.

> We here in British Columbia, are wont to believe that British Columbia lumber is the standard of the world, that everywhere this Province's name is known. It will doubtless be a keen disappointment to many to learn that so far as the lumber trade, at least, is concerned, by far the greater portion of our exports . . . were shipped through United States firms, billed as American lumber.[25]

A few days later MacMillan gave a public speech on his findings to a special meeting of the BC Lumber and Shingle Manufacturers Association. His theme was similar, but with much more detail than he had given the *Daily Colonist*. He told the province's lumbermen it was unlikely Europeans would come to the Pacific Northwest for lumber to rebuild after the war.

> To begin with, the cheap lumber which they use, will probably come from the Estate Forests of Europe, most of which will be capable of meeting even this unusual demand. Then these countries will all be very anxious to cut down their imports to the absolute minimum. Large numbers of soldiers and other munition workers who are now withdrawn from active life will be released, thousands of whom were interested in building either with brick or cement and the natural thing to do would be to organize these people when they are released from the war and put them back on the reconstruction of their own works, in the manner to which they have been accustomed and on which they have tremendous quantities of raw material, that is, cement and clay.[26]

He went on to explain that attitudes about the Pacific Northwest's most valuable species, Douglas fir, were poor in the countries he had visited. It could not compete with Scandinavian or Baltic timber, nor with pitch pine.

Apart from educating foreign consumers about the qualities of fir, the other advice he offered was to raise the price. "The profit on the Douglas fir [abroad] is more than the F.O.B price here. They say here is a cheap fir, the foolish people are willing and we may as well make all we can out of it." He urged the lumber producers to set up a co-operative trading organization that could accept orders from foreign buyers, fill them to specification from various BC mills, and deliver them on time, at a price competitive with those offered in the US. In great detail he listed the shortcomings of the BC lumber export business: the lack of an adequate assembly wharf where lumber could be delivered and sorted into individual orders, the dependence on ships owned by US firms with connections to US lumber companies, the lack of coaling facilities in Vancouver and the slowness of local longshoremen, all resulting in higher shipping costs.

From the moment he returned, MacMillan was faced with the choice he had been able to put off while out of the country: where he should work. He had many opportunities, but had to make the difficult choice between the public and private sectors. He had spent all his working life in public service, either federal or provincial; he was extremely sensitive to the public interest and fully cognizant of the need for dedicated public servants. MacMillan had tried to describe his struggle to Wesbrook in a letter he wrote from Australia:

> I think I am perhaps too enthusiastic. I have had moments when, a little discouraged with inefficiency in government administration, have felt like taking the advice of business friends and going into business. But as I have done nothing for the country at war feel I can only be fair to myself if I make sacrifices and serve her in the inevitable troubles of peace. Now that the Forest Branch has suffered so during war I cannot exactly determine where I can do most. I am inclined to think I can do most with you, but if you do not consider it prejudicial to your plans would prefer to look the ground over in B.C. before making definite decision. I fear the Forest Branch is in for some lean years. Many of the men whom I took there have been improperly treated and I don't care to abandon them. Nevertheless, forest administration needs university assistance in B.C.[27]

As well as the offers from UBC and the University of Toronto, at least two American universities wanted him to head their forest schools. Shortly after returning to Canada, Sir George Foster, a leading cabinet minister in Robert Borden's Conservative government, offered him a senior position in the Department of Trade and Commerce.

I was on the verge of accepting—it was quite a promotion for me—but I had the presence of mind to say that I supposed I would have the authority to get rid of any Trade Commissioner whom I could prove was a complete failure, and I mentioned two or three—one in Bristol and one in Manchester. He answered that such action could not be delegated by the Minister. I then thanked him and told him that I thought I would do better in the export lumber business, which I was planning to enter.[28]

He had decided to accept E.J. Palmer's offer of a job at Victoria Lumber & Manufacturing in Chemainus, just south of Nanaimo on Vancouver Island. In later years he was criticized, usually behind his back by business competitors, for going into private business after travelling the world at public expense to study the global lumber business. As his letter to Wesbrook indicates, he was acutely aware of his public duty in this regard. But it is also true that he considered the establishment of a BC-controlled lumber export business to be an urgent public need. A few months earlier he had addressed this issue in another context.

You are sufficiently well acquainted with lumber conditions on the West Coast to know that one of the first necessities of a forest service having its whole interests west of the Rocky Mountains is to create a value for stumpage and a stable timber industry. Unless success can be attained in this direction, it will be increasingly difficult to secure that public backing which is necessary to provide the Forest Service with the necessary funds and the necessary staff, and also—what is probably more important—the necessary influence to enable it to give the Forest country fire protection and such intensity of Forest management as is practicable.[29]

When MacMillan came to BC, he brought with him the most advanced ideas about forest management in North America. Much of the forest industry at that time was barely out of its pre-industrial stage. Steam-powered logging equipment was just coming into use, with the bigger logging camps using ground-lead yarding systems and the first steam railways. Most of the timber used in the mills, however, was still logged with oxen and horses, and the hard physical labour of human beings. In this context, many of his ideas about forest management placed him far ahead of his time, a situation he could not long ignore.

A few weeks in Victoria convinced him that working for the government was not necessarily the most effective means to advance the cause of forestry in BC. By now he knew that working for government could not necessarily be equated with working in the public interest. As well, his travels had given him a new outlook on some of the assumptions about forestry he had acquired from Fernow and his protégés, including those concerning the supposed virtues of public administration.

On September 12, MacMillan sent a coded telegram of resignation to Lands Minister Ross in Prince George.

AM TAKING UP PRIVATE WORK WISH NOW INFORM YOU OF MY RESIGNATION THANK YOU FOR YOUR PERSONAL SUPPORT DURING YEARS DEVELOPMENT OF FOREST BRANCH AND STRONGLY RECOMMEND CONFIRMATION GRAINGER PRESENT APPOINT-MENT AS CHIEF FORESTER DO NOT DESIRE PUBLIC ANNOUNCEMENT MY RESIGNATION BEFORE COMPLETION MY DOMINION WORK NEXT WEEK UNLESS YOU THINK NECES-SARY. MACMILLAN.[30]

The resignation was made public ten days later, the same day MacMillan appeared before a royal commission on trade under Sir George Foster. In his submission he elaborated on the descriptions of the BC lumber export trade he had made elsewhere since his return. He gave a lucid explanation of what had happened to the lumber export market since 1905.

While we were switching from the export trade to the supply of our domestic demands, the sailing vessel passed out of the export trade to a very large extent. When the transportation was furnished by sailing vessels the mill owner was able to become responsible for the chartering of the vessel without seeking the co-operation of any other mill, or without undertaking a contract which one mill might hesitate to fulfil. Sailing vessels only carried from 400 to 1,500 thousand feet, and they didn't load at a rate exceeding 75,000 feet per day; but when the steamer came in it carried one and one-half to four million feet, and loaded at the rate of 400,000 feet per day. The sailing vessel carried a cargo which one merchant in almost any port could buy, and it was possible for the mill owner to charter a vessel and load it without incurring demurrage, and sell it to a purchaser in a foreign country. No middle man was necessary. When the steamer came in, one mill could not charter the vessel. . . . The steamer cargo was too great for one purchaser; it needed distribution. Therefore the middle man came into the business. . . . The result of the transfer of the trade from sail to steam has been even more serious in that it has placed the full control of the overseas lumber export trade in the hands of the large companies operating or engaged in the merchandising business. . . . Most of the business is now in the hands of the [US] exporting companies; those companies grew up at a time when these changes were taking place, and when the Canadian mills were turning their attention to the domestic trade. . . . This is more serious when you remember that the United States has an operating mill capacity on the

Pacific coast now to take care of double the present export trade without any assistance from Canada whatever.[31]

His recommendation was to encourage British investment in the processing and marketing of BC timber, by taking advantage of BC's accessible
timber supplies. His argument was predicated on the assumption that British
Empire countries would consider it in their best interest to give preferential
treatment to BC producers, either through tariffs or other measures.

> We have the greatest quantity of accessible timber to be found in the world, and
> by accessible, I mean, accessible for the purposes of an export trade. . . . It should
> not be too great a scheme for the British Empire to place their resources behind
> the development of this region and as a result transfer most of the present trade
> from the [U.S.] to this one. . . . It is much beyond the resources of the people of
> this Province to do so. We do not have here the capital nor the experience. . . .
> There is particularly the question of transportation and the trading ability. That
> is where all the money is made in this export timber business. . . . Surely if we, as
> a people, possess the resources and furnish the market and supply the money to
> develop the demand for this timber, we should have the genius and the ability to
> build up a British connection between our raw material and our market; that is
> all that is necessary.

Significantly, he no longer proposed that the Forest Service act as broker to the BC lumber industry. Instead, he suggested that British money be
used to help build a commercial brokerage capability to take control of the
business away from the Americans.

The day after his presentation, MacMillan went to work for the
Victoria Lumber & Manufacturing Company in Chemainus—a decision he
would soon regret.

In the private sector

MacMillan and his family, including Joanna, left Victoria and moved to Chemainus, BC, in late September 1916. They acquired one of the town's older houses on the brow of the hill overlooking the sawmill, a few yards from the water wheel which had once powered the mill. They adjusted quickly to the small-town atmosphere of the community. For the three adults, it was reminiscent of their days in York County. Shortly after arriving, MacMillan was elected president of the Chemainus Recreation Club, and Joanna became active in a local church. Fifty years later MacMillan recalled his time in the town for a local historical publication: "It was a very friendly community and a beautiful and pleasant place to live. Costs were low—Chum salmon were 5 cents each. I had no car, no spare time and no holidays."[1]

As it had been since its founding and would be for another seventy-five years, Chemainus was dependent for its existence on the sawmill. Victoria Lumber & Manufacturing Company was founded in 1889 by a syndicate which included Frederick Weyerhaeuser among its members. John A. Humbird, a Wisconsin lumberman, was president of the company. It purchased from the Esquimalt & Nanaimo Railway 100,000 acres of prime forest land, covered with some of the Pacific Northwest's best stands of Douglas fir, as well as a small sawmill built about 1860 and expanded several times since. By the time MacMillan hired on as assistant manager, the VL&M mill was one of the largest in BC, supplying lumber to the local markets, shipping east by rail to company-owned lumber yards on the prairies, and competing in the export trade through San Francisco brokers. One of MacMillan's few sales of lumber in Australia had been a 350,000-board-foot order filled by VL&M.

When the Humbird interests set up VL&M they hired E.J. Palmer, a former railway conductor who had found Humbird's favour. Eventually Palmer became general manager of the company, and earned a highly contradictory reputation. A paternalistic and arbitrary administrator, he was revered by many local citizens dependent on the company for a living. Others—including, in the end, MacMillan—did not remember him so fondly, and he was known as "Old Hickory" for his tough and contrary nature. All agreed, however, that one of Palmer's strong points was his interest in technological innovation. Under him, VL&M was the first BC logging company to use steam-powered equipment in the woods. Decades before it became practical, for instance, Palmer predicted the use of aircraft for water bombing forest fires. The company was very successful under his management. He was in his sixties when he learned MacMillan wanted to leave government service and offered him a job as his assistant. It was understood that the bright young forester would succeed him.

The company's preeminence was due primarily to its superb stands of timber, which it did nothing to enhance or protect. MacMillan, who by this point in his career had examined a lot of timber, was impressed.

I have seldom seen such a beautiful fir forest as the Victoria Lumber & Manufacturing Company possessed. However, this is what one would expect, knowing that it was the first selection any timber cruisers had the opportunity to make out of about one and one-half million acres of Douglas fir forest between Shawnigan Lake and Campbell River. There was not a dollar a year spent on any form of forestry or fire protection.[2]

The company used one of BC's first logging railways to haul logs to the mill, which cut about 100,000 board feet in a ten-hour shift.

> The mill was probably as good as any in the Vancouver forest district at that time. Everything was done by man-power. . . . All lumber was pushed out of the mill from the headsaw by man-power, rolled down the timber deck with peavies to dollies and then pushed across the uneven wharf by Chinese—sometimes forty to a big stick on a dolly. . . . When within the reach of the ship's slings the ship did the rest. Many of the sailing ships at that time carried small wood-fired donkey boilers providing power for loading. The crews of ships without donkeys attached ship's rigging to the heavy sticks, hauled them in and stowed them in the hold—all by man-power, via capstans and peavies.[3]

MacMillan's duties encompassed every aspect of the operation, from the woods to the loading wharves. The logging operations were completely under his direction, and within weeks he was overseeing the building of one new railway and rebuilding another, both of them several miles from the mill in opposite directions. He also served as general superintendent of the mill, the shipping dock and lumber yard, as well as sales manager. "I was expected to put in a full seven day week. I took inventory on Sundays in the inventory season, finishing it on New Years day. No overtime was ever paid to anyone."[4]

Palmer, on the other hand, had little interest in the forest or mill operations. He went out in the woods about one day a year and spent only two or three hours a week walking around the mill yard, rarely entering the mill itself. He was the archetypal office-bound manager that MacMillan later came to distrust. He usually left Chemainus Thursday afternoons, driving his new Buick roadster to Victoria where he did business on Friday and spent the weekend golfing.

This and other aspects of the job began to grate on MacMillan. He watched in amusement on one occasion when Palmer clashed with the stevedore boss, Bill Horton, a veteran of Vancouver's Hastings Mill who was credited with saving the mill during the fire which destroyed the city in 1886. Many years later MacMillan wrote about the incident in a local history.

> The Chemainus stevedores included a large number with Indian blood. They lived within ten miles of Chemainus, and were very much inter-related. They were competent and proud of loading a sailing ship so well and carefully that almost without exception more lumber was put on any specific ship at Chemainus than at any other port at which she had loaded. The ship always carried a record of previous loads.

On one occasion they proved their terrific ability. The sailing ship *Mabel Brown* came into Chemainus late on the 22nd of May, 1917. She was overdue, the cargo all ready, and the wharf congested, delaying the loading of other waiting ships. EJ was determined that the loading would continue without a break or slow down until the ship was finished and the berth cleared ready for the next ship now "waiting in the stream". The Indians had planned a big holiday with other Indians at some down-Island point for May 24th, which they were equally determined to attend. EJ told boss Horton that they could not go. Horton replied "We are going, but we will load your ship first." They put a full cargo of over one and one-half million feet of lumber, almost all six-by-twelve for Australia, on the ship before daylight on the 24th of May, and then went to their party. It is doubtful if that record has ever been beaten. They got no over-time pay.[5]

On another occasion, MacMillan was caught between Palmer and the workers.

The Chinese could not move the dollies if it snowed a foot or more; therefore, during and after a snow storm, the mill shut down. I remember once it snowed all night and in the morning there was more than a foot of damp snow. The old man, EJ, came to the office in high boots and a bad temper. He growled at me and said "Mac, go up to Chinatown and cut the Chinamen's pay two cents an hour." This meant they would then get only eight cents an hour instead of ten. We had 200 to 300 Chinese living in two big "China Houses" where they had individual bunks and got their meals in community kettles; each kettle was between two and three feet across, and sat on a brick arch.

I found the Chinese boss and gave him the message. He showed no more signs of interest than if I had told him "It is going to snow again tomorrow."[6]

During MacMillan's tenure at VL&M the company was logging along the lower end of Cowichan Lake. A contract had been let to Matt Hemmingsen, a highly respected logger who had come to Vancouver Island from Wisconsin ten years earlier to clear out logjams caused by futile attempts to drive logs down Vancouver Island's rivers.

He, with his wife and young family, were living near the job in a two or three-room shake shack on a cedar log float. Although one of the best loggers I have met, he was making only a very thin living, probably only low wages. He asked me if I thought he should get an extra dollar per thousand feet B.M. for timber sticks sixty feet and longer, for which there was quite a demand and which to my inexperienced eye obviously cost more money to handle. I agreed and on returning to Chemainus I told EJ. He was so angry that he pretty nearly fired me, but probably knowing Matt was still underpaid, he supported my promise.[7]

Mechanized logging was just beginning at this time and was still not a very well-developed art. Hemmingsen was one of its most skilled practitioners. He and MacMillan became good friends, and the skilled logger taught the young forester the business of logging. Hemmingsen's son John eventually became one of the top men in MacMillan's company.

Throughout his term at VL&M, MacMillan continued to correspond with Fernow and the other professional foresters with whom he still identified, although they viewed his move to VL&M as an act of apostasy. Writing in *Forestry Quarterly*, a publication he edited, Fernow commented: "The provincial government and the cause of forestry are losers, although we dare say Mr. MacMillan will not forget his forestry training and will eventually be again a power for good."[8] MacMillan maintained his interest in developing a forestry school at the University of British Columbia, in spite of a change of government that held out little promise for Wesbrook's struggling institution. His new job convinced him more than ever of the need for the school.

> Even my short and congested experience in the lumber business shows me the absolute necessity for systematic training in every branch of the industry, particularly in logging and manufacture. . . . This is the most chaotic business I can imagine. . . . There is no wiser, cheaper, more effective way of assisting the lumber industry than by placing science & knowledge at its disposal, and a Forest School and Laboratory would do that. We must have it.[9]

He and Wesbrook spent a day together in Chemainus over the winter, during which Wesbrook appears to have heard about, or sensed, MacMillan's growing dissatisfaction with his job. In the spring, the president sent MacMillan a letter in which he stepped up his campaign to recruit him.

> I am writing you now just to inquire whether you are so tied up with your present position as to be entirely out of our calculations. I have been so depending upon the possibility, in fact, certainty of being able to have your services in the development of these projects that I cannot lightly give up the notion now when there seems to be some prospect of beginning the work. Of course, I do not know whether I am warranted in writing you now. Perhaps I should wait until after legislative action has been taken but I cannot run the risk of losing our chance with you in case there should be a chance. Please think the matter over and do not say "No" unless you feel either that it is to your best interests to go on with what you have now in mind or that you are so committed that it would be impossible to withdraw.
>
> In case we are allowed to go forward, I have been counting upon you as Dean

and that you would be willing to assume responsibility for the development not only of the school but of the other features. . . . I think myself that you can do more for the province, the country and the work in which you are interested with the University than in a private capacity.[10]

MacMillan had little faith, however, in the government's willingness to fund the proposed forestry school. It appears he never did give Wesbrook a definite answer, but kept his options open while he continued to work elsewhere.

By the summer of 1917 the relationship between MacMillan and Palmer was strained. In early August, MacMillan wrote a long letter to Tom Humbird, who had taken control of the family interests in the company, outlining his problems, which included deteriorating health. Humbird replied with a warm letter, suggesting that when the fire hazard lessened he would come to Chemainus and meet with the two managers in an attempt to resolve their differences. But he was too late. By the time MacMillan received Humbird's response, he had written a long letter to Palmer.

The doctor's instructions have brought to a head what has been going through my own mind for some time—and that is that I have been required to attempt too much, either for the good of the Company or for my own good. . . . I am afraid that you will decide, upon reading this, that I am afraid of work. When I came here I understood from you that you contemplated building up an organization. As time went on, I learned that you expected me to handle all the details, single handed. I have tried this for several months and I am sure that the result has been a loss of money to the company.

I am not lazy, am not afraid of work. I think you will admit I have worked hard, staying on the job every week, seven days a week and counting neither hours nor holidays. I am not satisfied at all with the grade of work I have accomplished.[11]

He recommended that Palmer hire two men to replace him, one to be in charge of manufacturing, shipping and sales, and the other to oversee logging. His suggestions made it clear he did not see himself working in a reorganized company.

I want to repeat I am not kicking—I am simply pointing out it is a losing proposition for the company and consequently for myself. Furthermore I am afraid to risk my health on it—health is the only capital I have. Your ways of working and mine are different—I have the greatest friendship for you, but I have been terribly worried constantly, while trying to lay out my work by being held responsible continuously for details, instead of for results. . . . I therefore recommend you set out to get these two men and I will stay with you long enough to keep things going, while you get them. . . . I have nothing definite in mind to do myself—have not had time to think much of it yet.[12]

Palmer did not take well to MacMillan's letter. Not only did he dislike the recommendations for reorganizing the company's management structure, he also wanted to see the end of his assistant manager. Humbird came up from Idaho in an attempt to smooth things over, but in the end he and Palmer asked MacMillan to leave in early September. As he walked out the door, MacMillan turned to Humbird and, as legend has it, said: "The next time I walk through this door, Mr. Humbird, I am going to own this mill."

MacMillan's abrupt departure from VL&M stirred up animosities that coloured relations among the three families for years to come. On Christmas Eve 1923, shortly after the Chemainus mill burned to the ground, Palmer died of a heart attack. His wife, who never forgave MacMillan for what she considered an act of desertion, some years later spotted his picture on the wall of a Canada Trust office in Vancouver. Pointing at the portrait of one of the trust company's newest directors, she asked a companion: "Would you ever trust your money with that man?"[13] And although MacMillan's relations with Tom Humbird had been cordial, his encounters with the next generation of Humbirds would one day rock the forest industry.

MacMillan clearly anticipated his break with VL&M some weeks before his final encounter with Palmer and Humbird. Some months after his departure, he received a letter from Calcutta written by the head of an Indian forest company, in response to a letter he had sent at about the same time he wrote his long missive to Palmer.

My Dear MacMillan. I was very glad to get your letter of August 14th. It seems to have been a long time en route—in fact so long that I expect you will by now have fixed up with the Aeroplane spruce production proposition; but will you please write me fully for what period you could come to India on a general roving commission for us. Say 6 months; what would your fee be, we paying all expenses.[14]

By the time MacMillan received this letter, his course had long since been set. News spread quickly of his termination at Chemainus. On September 27, Fernow sent a note to University of Toronto president Robert Falconer:

I consider it my duty to advise you that there is a likelihood of securing a successor for my position, from every point of view satisfactory. I learn on my return, that Mr. H.R. MacMillan, who, not quite a year ago, left the British Columbia Forest Service to accept lucrative employment with a private concern, has resigned this position. I am told that he has made a bid for the directorship of the B.C. Forest School if satisfactory budgets can be guaranteed. At the same time, I have seen in the papers the statement that the organization of this school will be postponed for financial reasons. . . . Mr. MacMillan could however not be had at a low salary and would probably be unwilling to work with as niggardly appropriations as I have been willing to work. Incidentally I may add that the best forest schools in the States dispose of $20 to $30,000 budgets. I believe to secure Mr. MacMillan's services, it will be necessary to act very promptly.[15]

Fernow also sent a short note to MacMillan expressing his "hope you will look favourably on this proposition to take place when I retire next year to make place for a more enterprising genius."

But by the time MacMillan received Fernow's letter, he had already made up his mind what to do in the immediate future. His demanding sojourn at VL&M had allowed him to put the war in Europe to the back of his mind for almost a year. During that year, the nature of the war changed. In the early stages of the conflict, aircraft played a limited role and were used primarily for reconaissance and observation. Pilots and passengers were armed only with handguns or shotguns for close-range protection. By 1917, machine guns were mounted on the aircraft, leading to a heavy loss of planes and an accelerated demand for their replacement. The material critical to the building of aircraft at this time was Sitka spruce. It was light, strong and flexible. The largest and best supply of Sitka spruce readily available to the British was in British Columbia. In 1917, the Imperial Munitions Board, Britain's wartime military supply agency, established a Department of Aeronautical Supplies in Vancouver, and appointed Austin Taylor, a former executive of the Montreal Locomotive Works, director. His assigned task was to provide the British aircraft industry with unlimited supplies of aircraft-grade spruce. He arrived in Vancouver on November 7 to set up his office.

MacMillan, as his correspondence with India indicates, was aware of this project well before his break with Palmer, and it is likely the call to return to public service influenced his decision to leave the private sector. Fernow's appeals, backed up with a flurry of telegrams and letters from James H. White, a U of T forestry professor, trying to persuade him to take the Toronto position, fell on deaf ears. He already had an offer from Taylor giving him the opportunity to play a role in the war he had yearned for while travelling around the world as a trade commissioner.

After accepting an appointment as assistant director, MacMillan's first move was to relocate the family to Vancouver. Since moving out of the company house in Chemainus, they had been living in an unheated summer house at Cowichan Lake and were anxious to move into more comfortable quarters.

MacMillan's daughters, Jean (left) and Marion, at the time the family moved to Vancouver, 1918.

The city was bursting with wartime activity and housing was scarce. MacMillan obtained temporary accommodation in the fashionable Glencoe Lodge, a downtown boardinghouse. On the eve of their departure from Victoria, both his daughters came down with whooping cough and had to be hidden outside the lodge with Edna and Joanna while MacMillan registered. A few weeks later he found them a house at 2516 York Street, in Kitsilano.

Before he was able to devote himself fully to his new job, MacMillan ended up in hospital. His previous year's exertions at Chemainus, followed by the scramble to get started in his new task and the move to Vancouver, left him exhausted. On December 3, Wesbrook wrote to him from a train en route to Winnipeg, urging him to take it easy: "Do not fret as you are saving time for yourself and storing up energy upon which even the best meaning of us will be tempted later to draw unduly."[16] Calling upon habits learned in the sanatorium, MacMillan put his time in hospital to good use. He caught up on his correspondence with foresters throughout North America and those serving in the armed forces. Clyde Leavitt, chief forester of the Commission of Conservation, wrote to prod him into finishing a piece on timber markets. And he prepared reports and made plans for the massive onslaught on the province's coastal spruce forests. By Christmas he was out of hospital and hard at work.

Austin Taylor and his group faced a formidable task. First they had to determine exactly where the highest quality spruce was located. They were inundated with speculators and loggers who held timber licences, insisted their claims were covered with unimaginable volumes of the highest grade timber, and offered to sell them to the government. MacMillan rejected this approach. The concentrations of spruce were too low on these licences and would have required logging large volumes of other, undesirable species, which would have slowed the entire operation. Instead, he drew on Roland Craig's work for the Conservation Commission and called on Grainger to provide whatever inventory information the Forest Service had. It was quickly determined that apart from a relatively small amount on the west coast of Vancouver Island, the best and largest spruce forests were on the Queen Charlotte Islands. The BC government immediately issued an order in council making all spruce on Crown land available for logging, and instructing private timber owners to proceed immediately with logging all airplane-grade spruce, with provision for a fixed stumpage rate of $6 a thousand board feet.

Drawing on his contacts in the industry, MacMillan assembled a team of capable people. He placed Frank Pendleton, an Everett, Washington, logger and sawmill operator with interests in BC, in charge of sawmills. Fenwick Riley, Bloedel Stewart & Welch's logging superintendent, performed the same function for the Munitions Board. The responsibility for transporting equipment into the Charlottes and bringing logs and lumber out was assigned to James Coyle, a partner in Greer and Coyle Towing. G.G. Davis, a logging superintendent from Port Renfrew who had devised a type of raft for towing logs across open ocean, was hired to put his invention to work moving logs. And, once again, MacMillan called on VanDusen, who was appointed superintendent of operations and stationed at Port Clements on Graham Island. L.R. Scott, a friend of Taylor who would later perform a major role in MacMillan's business ventures, helped run the Vancouver office.

The next task was to organize from scratch the biggest logging show anyone ever attempted to put together. In 1918 virtually no settlement or commercial timber operations existed on the Queen Charlotte Islands. Although loggers had long known about the magnificent stands of timber, markets had never been sufficiently strong to warrant the formidable task of moving logs from the islands 400 miles south to the mills in Vancouver. Until the development of log barges, the major obstacle to moving logs was crossing storm-prone Hecate Strait with a tugboat towing a boom. The population on the islands was not large enough to provide labour for on-site mills. On MacMillan's advice, Taylor decided the best strategy would be to issue logging contracts to about 300 small operators. Some were handloggers, who felled

MacMillan and W.J. VanDusen aboard a Forest Service launch, 1918.

trees along the steep shorelines of Graham and Moresby islands and wrestled them into the water with jacks and human muscle. Other contracts were issued to riving crews, who felled the trees in the woods, bucked them into fourteen-foot logs and split the logs into cants of aircraft-grade timber, which were loaded onto barges and shipped to cut-up plants in Vancouver and Prince Rupert. This method was also employed by the Gibson family along the west coast of Vancouver Island, who used a riving contract to pioneer logging there.

As well, a large number of mechanized logging camps, utilizing ground-lead steam donkeys, were established on the Queen Charlottes.

It was decided to log the spruce selectively, which turned out to be the key to the success of the undertaking. MacMillan realized that the usual method of logging all the spruce and milling it would produce less than 2 percent of aircraft-grade lumber. By carefully selecting the best trees for falling and processing, an average 30 percent recovery of aircraft-grade lumber was achieved. Several sawmills were built at various locations in the Queen Charlottes, and existing mills at Vancouver and Prince Rupert were modified and pressed into service to handle logs and the cants from riving operations. During December 1917, Coyle assembled a fleet of tugs, barges and scows. By January he began moving a mountain of logging and sawmill equipment, construction crews and loggers through the hazardous winter storms into various locations in the Charlottes.

Logging began in mid-January, with production rising steadily through the late winter and early spring. In the Queen Charlottes, logs were assembled at two locations to be sorted for delivery to the local mills or built into Davis rafts for towing to Vancouver and Prince Rupert. Grand Trunk Pacific express trains carried the bulk of the finished product—clear, straight Sitka spruce—out of Prince Rupert to the Atlantic coast, where it was quickly loaded on ships bound for England.

While Taylor maintained operations in Vancouver, MacMillan spent the first nine months of 1918 roving up and down the coast, catching rides in whatever way he could. Half a century later, he noted in his yacht log:

> Yesterday we passed several tow boats plugging along South with large rafts of logs, amongst them my old friend the *Commodore*, built by Hastings Mill 50 or more years ago on which I lived and travelled in waters around Queen Charlottes in 1918—some very stormy passages. Captain Bjerre, an old Dane, was her master, a staunch old man.[17]

MacMillan was in his element. He seemed to have an instinctive flair for working and leading people in this environment, and he was constantly in the thick of the action. The formidable reputation he later developed among coastal loggers had its origins in this period.

MacMillan (left) with officials of the Imperial Munitions Board at Masset Timber Company mill, 1918.

One of his most delicate negotiations was with the maritime unions, which were concerned with keeping their members working, and for their safety on the stormy waters of Hecate Strait and Queen Charlotte Sound. In the end, because he travelled on the boats himself, MacMillan prevailed, after agreeing that no tug would sail with an inadequate crew. In order to maintain production, he would not let the crews leave their ships when they brought in a tow, but sent them right out again for another. "We kept those poor fellows afloat within sight of shore sometimes four months at a time. Damn it, we had to. Delivery of that spruce was vital to the war effort."[18]

By 1918 the BC public was far from united on the war. More and more people opposed it, some because they were pacifists, others for political reasons. Although jobs were more plentiful toward the end of the war, wages were fixed and inflation was eroding the workers' spending power. Labour unrest was increasing throughout North America, and the Industrial Workers of the World—the Wobblies—were making inroads in the Pacific Northwest.

The spectre of class warfare was a real one to many British Columbians. Less than a year earlier, the Bolsheviks had taken power in Russia. Closer to home, two years previously, several US Wobblies had been killed and a score wounded in an ambush at Everett, just south of the border. Strikes were proliferating in BC. In the Crowsnest area, the federal government stepped in to end a coal strike and assumed control of the mines, forcing the miners back to work. And in Vancouver several unions went out on strike. Events came to a head with the shooting of union organizer Ginger Goodwin in July 1918 on Vancouver Island. His death precipitated a twenty-four-hour general strike in the city, amid great fears it would explode into a larger conflict.

MacMillan refrained from commenting on these events, at least publicly. While he had no use for socialists, leftists or politically inspired labour organizations, he had a profound respect for people who did their jobs well and took pride in their work, whether they were whistle punks or mill owners. In June 1918, he was appointed a special police constable by the federal government, but there is no record of him exercising the powers granted to him; he usually managed difficulties with more direct measures.

One night in Prince Rupert, a tugboat crewman slipped over the side and swam to shore for a night of carousing. The captain sent word to MacMillan, who was in town, that he would stay put until the crew was brought up to full complement. This ultimatum sent MacMillan out onto the streets of Prince Rupert, scouring the bars, restaurants and hotels, looking for someone to fill the position. But in the late stages of the war, unemployed men were scarce in the port city. Finally, in desperation, he went to the local jail and asked the chief of police if he had anyone who could be "borrowed" for a couple of weeks. The chief checked his inmates and produced one who was dead drunk and snoring peacefully. Undaunted, MacMillan dragged him down to the dock, dumped him in a skiff and rowed out to the tug anchored in the harbour. When the crew hoisted the drunk up on deck they discovered he was their missing sailor. Marvelling at how efficiently MacMillan got things done, they hauled up their anchor and headed back to the Charlottes for another boom of spruce.

By July 1918 the optimistic production goal of 3 million board feet a month, set by the Munitions Board, was passed, and production continued to rise until it reached the phenomenal rate of almost 9 million feet a month by the time the war ended in November. Those most impressed by this accomplishment were coastal loggers; more than anyone else, they knew what it represented. They also knew who was primarily responsible. MacMillan had been a familiar figure in virtually every logging camp and sawmill on the coast throughout the nine-month undertaking. By the time the war ended, he was on a first-name basis with practically every owner and manager in the industry, and he could recognize hundreds of loggers, millworkers and towboat workers. What earned their respect was the fact that as the man in charge, he spent most of his time in the woods, in the mills and on the boats, not in a downtown office. By November 1918, the legend of H.R. MacMillan had taken root along the BC coast.

With the war over, MacMillan once again faced the decision of what to do for a living. He had several months in which to make his choice while he wound down the spruce operation, closing the camps and mills, paying off contractors and disposing of the board's assets. The long-discussed forestry school at UBC was by now a dead issue. The government had refused to fund it, and its main supporter, Wesbrook, died three weeks before the war ended. Throughout the war, Fernow had continued his campaign from Toronto, intent upon luring MacMillan into the dean's office. Fernow's health was deteriorating and he was anxious to find a successor, preferably MacMillan. "My dear Mr. MacMillan," he had written in March 1918,

> Your apologetic letter of Dec. 14 deserved a more prompt acknowledgement and would have had it, if my physical condition had not made me delay it. I have a curious nervous and muscular complaint unknown to doctors, which when I write, even only to sign my name produces a muscular tension in my leg. I had hoped to get rid of it by massage and baking, but have so far had no success, and I feel considerably handicapped, for it also acts if I merely think hard. This, therefore, only to tell you that I never laid it up to you that you did not reply more promptly, that you are in all respects forgiven, and that the situation here is the same as it was when I wrote to you about the opening.[19]

The two men continued to correspond until the end of the war, after which Fernow pressed MacMillan hard to come to Toronto and meet President Falconer to discuss his appointment as dean. Coming as they did from the most prominent forester in North America, these pleas finally persuaded MacMillan to visit Toronto. In February he cabled Fernow: "WILL

MacMillan with his lifelong friend Austin Taylor (at right), on the steps of the Imperial Munitions Board office, Port Clements, 1918.

ARRANGE VISIT TORONTO IN APRIL IF YOU WILL KINDLY INDICATE MOST SUITABLE DATE STOP PRESUME UNIVERSITY WILL MAKE ALLOWANCE TRAVELLING EXPENSES."[20] A series of telegrams finally established April 10 as the date MacMillan would meet Falconer in Toronto. On April 7, MacMillan cabled again. "I REGRET I MUST ADVISE YOU THAT IMPORTANT WORK OF SETTLING CONTRACTS HAS BEEN SO DELAYED THAT I FIND IT IMPRACTICABLE TO LEAVE THIS MONTH SO AM WRITING EXPRESSING MY PERSONAL DISAPPOINTMENT AND BEST REGARDS FOR YOURSELF."[21]

MacMillan's vacillation over this offer, reminiscent of his indecision at the time he was hired as BC's chief forester, is again out of character. He may have been keeping all his options open while looking for something better, or trying to negotiate better terms. Or, as a former acquaintance suggested, he may have been influenced by Edna, who did not want him to return to a public service job.[22]

MacMillan never made the trip to meet Fernow and Falconer. At about this time, he learned that a delegation of British timber dealers was planning a visit to Vancouver. One of them was Montague Meyer, whom MacMillan had met in London four years earlier. In his capacity as British timber controller during the war, Meyer had been in as good a position as coastal loggers to recognize MacMillan's crucial role in the aircraft spruce program. The British industrialist had an interesting proposition for the about-to-be unemployed forester.

founding the export company

British Columbia's economic outlook in early 1919 was mixed. As the province emerged from the aftermath of war and the annual winter slowdown of a resource-based economy, its half-million residents took stock of their prospects. Demobilization of the armed forces had dumped thousands of workers on a sluggish job market. Pent-up labour discontent was expressing itself in a politicized union movement talking openly of violently overthrowing the established political and economic order. In May, the Winnipeg General Strike spread west, tying up parts of BC. Most sectors of the provincial economy were stagnant after focussing intently for the previous four years on providing materials for a distant war. Mining was down. The fishing industry was just as bad, a victim of declining markets and the destruction of the Fraser River salmon runs by the 1913 Hell's Gate slide.

The only bright spot was the lumber industry. Wartime export markets held up as Britain and Europe began the process of rebuilding, and sales actually increased slightly during 1919. Perceptive people could see possibilities in this sector that did not exist elsewhere.

One of those in the best position to gauge the opportunities was Montague Meyer. In May he accompanied Sir James Ball, the new British timber controller, on a visit to BC to assess how much lumber the forest industry could supply for British reconstruction. In anticipation of this visit, the large exporting sawmills hastily organized themselves into the Associated Timber Exporters of British Columbia (Astexo). It was the type of organization MacMillan had recommended following his global tour during the war, and was conceived as a vehicle for assembling large timber export orders. Astexo members and the provincial government, under Premier John Oliver, took Ball and Meyer on a tour of the coastal forest industry, walking them through the best stands of untouched timber and showing them mills and shipping facilities, in an all-out attempt to convince them that BC was a reliable source of timber. The two were sufficiently impressed to place an order for 70 million feet of railway ties and crossarms—an order of unprecedented size which would require the participation of every mill on the coast. It was a daunting challenge for some mill owners. In a Vancouver speech, Ball warned them they had to improve their marketing capabilities if they expected to compete in the future with American, Baltic and Russian suppliers.

The tour travelled to mills outside the Vancouver area by boat, and when it stopped at the newsprint mill in Powell River, Meyer stayed on board to nurse a sore leg. MacMillan stayed with him while industry and government officials went ashore with Ball to inspect the mill. During their time alone, the two men discussed the idea of setting up a timber export company in BC to sell lumber in Britain and Europe. Meyer pointed out that the Panama Canal, by then functioning efficiently after a series of mishaps since its 1914 opening, had cut 10,000 miles off the Vancouver-to-London route. He also noted that the established American brokers who controlled the export business had not adjusted their operations to the realities of this new route, and expressed the view that an aggressive new export company could drive them from the field. Meyer proposed that he finance a Vancouver-based export company, run by MacMillan, which would ship lumber from BC to Europe to fill orders sold by Meyer. MacMillan asked for some time to consider the proposal, and they agreed to meet in New York a few weeks later.

After his experience at Victoria Lumber & Manufacturing, MacMillan was not interested in working for a company controlled by someone else. Nor did he see the advantage of replacing the American brokers who dominated the export business with a British-owned firm. Furthermore, he had other, related irons in the fire. At the end of the war he had purchased a share of Big Bay Lumber, a spruce mill near Prince Rupert, that cut shooks—thin, short boards used for making boxes—and aircraft lumber. He had also bought shares in Percy Sills's Premier Lumber Company, a wholesale operation. Except for his unfortunate stint with Palmer at VL&M, and his experience as a travelling salesman while he was a student, MacMillan had never worked in private business. His association with Austin Taylor during the war had introduced him to the business world, but the special circumstances of that period were of limited value in peacetime. His years of public service had not taught him the basics of business management, and, while he was employed getting out spruce during the war, he filled in the long hours on coastal tugs and steamers studying "The Modern Business Course," a set of correspondence lessons from the Alexander Hamilton Institute in New York. Thus he had already taken the first steps to establish himself in the lumber business when Meyer made his offer.

In June MacMillan travelled by train to New York and on June 12 he and Meyer signed an agreement establishing the company—the first locally owned lumber export company in BC. It was called H.R. MacMillan Export Company because of a convention in the British timber industry that a principal attached his name to a company to demonstrate his personal commitment.[1] At MacMillan's insistence, he and Meyer each put up $10,000 capital. Meyer had offered to provide all the investment money, but MacMillan would not accept anything more from him than what he could raise himself. His own share turned out to be $10,000 he borrowed on a house he had purchased recently on West 16th Avenue in Vancouver. Meyer would be president, and MacMillan would serve as manager at a salary of $5,000 a year, plus a bonus to a maximum of $2,500 a year from profits. It was agreed that Meyer would finance operations when necessary, and would receive stipulated discounts and commissions. The new company would be Meyer's sole supplier from the Pacific Northwest and MacMillan's only export vehicle.

The company was incorporated in July, a stenographer was hired, and

MacMillan opened for business in one of Premier Lumber's two rooms on the seventh floor of the Metropolitan Building, two floors below the recently opened Astexo office. This building, with the Terminal City Club on the ground floor, was for several decades the hub of the BC forest industry.

Orders began to arrive from Meyer's London office almost immediately. MacMillan's role was to fill them at the best price and to arrange for shipping and insurance. By fall, he had more work on his desk than he could handle. He got in touch with VanDusen, who was back at his job as district forester in Vancouver, and hired him to start in November as manager, at $325 a month. They crowded another desk into MacMillan's small office, and the two men sat back to back. The stenographer moved into a room across the hall, which she soon shared with Harold Wallace, an accountant who was a boyhood friend of Aird Flavelle. Also hired in November was L.R. (Leon) Scott, the young lumber buyer from Ontario who had come to Vancouver a couple of years earlier with Austin Taylor.

When the H.R. MacMillan Export Company began operations, every stick of BC lumber sold on the export market was handled by US brokers. MacMillan and Meyer's method of penetrating this market was simple, if risky. The lumber export business utilized two basic systems, commonly known as "FAS" and "CIF." The first, "Free Along Side," involved a mill selling a parcel of lumber and delivering it to the wharf to be loaded on a ship for transport abroad. Under this system, the buyer or his agent took delivery on the wharf, paying the seller his price, and then arranging for shipment. It was the system used by all BC mills, none of which was sufficiently big or well established to underwrite shipping costs. Even the formation of Astexo did not end this practice, as its members continued to sell FAS. Because foreign buyers were reluctant to purchase lumber sitting on a wharf in Vancouver or Chemainus, a number of brokerage businesses had developed in San Francisco, Portland and Seattle. They made purchases from BC mills on behalf of buyers overseas, for fees or commissions.

In the second system, "Cost, Insurance and Freight," lumber was sold at its final destination for a price that included the cost of buying it at the mill, insurance on shipping it to the buyer, and the cost of shipping it to a designated port. In this system, exporters accepted orders from abroad, purchased the lumber, arranged to have it shipped and collected payment when the lumber was delivered to its final destination. They did not charge a fee or commission, but made their money by buying at a lower price than they sold and by obtaining the best possible shipping rates. Since all three of these factors—cost price, sale price and shipping rates—fluctuated daily, it was a risky but potentially profitable business.

In both systems, a buyer in a distant city cabled an inquiry to brokers and suppliers in Vancouver, asking for delivery of a specified volume and grade of lumber on a certain date at a designated location. If the broker wanted the order, he cabled a response. From the outset MacMillan laid down a hard and fast rule that all inquiries should get a response within twenty-four hours, an unusual practice in the business at that time. Meyer and MacMillan's method was to offer firm CIF prices to the largest one hundred dealers in cities across Britain, as well as to railways and government departments. Whereas the American brokers sold only full shipload lots to the five or six largest timber merchants, who resold at high prices, the H.R. MacMillan Export Company (HRME) sold small parcels to almost anyone. Because of the high price of Douglas fir demanded for US-brokered lumber, only high-grade products such as decking and timbers were handled, but because it sold at lower rates, HRME was able to market fir railway ties and lower grades of fir lumber as well.

The keys to success in the lumber export business were, and still are, simple and straightforward. A broker must be able to obtain lumber cut to the precise specifications and in the grades and species the buyer orders, at a lower price than competitors. He must be able to secure shipping space within the required time and at a reasonable cost. And he must develop and maintain confidence among buyers that he can be relied upon to complete the agreement. All aspects require rigid fulfillment of contractual obligations, combined with trust and gentlemanly conduct. MacMillan, with VanDusen's and Scott's assistance, was quite capable of providing the first two, while Meyer, who already occupied an eminent position in the British and European lumber business, provided the third. Together the partners were an effective combination.

In London, the immediate success of the export company created a slight problem for Meyer. The British government had launched a major reconstruction program and would not release him from his public service position, creating a potential for a conflict of interest in his dealings with MacMillan. Early in the new year, Meyer informed MacMillan that he had set up another company, Canusa Trading, to serve as the selling agency for HRME. It would perform this function long after Meyer was freed of his public service.

The activities of the partnership were complicated by the communications technology of the period. The primary way to exchange information in 1920 was the written letter, but it took two weeks for mail to travel between Vancouver and London. The lumber business was sufficiently complicated that MacMillan and Meyer wrote five- to ten-page letters to each other daily, sometimes twice a day. By the time a letter had crossed the continent and the Atlantic, and an answer had been received, a month might have passed. The men numbered their letters, but keeping matters straight required enormous competence on both sides. It also required clear, concise writing, of which both partners were quite capable, and which was no insignificant factor in their success. Telegraphic cables, the only other means of communication available, were used frequently in placing and accepting lumber orders. But cable rates were very high. To save expense and maintain confidentiality, businesses used abbreviated code books to write terse, cryptic messages— almost an art form in itself. Early on in the business, MacMillan became a master of the form and developed an uncanny ability to write lucid, brief, forceful cables.

While Meyer was making his own adjustments in London, MacMillan had no intention of sitting quietly in Vancouver, dispatching shipments of lumber and leaving the sales end to Meyer's organization. With his small office staff in place, he spent most of 1920 retracing his 1915–16 trip as timber commissioner. He and Edna left Vancouver January 15 for New Zealand and Australia, where he re-established old connections and laid the groundwork for future export sales. After two months in Australia, he received a cable from VanDusen, advising him of rumours that two Indian railways were looking for someone to fill a large order for railway ties—sleepers, as they were called in the trade—because their regular Russian suppliers had succumbed to the chaos of the Bolshevik revolution. MacMillan cut short his business in Australia, and he and his wife took a ship to Calcutta. After several weeks of meetings with Indian railway officials, whom he impressed with his arguments for the virtues of Douglas fir sleepers, he still was unable to find anyone with the authority to sign a contract. Mystified, he cabled Meyer for advice and received a reply informing him that a London brokerage firm had a contract to purchase the ties. MacMillan decided to leave immediately for London.

When MacMillan and Edna arrived in Britain, a long, enthusiastic, handwritten letter from VanDusen awaited them, full of news from Vancouver. VanDusen described his attempts to secure sleepers for the Indian railways, and his difficulties in having the wood creosoted, in anticipation of MacMillan obtaining the order. This transaction was an early example of how MacMillan and VanDusen worked as a team—MacMillan in hot pursuit of the current objective, with VanDusen carefully executing the details to fulfill it. With the assurance that VanDusen was handling lumber, creosoting and shipping in Vancouver, MacMillan went in search of the broker with the contract to purchase the sleepers. Eventually he and Meyer tracked down an elderly gentleman in a cluttered, musty office who, after much searching, produced the documents authorizing him to purchase three shiploads of creosoted ties.

Filling the order turned out to be just as harrowing as winning it. As the sleepers were being prepared for shipping in the fall of 1920, the price of lumber began to drop. MacMillan and VanDusen realized that if they were unable to load and dispatch the final shipload before the terms of credit expired, the order would probably be cancelled and given to a competitor at a lower price. They exerted as much pressure as they could muster on the mills, the creosoting plant and even the longshoremen loading the ship. Late in the evening of the final day of the contract, the last ties were loaded. The men cleared the paperwork by persuading the bank's foreign exchange section to stay open until 11:00 that night. Typical of the luck that had blessed the company since its formation, a last-minute shift in exchange rates produced an unforeseen and substantial 23-percent bonus.

The profits earned in the first year on MacMillan and Meyer's $20,000 investment were enormous—close to $250,000 by the end of August.[2] Far beyond their expectations, the success was due to a combination of hard work and sheer luck. Both partners recognized that the addition of VanDusen to the company was another significant factor. While MacMillan had been on his world tour, Meyer had visited Vancouver and talked to VanDusen about his future in the company. VanDusen wrote to MacMillan in Bombay on July 5: "I was glad of an opportunity to explain my viewpoint. [Meyer] was very heartily in accord with the suggestion that I should take an interest in the company."[3]

While MacMillan was in London, he and Meyer agreed to pay VanDusen $5,000 a year plus 10 percent of profits. MacMillan informed

VanDusen of their decision on his return to Vancouver in September. As MacMillan explained in a letter to Meyer, the offer did not sit well.

> My proposals disappointed him deeply. He had in my absence been offered a straight salary in the neighbourhood of $1,000 per month and had turned it down. He is ambitious and felt that his ability in buying and negotiating had enabled us to profit considerably, that he liked the association and wanted to feel that we could allow him a larger share in the Company's profits now and henceforward. . . . I believe that he is too valuable for us to lose. I have in the meantime looked around amongst my business acquaintances to see on what basis really good men were paid and he is a really good man. I arranged with him that he should draw $350 per month salary, 10 percent on the profits of this Company and should purchase at par seven and one-half percent interest in the Company putting up with us a proportionate share of any new capital subscribed or loans made to the Company.[4]

MacMillan concluded by explaining that VanDusen was planning to go to South America in October and would travel via London to discuss the matter with Meyer. VanDusen would also stop off in Montreal to look into setting up an office.

Meyer was appalled by MacMillan's offer, and his response reveals the tensions inherent in operating an export company stretched between London and Vancouver in an era of delayed communications. He wrote back to MacMillan on October 12.

> I cannot yet consider him our equal or anything like it, and when you come to think the matter over the majority of the profits which have been earned up to now have not really been earned through VanDusen's ability but through our very good fortune on this side and the financial backing which we have unstintingly given to the H.R. MacMillan Export Co. Ltd. . . . I would have been pleased for him to have [10 percent of profits] and thus to share in our prosperity, but I cannot entertain the suggestion that he should, in addition, be allowed to purchase seven and one-half percent of the Company's shares at par. These shares are already worth very much in excess of their par value. . . . I do not think it a good suggestion that we should have a third party coming in at the present time as a shareholder in the Company and I think that whatever VanDusen gets should be as an employee of the Company, for the time being at any rate.[5]

In a reference to the proposed Montreal office, Meyer allowed MacMillan a glimpse of the iron fist he usually kept covered by the velvet glove of British manners.

Any action of taking in somebody else is bound to affect my policy in regard to Eastern Canadian business and I think we shall have to start a separate business over there rather than amalgamate it with the Export Co. in the West.

Before VanDusen left Vancouver, tension between the partners escalated further. MacMillan received a telegram from Meyer chastising him for chartering ships carrying sleepers to India on a prepaid basis, rather than collect-on-delivery. Under the terms of the agreement establishing the company, it was Meyer's responsibility to provide whatever up-front financing was needed. He objected to having to tie up $35,000 per ship for four months and ordered, in blunt language, that charters be paid collect-on-delivery. In a subsequent letter he apologized for the telegram, but by then VanDusen had left for London. As soon as he arrived at Meyer's office, the two men got into a fierce argument. VanDusen wrote to MacMillan:

> About the first thing Monty said to me when I first landed in his office was "Why the devil didn't you charter collect even at 45 rather than 43 prepaid." His remarks were in such a tone that I saw red immediately and thumped the table and asked him what kind of a bunch of ninnies did he think we were. That we didn't deal with pious hopes but blunt facts. . . . He started out to boss me the same as he does Bourne [the Canusa manager] who can't call his soul his own & he got rudely awakened and shocked I guess. We soon got so we were talking on an equal footing and things have gone better since.[6]

In fact, things went very well from that point on. Meyer capitulated, agreeing to VanDusen's requests and going so far as to sweeten the offer. He wired his agreement to MacMillan and followed with a letter that began:

> I thought I would just drop you a line to tell you how much we all like Van & that we hold a high opinion of his business ability . . . everything is all right now & I think Van will be a great asset. As the Company has made a lot of money I think if you agree that it might pay the travelling expenses of Van's wife.[7]

The final agreement gave VanDusen a 17-percent share in the company, with the stipulation that he reinvest his share of the dividends at the same rate as Meyer and MacMillan. This was the first major disagreement between the partners and there would be more, frequently over MacMillan's choice of staff or business associates. The urbane Meyer, with a bit of the archetypal stuffy Englishman in his manner, occupied a much different world than his Canadian partner, who dealt with and enjoyed the company of more practical, down-to-earth people. When circumstances caused the extremes of their personalities to connect, conflict was frequently the result.

Interspersed with the correspondence about VanDusen's position was a great deal of information on company affairs. Some of it had to do with other products the firm dabbled in during its first two or three years in business. Meyer was particularly interested in the wheat business because of the large amount of Canadian grain imported into Britain at that time, and he queried MacMillan regarding salmon at least once. One of the few commodities that was actually shipped by HRME was cement, but it was not a successful venture and was not repeated.

By the end of 1920, lumber prices had dropped to half what they had been a few months earlier. Given the company's accumulated profits and some fortuitous events, MacMillan was able to take advantage of the situation. After the war the Canadian government established the Canadian Government Merchant Marine, to maintain the shipping capability developed with the cargo vessels built by the wartime Imperial Munitions Board. For a few years the Merchant Marine ran a fleet of ships out of Vancouver, operating coastwise along the US and South American seaboard as well as on regular routes to Australia. It offered the MacMillan company an alternative to the US-owned ships that had dominated the trade until then.

Shipping was so easily available during 1921 that freight rates were driven down dramatically. These low rates, combined with upheavals in the lumber export business because of falling prices, prompted MacMillan to purchase timber in Washington and Oregon, in direct competition with the same firms that had once controlled the BC export business. In March, during an extended trip down the west coast, he wrote to Meyer from San Francisco: "I have spent the last two or three days in Portland and here looking into market connections and conditions. Trade in this city is very depressed, several of our competitors being on the rocks or in the breakers."[8] He did, however, find a silver lining in the lowering clouds of depression. He signed an agreement with a US exporter to share control of the Pacific Northwest sleeper trade to India. It was, at a time when such agreements were quite legal, a cartel.

Through the 1920s, railway sleepers were the main item of business for the company, and by the end of the decade it had shipped 215 million board feet of Douglas fir sleepers, more than any other North American exporter. General economic growth and problems facing Russian timber exporters contributed to the growth of this business. But persuading railway companies to use Douglas fir was still a hard sell. This was mostly because it was a species unknown to the railway engineers and purchasing agents who made the decisions, but also because many railway companies, including the British, preferred to do the creosoting themselves to save on shipping costs. The railways' creosoting processes, which worked well on species such as Baltic fir and pine, were not effective on Douglas fir. To deal with this problem, MacMillan persuaded George Herrmann, owner of Vancouver Creosoting Ltd., to make an extended trip to the UK in 1924 to inspect the British creosoting plants and recommend better methods, and then to visit Indian railways to determine the type of creosoting needed on sleepers used there. This initiative, combined with a marketing campaign extolling the virtues of Douglas fir, increased the use of the Pacific Northwest species in Asia, Africa and Europe. MacMillan knew from his trade commission trip that the superior properties of fir were not recognized outside the Pacific Northwest. Beginning with sleepers, one of his major campaigns as an exporter was to preach the virtues of this valuable BC coastal softwood.

In August 1921, MacMillan and Percy Sills went to Aberdeen, Washington, to examine a loaded lumber ship, the *Canadian Exporter*, that had run aground on the Grays Harbour bar. They decided there was a good chance of recovering the cargo and ship's machinery, and bid $2,000 for the salvage rights. By the time early winter storms put an end to the operation, Sills had managed to get off almost $18,000 worth of lumber and equipment. Although the exercise cost them more than $20,000, they were both happy with it; they had learned they had no future in the marine salvage business, and had enhanced their reputations throughout the Pacific Northwest as energetic, daring businessmen.

That same summer, MacMillan purchased a new house for his family at 1326 West 13th Avenue in Vancouver, precipitating a major confrontation with his mother. Ever since his release from the sanatorium a decade earlier, he had provided for Joanna as if in repayment for the years of hard, domestic labour she had put into raising and educating him, and paying for his convalescence. She had lived with her son and his growing family ever since their move to BC, but now her greatest desire was to move into a nearby apartment where she could lead a life of her own and devote her time to church missionary work. MacMillan would have none of it. He was determined to care for her and insisted that if she moved out it would appear as though he were forcing her to leave. Over the previous ten years he had developed a forceful, domineering personality; when, on occasions such as this, he exercised it within his family, he left the household of four females in tears. They were relieved when he departed on one of his long trips.

Although the domestic scene was much more peaceful when he was absent, MacMillan the father, husband and devoted son could also be a source of great delight. He hated inactivity and looked upon holidays and Sundays as heaven-sent opportunities for fun. His favourite Sunday occupation was to rise early, before anyone else, and drive to a neighbourhood green grocer store to purchase the ingredients for an elaborate picnic, which he then prepared himself. He would load the baskets, the children, Edna and Joanna into the car and drive them to the beach, a farm in Richmond or one of many favourite sites around the Lower Mainland. During the late 1920s and early 1930s, the MacMillans often entertained visiting Japanese timber buyers, for whom picnics were a cultural tradition, on Sunday afternoons at Spanish Banks, a Vancouver beach with superb views of the city and North Shore mountains. If the weather was poor, a family outing consisted of a long drive around the rapidly developing metropolitan area and lunch with the Flavelles in Port Moody, the Taylors or Graingers in Vancouver, or the Robsons or MacDonalds in New Westminster—all of them active in the lumber business. The centre of MacMillan's social life was his family; he was definitely not one of the "beautiful people" of the era.

Vancouver in the 1920s was, by the standards of the day, a pretty wild place. Prohibition in the US had led to the establishment of a large liquor business in the city, which helped lubricate changing moral and social standards. MacMillan was not a part of the fast life. He had no strong feelings about the behaviour of others, but appears to have had little interest in an active social life himself. Dinner parties with business associates and the occasional social event at the Hotel Vancouver were enough for him, and one of his greatest pleasures was a few weeks of hunting and fishing each year. On a trip to Portland in 1921, he met one of the partners in an insurance firm engaged to underwrite a cargo of lumber. Don Bates, an avid sportsman, soon became one of his closest friends. Bates had discovered the northern Cascades wilderness between Hope and Princeton in southern BC, and in the last half of the

1920s he and MacMillan began taking regular trips into the area with pack-horses, to hunt and fish. As his daughters, Marion and Jean, grew older, MacMillan organized annual family expeditions by horse along the trail from Princeton to Hope. On these trips he collected plants, which he carefully pressed and later took to botanists at the University of British Columbia for identification. With his time now consumed by the affairs of business, and his interest in forestry a receding memory, his recreational visits to the country were the only outlet for his scientific training and interests.

In a few short years MacMillan, the dedicated forester, had become completely absorbed in the commercial end of forestry. The buying and selling of lumber was as far removed as he could get from the ideas and issues with which he had concerned himself for more than fifteen years. The ardent conservationist had become a key player in a complex industry which existed by outright exploitation of forests. In February 1923, Fernow, one of MacMillan's few remaining ties to the profession and ideals of forestry, the teacher and leader who had selected MacMillan as his replacement, died.

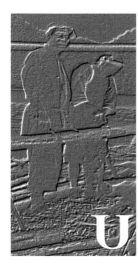

Uneasy partnership

As business began to pick up in 1922, the H.R. MacMillan Export Company was positioned to expand with it. The exceptionally high profits earned and the decision to leave them in the company left HRME well situated in regard to its competitors. Leon Scott opened up a lumber purchasing office in Seattle that year, and at about the same time MacMillan bought a small sawmill at Qualicum on Vancouver Island, the Bluebird Lumber Company. Percy Sills, whose relationship with MacMillan alternated between associate and employee, took over as manager of the sawmill and ran it until it closed two years later.

At dawn one morning in early September 1923, Sills was dragged from his bed by a phone call from MacMillan. A few days earlier a powerful earthquake had levelled Yokohama and most of Tokyo. Arthur

Bryan, the Canadian trade commissioner in Yokohama, had stopped in Vancouver on his way to Ottawa and given MacMillan an order for 40 million feet of lumber to help deal with the catastrophe. Bryan had been trying unsuccessfully for years to obtain orders for BC mills; the disaster was the opportunity he had been waiting for. Not bothering to take the order to Ottawa and pass it on through normal channels, he simply turned it over to MacMillan. In his usual peremptory manner, MacMillan told Sills to catch a ferry leaving Nanaimo in two hours, allowing him to board the *Empress of Russia*, which was leaving Vancouver for Japan at noon. His instructions were to help facilitate the arrival of the lumber and to obtain more orders.

In Vancouver, MacMillan and VanDusen scrambled to divert lumber to Japan so that they could take advantage of the fortuitous opening into the Japanese market. They redirected two ships loaded with MacMillan-exported timber to Yokohama, and placed large orders with practically every BC mill. By the beginning of 1924, confusion reigned. The Japanese agency established to handle the emergency purchases, the Fukoin, had overestimated its lumber needs and began cancelling orders. MacMillan decided in March that he had

Loading Powell River Company newsprint in Japan, 1924. MacMillan took this photo.

better go to Japan to protect the company interests. He and Edna sailed on the *Empress of Australia* and arrived in Yokohama on April 2. Superfluous lumber was stacked on the docks as far as the eye could see. The Fukoin was using every possible pretext to withhold or delay payment for lumber shipped.

> The Japanese companies in the timber trade set out to hurt us, their policy to use influence to delay payment. Our lawyer, E.C. Mayers, told us if we got in trouble on the official contract we had no chance, payments ceased. I went to Tokyo and placed my case for 3 continuous days before the Fukoin bench of three—one Mitsubishi man, the chairman an official of the Japanese government railway system, and one other, forgotten. The interminable conversations, I in English via an interpreter to the judges and back to me through Canadian government interpreter, got nowhere.
>
> On Sunday at the Imperial Hotel I was mooning around. The head of American Trading Company, an American Irishman, remarked on my disconsolate appearance. I told him my troubles with the Fukoin who spoke no English and seemed not to understand I was within my contract and should be paid. He said "No English? Why, the chairman is a graduate of Lehigh University and is proficient in English. I know him well, he lives near me." I arranged to meet him that day and got a decision in full the next day, to ship 10 percent extra and be paid in full quickly.[1]

There are several versions of the dramatic aftermath of this incident. When the MacMillan company began shipments to Japan, it engaged a Japanese agent named Watarai on commission to help with sales and, primarily, to protect payments to the company. Years later MacMillan described what occurred when he refused to pay the agent because of his failure to protect the company in its dealings with the Fukoin.

> I refused to pay him (Watarai). When I returned to Japan in 1927 he was on the warpath. He had a hired ruffian—a known type, something sounds like Feroski—German type helmet, glazed jackboots, carried a club. He threatened my safety, tried to take me out of my hotel room, also attacked me in public hall at entrance dining saloon on Empress. Also tried to pull me out of my room onto the gangway from ship to wharfside, but was each time ejected by a Japanese "gens d'armes" who while foreign ship in port maintain law & order. The next time I heard of Watarai he was in prison for conspiring to defraud Japanese government in importations of wood from Philippines.[2]

More dramatic versions of this story have MacMillan grappling with the intruder, who was attempting to assassinate him in his hotel room, and eventually throwing him through a window. There are various accounts of a gang of up to fifty thugs chasing MacMillan and Bryan around the ship, and almost succeeding in dragging MacMillan to shore. Whatever the details of the fracas, it was not enough to put him off doing business in Japan.

MacMillan's abrupt ordering of Sills to Japan was typical of the manner in which he dominated company employees, in much the same way as he dominated his family. Physically he could be very intimidating: in his prime, he stood six feet, four inches, and weighed just under two hundred pounds. His intense blue eyes glinted from under a pair of huge eyebrows. In anger, he inspired fear and obedience. Aware of his effect on others, he was not above using his powers of intimidation to get his own way. He liked to be in control of events, down to the most minute details. One of his favourite management principles was summed up in the phrase: Analyze, Organize, Deputize and Supervise. When he felt tough-minded, he often added "excise" to this dictum.

The detailed supervision of deputies was one of his best-known tactics. A youthful Henry Bell-Irving, who later became lieutenant governor of British Columbia, worked as a document clerk in the HRME office. He made a fatal decision to ship an order of airplane spruce to Italy without first obtaining an inspection certificate. When MacMillan noticed the omission, he called in Bell-Irving and fired him. He then proceeded to give him a stern lecture, during which Bell-Irving said in his defence it had not been possible to get a certificate. In tones he reserved for towering pronouncements, MacMillan shouted: "Young man! You must understand that nothing is impossible." He then rehired the young man, who stayed with the company until World War Two. Years later, when Bell-Irving was given a grant of arms, he adopted as his family motto the phrase, "Rien n'est impossible."[3]

On the other hand, MacMillan had a mischievous, occasionally cruel sense of humour, and it was often difficult for inexperienced employees to know whether he was serious or joking. One of his favourite tricks was to turn to a new employee while riding up in the cramped Metropolitan Building elevator and, peering down from his great height from under his shaggy eyebrows, inquire:

"And how long have you worked for the company?"

"Three weeks, Mr. MacMillan," the nervous young man might respond.

After a long moment of thoughtful silence, MacMillan would enquire: "And how much are we paying you?"

"Seventy-five dollars a month, Mr. MacMillan," the by now very nervous young man would answer.

MacMillan's timing was always impeccable. Just as the elevator doors opened he would shake his head gravely and say: "No one should be paid that much," stalking off down the hall with a hidden grin on his face, leaving the stricken employee quaking in his wake.[4]

As the company's business expanded steadily, the rented parcel space on Canadian Government Merchant Marine ships and other shipping lines was no longer adequate. The next step was to work through a British shipping company, the Furness Withy Line, which chartered entire ships for a cargo of MacMillan timber. The company tried a couple of time charters on its own in 1923—leasing ships for a set period of time, during which they would make a number of trips. This arrangement gave the company a definite edge over its competitors and, thus encouraged, a subsidiary shipping company, the Canadian Transport Company, was formed in June 1924. The new company's task was to hire ships on long-term leases and, when they were not in use carrying lumber from the Pacific Northwest for HRME, to book other cargoes at competitive commercial rates.

The time-charter business, while potentially lucrative, is a hazardous one, as MacMillan's successors learned almost fifty years later. In the 1920s it was a glamorous activity that appealed to MacMillan's sense of adventure, with ships coming and going to all the ports of the world, under the command of colourful masters who, when they visited the company offices in the Metropolitan Building, added an element of excitement and immediacy to the business. Now, MacMillan, VanDusen and others were apt to find themselves sitting up late into the night, following the course of one of their ships, loaded with a cargo of lumber, battering its way through a storm in the mid-Pacific or north Atlantic.

Initially, a British firm, Dingwall Cotts and Company, was hired to manage the shipping operation, but that company's Vancouver agent, H.S. Stevenson, fit so well into the MacMillan organization that within two years he took over as manager and ran the company until the end of World War Two.

[Stevenson] came in with the idea of loading fast and discharging fast which enabled us to make a really good deal at a very cheap freight rate. We organized a cargo of logs in Vancouver for discharging in Yokohama—a full cargo. We put it all in one spot, a berth down at Lapointe Pier, and worked around the clock until the time charter vessel was loaded. We turned it out in two days, or something like that. . . . We had the ship delivered back to the owners in Yokohama before they thought it would be sailing from Vancouver, and that cut the rate practically in half. We made a bucket with that.[5]

By the end of the decade Stevenson had between forty and fifty ships under charter at any given time and had opened up scheduled shipping lines to the West Indies and Montreal, the latter in successful competition with Canadian railways.

Many of the chartered ships were owned by Norwegians, and the nature of their businesses intrigued MacMillan. On a European trip in 1931, he visited some of the owners in their home ports, small towns and villages tucked away in fjords along the Norwegian coast, a geography similar to that of maritime BC. MacMillan was impressed with the "snugness" of life in these communities, and saw much he thought relevant for the development of BC's coastal regions. Occasionally these well-appointed, deep-sea ships returned to their home ports, and while the crews spent time with their families, other villagers overhauled the vessels and readied them for more months or years at sea. This, he realized, was the source of the Norwegian merchant marine's great strength, and was also a contributing factor to the kind of community stability missing in frontier-minded BC.

At about the same time that he set up Canadian Transport, MacMillan decided to broaden the base of the lumber export business. One of the problems created by the success of the fir sleeper business was that there was no market for the large amount of other grades and types of lumber cut out of fir logs. The use of a high-quality wood like Douglas fir for railway ties not only offended everything MacMillan had learned as a forestry student, in the long run it was simply bad business.

By early 1924, Meyer had had some success in securing an order for a full range of fir products for the British Admiralty. Thus encouraged, MacMillan decided to move into the high-grade spruce trade as well. Although he agreed to this decision, Meyer was not enthusiastic about it, and the message had got through to MacMillan in Vancouver. In June he wrote and told Meyer he had arranged for Percy Sills to take charge of purchasing spruce on the BC and US west coast, and in a series of letters he attempted to get Meyer and the Canusa organization to take a more aggressive approach to marketing aircraft spruce. What he had observed, during the war and through his partnership in Bay Lumber, was that aircraft-grade spruce commanded exceptionally high prices, and other exporters were shipping large volumes to Europe, Japan and the UK. Spruce was still one of the critical materials in aircraft construction and, as these countries recovered economically from the war, they were beginning to construct military aircraft on a large scale. Although it was not apparent to many at the time, it was one of the early indications another war was approaching.

In July, MacMillan wrote to Meyer again, informing him that Sills would be visiting England to feel out the British market for spruce. In case Meyer had misread his feelings, he wrote:

> As an instance of the fact that your organization has some work to do before they get in good shape to handle this business I would mention that Bamberger [a major British aero spruce buyer] is working here on an enquiry of 2,000,000 ft. concerning which we have heard nothing. The Ocean have been shipping at least 500,000 ft. per month and they have probably 2,000,000 ft. brought forward although we received no enquiries from you to indicate that such a volume of business was being done. Competitors are buying all the time. We hear of their activities continually, nevertheless in the four weeks since we advised you we were starting this department we have had no enquiries.[6]

It appears that Sills's mission was broader than merely investigating the spruce market. In August, on board the *Empress of France* in the North Atlantic, he received a telegram from MacMillan.

> HAVE NOTICED LAST MONTHS COMPETITORS DOING MORE UKAY BUSINESS THAN WE THOUGH OUR PRICES CERTAINLY COMPETITIVE ALSO THAT SEVERAL LARGE SLEEPER CONTRACTS PLACED WITH COMPETITORS WHICH CANUSA EVIDENTLY NEVER HEARD OF STOP I ATTRIBUTE THIS EITHER TO GENERAL SLACKNESS ON PART OF CANUSA OR THEIR PREOCCUPATION OTHER THINGS STOP I AM CABLING THEM URGING GREATER ACTIVITY BUT WE CANNOT CONTINUALLY ENERGIZE THEM SO PLEASE CONVEY MY IDEAS TO MONTY DIPLOMATICALLY AND FORM YOUR OWN OPINION CONCERNING THEIR CONTINUED DEVOTION OF ENERGY AND ABILITY TO OUR BUSINESS AND CABLE ME SO THAT ANY NECESSARY ACTION MAY BE PROMPTLY TAKEN STOP REMEMBER MALLINSON BAMBERGER LAMB BROTHERS VINCENT MURPHY HOLME THE ADMIRALTY AND THE GLASGOW CROWD [all purchasers of spruce] ARE YOUR MEAT[7]

After two months in England, Sills returned on the *Empress of Scotland*, from where he sent MacMillan a long handwritten letter assessing the Canusa operation.

This Canusa outfit appears to me very much like a watch—wheels within wheels with Monty Meyer as the main spring and the [one] which you will have to keep on winding up. It is not a clear cut business organization like yours is—solidly grounded, with its parts functioning according to and solely by their energy & ability and sent into the discard if they don't—but all infected with a sort of cancerous growth of intrigue and fear and bootlicking, each one seems to have something on everyone else.

The organization itself is rotten (I will make no reference anywhere to finance and financing) chiefly through overlapping of the three main executives. . . . God damn it, they don't know how to work. I'd get down at 9:15 to find a clerk just straggling in—it is 10:30 before they get to work, and by God any time I got a cable translated before 10:15 it was a lucky day. And as you know they only claim to work from Monday noon to Friday night—they talk custom of the country and English mentality—I think this is largely bunk as I believe in other offices they work more or less like we do.[8]

When, after a couple of months, there were still no inquiries from Canusa for spruce, MacMillan sent Sills back to England, where he concluded a sale with one of the large aero spruce buyers. Meyer was furious, and in February sent MacMillan a twelve-page letter describing in minute detail his reactions to Sills's performance and character.

As regards Sills himself, I took rather a fancy to him. He is an amusing fellow, with a certain amount of technical Timber knowledge. I found him, however, what I would call, somewhat unstable commercially and a person who takes every opportunity of talking. . . . I do not think that his mentality is one which would make him a permanent servant of the MacMillan Co. and I would, therefore, suggest to you, even if it is difficult or expensive, that you should get somebody else, on whom you can permanently rely, to make himself acquainted with this very special and difficult trade so that in case Sills decides to suddenly retire from his association with the firm, you will not be in the position of having everything chaotic, but will be able to carry on without his help almost as well as with it.

I think you know me so well that you will realise that there is nothing vindictive in my nature. I am praising Sills in this letter, and I am blaming him also. He is a man who has mixed up in him merit and stupidity. He is keen, energetic and knows certain things about wood. He is not a business man, he talks too much, he is too conceited, and is generally somewhat unstable.[9]

MacMillan responded with a brief letter, saying "I am now in the position of reserving judgement as to what we should do until I receive the whole history . . . from Sills."[10] There is no record of his conversations with Sills on the latter's return to Vancouver, but within two years MacMillan had drastically altered his business relationship with Meyer.

By the mid-1920s, with BC enjoying one of the strongest economic booms in its history, MacMillan was expanding his business and personal interests in many directions, and on several planes. By the time he celebrated his fortieth birthday at a Chinese restaurant in London with Edna and Loren Brown, BC's lumber commissioner in the UK, he had reached his full stride. The export company had established itself as a respected part of the British timber business and was expanding rapidly into Europe, Africa, India, Australia and Japan. Offices had been opened in Montreal, Seattle, Portland, Japan, Australia and New York. The growing markets were being serviced with a company-owned shipping agency which, within two or three years, had become a highly profitable business in its own right.

Having made a serious dent in the US brokers' domination of the BC export market, as well as having established a purchasing capability in their back yard, MacMillan proceeded to attack them directly on their position in BC. In 1925 he initiated a lobby of federal politicians, through Vancouver MP Henry Stevens, aimed at undercutting the Americans' business in BC.

> British Columbia exports each year about six hundred million feet of lumber with a CIF value of something over $20,000,000. . . . Under present conditions about one-third of this business is conducted by Canadian companies with headquarters in Canada operating under Canadian laws, and the balance of it is conducted by foreign companies operating almost altogether with headquarters in San Francisco, Portland and Seattle, and not subject to Canadian laws. . . . So far as I can learn, and I have gone into the matter with all the facilities at my disposal, the American companies resident in Seattle conduct this merchant business in British Columbia with the same freedom as we conduct it and pay no taxes whatever either on corporation profits or on individual incomes to either the Dominion Government or the Provincial Government. . . . The remedy for this situation appears to me to lie in having the Dominion and Provincial Income Tax Departments collect the same taxes from foreign merchants who buy and sell Canadian products as they collect from resident merchants.[11]

Stevens and MacMillan maintained a lengthy correspondence over the next decade or so, which, in light of Stevens's controversial political career, raises the question of MacMillan's influence on federal politics between 1925 and the outbreak of World War Two. Stevens was a longtime Conservative who served in Arthur Meighen's cabinet during the 1920s. In the 1930s he headed a Royal Commission on prices and embarrassed the government with his strong attack on the practices of some big businesses. He resigned from the commission and the caucus and started his own Reconstruction Party, which took 10 percent of the federal vote in the 1935 election. For a brief period, during MacMillan's initial dabbling in national political issues, he and Stevens shared several political opinions.

MacMillan aimed another shot in his campaign against American timber brokers at R.B. Teakle, head of Canadian Government Merchant Marine.

> Although we, a Vancouver firm, had tried unsuccessfully for eight months to secure space on the Isthmian Line, which is the largest carrier of lumber from British Columbia to New York, nevertheless some of our competitors, with their head office in Seattle and their bank accounts in Seattle, were able to get space both on these American ships, from which we are debarred, and also on the Canadian Government Merchant Marine ships. I mention this, not as an indication of any bad treatment we were receiving, from the Canadian Government Merchant Marine, but as an indication of the freedom with which we in this country allow all people to trade as opposed to the manner in which American ship owners and American merchants work together whenever feasible to exclude foreigners, and to develop American trade and keep it in American hands, even to the point of discrimination against Canadians. Another instance of this same practice might be seen in the reported arrangement made by the Hearst Newspapers to discontinue buying Canadian paper CIF so long as it is shipped in Canadian ships, and to adopt the policy of buying it FOB plant, so that they can make arrangements to ship it in American ships. This policy of working together to extend American trade may be very well commended by Americans, but I have noticed in recent years in Vancouver several instances where it is undoubtedly working against Canadians.[12]

This issue and these letters represent MacMillan's initial foray into the public policy arena. As chief forester and trade commissioner he had enjoyed a certain level of public esteem and attention from the media, but this had declined with his move into private business. Now, having established a local reputation in business, he again began to express his views on public issues—particularly if they affected his own interests.

One issue over which MacMillan and his partner Meyer disagreed was whether to expand the business into the lumber manufacturing sector. Meyer was a timber dealer, the archetypal middleman; he had little interest in the sawmill business. MacMillan, on the other hand, had the instincts and habits of George Weyerhaeuser, who built his timber empire by expanding toward the source of supply. In 1925, MacMillan developed his interests independently by taking a share of Chehalis Logging with Aird Flavelle. This old friend had continued to develop his business in the province and, among his other activities, was a partner in a timber management firm with Frank Pendleton, the US logger MacMillan had brought in to manage the wartime spruce-logging exercise. Martin Grainger and John Lafon, both of whom left the BC Forest Service, also joined the firm.

The following year the H.R. MacMillan Export Company, acting through its subsidiary, the Bluebird Lumber Company, purchased the bankrupt Pacific Cedar Company on the north arm of the Fraser River in south Vancouver. Shortly thereafter, it was renamed the Canadian White Pine Company (CWP) and, under the management of Sills, absorbed Bluebird Lumber. The purpose of CWP was to mill western white pine, a relatively scarce wood on the West Coast, but one for which an established market existed because of the widespread use of eastern white pine, upon which the forest products industries of Ontario, Quebec and the Maritimes had been founded. In the past, most white pine from BC had been sold in the form of hand-hewn cants, known as "waney pine," that were used in industrial pattern making in the UK and Europe.

Within a year of this purchase, MacMillan and Meyer terminated their partnership. Although no account of the circumstances surrounding the separation exists, it appears to have taken place on MacMillan's initiative. MacMillan purchased Meyer's shares, with payments spread over a year. Canusa continued as the MacMillan company's UK agent, and the two principals remained friends until Meyer's death in 1961.

However, the men's friendship was severely tested a year after they dissolved their partnership. Under the terms of the transaction, MacMillan agreed to pay Meyer the relatively small sum of $500 a year in exchange for Meyer's agreeing to spread payment for the shares over one year. A year later, after making the final payment for the shares, MacMillan terminated payment of the $500, and Meyer was furious. He expressed his feelings in a letter.

I tell you perfectly frankly that I felt extremely hurt that you should suddenly decide that business was too bad to pay me the 500 pounds per year, which you had agreed to do. I liked receiving this amount, and in view of the fact that we started this business together and that my association with it for some years must have been of inestimable value to it, I should not have considered—if I had been in your position—that it would be paying too high a price to retain by this payment the permanent interests of your original partner in a business which he had certainly helped you to create. . . . I need scarcely tell you that the financial difference to me is comparatively small, but the sentimental effect is consider-able. I feel now that I am nothing in the business, that I have nothing to do with it in any way whatsoever, and that I am simply classed with Martin Olsson, Foy Morgan and Denny Mott, as an agent or customer.[13]

MacMillan wrote back a letter of profuse apology—although he did not reinstate the £500 payment—and his friendship with Meyer was restored. The relationship between their firms changed considerably, however, as the MacMillan company continued to diversify and expand, eventually dwarfing Meyer's company, Canusa.

By early 1927, MacMillan had not paid any attention to the forestry profession for almost a decade. He had very little contact with foresters, did not write about the subject as he once had, and appears to have lost all inter-est in the subject. Fernow's gibe concerning his employment at VL&M appears to have stuck in his mind, and he may have felt a few pangs of guilt. In response to a request for information from the Yale Forest School Alumni Association, he wrote: "I have strayed very far from the field of professional forestry, and have got so far away that I hardly ever see any foresters. My work has been entirely in forest destruction in recent years."[14]

The request may have stirred his memories: it was at about this time that he resumed correspondence with his old Yale roommate, W.N. Millar, now an associate professor of forestry at the University of Toronto.

I am out of touch with things but I fancy that the various organizations are in just about the same shape they were when I left the profession. The public is still in the same mental condition, that there should be some forestry practised some-where, preferably some place else. The public, so far as I have seen, has very lit-tle understanding of how the Government, to whom the duty of forestry is assigned, should practice forestry on public lands. The general assumption is that the practice of forestry by the Government on public lands is never going to interfere with the profit or comfort of the person who is thinking about it. . . . It will be quite a while yet before concrete thinking will be done by the foresters.[15]

As it turned out, his dissociation from forests and forestry would not continue for long.

MacMillan and Meyer's success in persuading British railways to use Douglas fir sleepers had some unforeseen consequences. The railways drew up tough, new specifications for fir sleepers in 1927, precipitating a conflict between MacMillan and most of the mills whose lumber he sold.

The Astexo group of companies, which banded together the same year MacMillan went into the export business, had profited enormously from his success. They were by far his largest source of supply. The relationship between his company and the thirty-odd Astexo mills was an uneasy one, although mutually profitable. While the mill owners did not want to be directly engaged in the lumber-export business themselves, preferring to sell FAS, they came to resent MacMillan's success. Many considered him an upstart and an outsider. Others thought, not entirely without cause, that he was making unreasonably high profits at their expense. A common accusation was that he accepted orders for lumber and sat on them, waiting until the mills accumulated excess inventories which drove prices down, when he would then buy, increasing the spread between his buying and selling price to his advantage. He did engage in this practice, as he admitted in a letter to Meyer: "I will tell you in confidence that I did not buy all our goods as soon as I sold them and benefited somewhat thereby."[16] Mill owners were further irritated when MacMillan established the Canadian Transport Co.; they resented the fact that he made a further profit on shipping which he did not pass on to them. The final straw was his establishment in 1925 of a sales office in New York to sell into the US east coast market, which the Astexo group viewed as their turf. At the end of 1926, Charles Grinnell, a highly respected American lumber dealer from Tacoma, became manager of MacMillan's New York office.

When the British railways put out orders for fir sleepers under the revised specifications, MacMillan accepted their orders and shopped them out to the mills. The Astexo group declined to quote on them, indicating their refusal to sell to MacMillan. Having accepted the orders, MacMillan found himself in the untenable situation of having no source of supply. Instead of attempting to compromise with the mills, which had been their objective in refusing to quote, MacMillan took direct and drastic action. Late that fall he went to southern Vancouver Island in search of his old logging friend, Matt Hemmingsen. He learned that Hemmingsen was on his way to Victoria the next morning to sell a large tract of timber he owned in the Malahat district, north of Victoria, so MacMillan spent the night in his car on the side of the road. Early the next morning he stopped Hemmingsen and negotiated the purchase of the tract, containing about 32 million feet of timber, mostly fir. This purchase also gave him access to a smaller, adjacent stand which he later obtained from the Esquimalt & Nanaimo Railway Company.

By the new year, ten portable sawmills and a score of small logging contractors were operating in the hills above Saanich Inlet. The first lumber from these mills was used to build housing for the workers and their families, many of them refugees from failed prairie farms. Once the mills were established, they began cutting sleepers at a phenomenal rate. Teams of horses hauled wagonloads of ties to hastily erected chutes that carried them down to scows anchored in Mill Bay, from where they were towed to ships anchored across the inlet in Brentwood Bay. In less then two years, this classic gyppo operation, reminiscent of the Queen Charlotte Islands spruce endeavour, shipped 22 million feet of sleepers. When it became impossible to fill the orders by the contracted date, Meyer persuaded the railways to grant an extension. MacMillan had won the first skirmish in the developing battle with Astexo.

An incidental aspect of these events rated coverage in Vancouver newspapers. At one point during the discussions, MacMillan made use of the recently installed trans-Atlantic telephone system. One day he placed a call to Meyer for early the next morning, and at 4:00 a.m. was notified at home that his call was ready. His account of the event was fully described in the press. "The clicking of the keys of a typewriter was the first sound I heard from London. This was followed by a voice saying, 'Stop that typewriter,' and then Mr. Meyer came on the line."[17] The two-minute call cost $57, and signalled a major change in the way the lumber export business was conducted.

A few months later, in June 1928, the Astexo companies made their next move with the creation of Seaboard Lumber Sales, a co-operatively owned lumber exporting firm initially established to corner the eastern US market. The industry was surprised to learn that Charles Grinnell, MacMillan's recently hired New York manager, was to be Seaboard's manager. Any solid evidence has long disappeared—perhaps consigned to shredding

MacMillan, the successful businessman, 1930.

machines—but many in the business thought Grinnell had moved to the MacMillan company only to acquire knowledge of its operations in anticipation of the formation of Seaboard.

MacMillan's response, again, was to seek an alternative source of lumber. To avoid the possibility of being refused a sale, he engaged Aird Flavelle to make inquiries and initiate discussions with the owners of mills and timber that might be for sale. Flavelle quickly focussed his attentions on the Alberni Pacific Lumber Company operation at the head of Alberni Inlet on the west coast of Vancouver Island. It consisted of a large, well-run, profitable sawmill and shingle mill, with enough privately owned timber for almost fifteen years' operation. The company, which had 650 employees and produced between 80 and 100 million feet of lumber a year, was owned by Denny, Mott & Dickson, a prestigious British firm. It was managed by Ross Pendleton, a son of MacMillan's old associate and Flavelle's partner, Frank Pendleton, who by this point was a central player in the Astexo–Seaboard group. At the same time they were trying to arrange a sale to MacMillan, Flavelle and Martin Grainger were engaged as agents of the owners to help manage its affairs. Given the number of hidden agendas, overlapping interests and personal connections involved, it was a complicated situation, but by the spring of 1929, Flavelle had negotiated the basics of a deal with the Alberni Pacific owners. MacMillan assumed a controlling interest in the company, while Denny, Mott & Dickson retained a smaller interest and the remaining shares were sold in a public offering.

The long-term health of Alberni Pacific, however, required the acquisition of additional timber, and Flavelle concentrated his efforts on discussions with Sid Smith, timber manager for the US-owned firm Bloedel, Stewart & Welch, by then the largest and most aggressive logging and sawmilling company on the coast.

Since the Bloedel company was known for its huge appetite for timber holdings, it is surprising that Smith even entertained the possibility of selling some to MacMillan, particularly in view of subsequent events. What is even more surprising, more than two decades before it was eventually consummated, the MacMillan and Bloedel companies were discussing a merger. In April, Flavelle cabled MacMillan in Toronto:

SAW SMITH WITHIN MONTH HE INDICATED THEY WOULD PROBABLY CONSIDER FAVOURABLY OFFER FOR TIMBER ON BASIS FIVE HUNDRED CASH IF MADE IN REASON-

ABLY NEAR FUTURE STOP HAD LONG TALK VAN YESTERDAY HE IS VERY ENTHUSIASTIC
STOP THINKS WORTH CONSIDERING PUT YOUR EXPORT COMPANY ALBERNI AND
BLOEDEL TIMBERS ALL INTO ONE HOLDING COMPANY AND REFINANCE. . . . I WOULD
PERSONALLY RECOMMEND GOING AHEAD WITHOUT ADDITIONAL TIMBER STOP WE
WOULD BE ASSURED TWELVE TO FIFTEEN YEARS SUPPLY AT PRESENT RATE OF CUTTING
FROM PRESENT HOLDINGS STOP PRESENT RATE DEPRECIATION AND DEPLETION
WOULD PAY OFF ALBERNI INDEBTEDNESS AND COST OF PROPERTY IN TEN TO TWELVE
YEARS LEAVING ALL NET EARNINGS AND THE PLANT AND EQUIPMENT AS VELVET
WHILE WE COULD CERTAINLY PROLONG THE LIFE OF THE OPERATION CONSIDERABLY
DURING THE NEXT TEN YEARS BY OTHER PURCHASES OF TIMBERS WHICH WHILE NOT
PERHAPS EQUAL BLOEDEL IN QUALITY OR CHEAPNESS OF OPERATION WOULD STILL BE
VERY USEFUL. AIRD[18]

Before the proposed transaction could be concluded, circumstances
in the wider world overshadowed the immediate concerns of the H.R.
MacMillan Export Company. Early in 1929, the US lumber industry began calling for increased tariffs on Canadian lumber. This was precipitated in large
part by the rapidly escalating volume of BC lumber—much of it MacMillan's—
going into the Atlantic coast market. The formation of Seaboard, for the
express purpose of feeding that market, added fuel to the flames, and
MacMillan blamed Seaboard for upsetting the Americans.

> Although when the business was first developed by the Seaboard Sales in the
> Atlantic Coast market, the dealings were almost entirely with wholesalers, the
> policy was gradually changed to a point where wholesalers were eliminated to a
> great degree and sales were made direct to most of the wholesalers' good cus-
> tomers. This finally earned for the Seaboard Sales the hostility of the wholesalers.
> As the strength of the Seaboard Sales developed, the management discovered in
> their hands a great power in control over freight, which power they used almost
> entirely against the interests of [US shippers] . . . thereby bringing the American
> Shipowners' Association into the field as strong and aggressive enemies. The
> combination of these circumstances . . . resulted in the imposition of the present
> prohibitive tariff on lumber into the United States.[19]

While he was in the east in April, MacMillan went to Baltimore to
address the US National Trade Council with his first significant public speech
in more than ten years. In it he described clearly the Canadian position, in an
attempt to quell the rising emotional tide in the US calling for tariffs. Behind
the scenes in Canada, he was supporting stronger tactics, as he wrote to the
president of the Vancouver Board of Trade:

The business men of Canada should be preparing for unfavorable tariff action by the United States. Any policies that it may be wise to adopt, should be based on sound study and not on instincts of anger, retaliation or other emotional reactions. Possible new fiscal policies of self-help by Canada might include, amongst other things, export taxes on certain raw materials as are returned to us in manufactured form.[20]

Then, in October, the stock market crash brought down the curtain on the decade-long period of economic growth that had vaulted the BC economy into the midstream of the modern world. It took the lumber industry a few months longer to slow down than most economic sectors, although it was quite clear to those in the industry that the days of quick and generous prosperity were over. MacMillan shared, or adopted, the view of Sir Joseph Flavelle that common sense in investment planning had been replaced by a speculative fever.

During the 1920s, enormous economic changes had been wrought in BC, one of the most significant being that a number of Vancouver businessmen acquired control over major sections of the economy. Before World War One, the business life of the province had been controlled from elsewhere—San Francisco, Toronto, New York and London. Ten years after the war ended, much of BC's business life was in the hands of young entrepreneurs like MacMillan. The locally owned *Vancouver Sun* newspaper applauded the success of the hometown business elite and used it to attack its chief competitor, the *Province*, owned by Montreal interests.

The Vancouver Sun prefers to help and promote the activities of local people. When the Austin Taylors, the W.H. MacMillans [sic], the Odlums, the Victor Spencers, the Stanley Burkes, the Phil Malkins, the Woodwards, the Buckerfields, and scores of others of that large growing circle of active young business men who compose the Vancouver group, make big money, that money stays in Vancouver.[21]

For a decade, MacMillan had focussed his energy on creating the H.R. MacMillan Export Company. From its original capitalization of $20,000, the company had grown to the point where it was doing more than $15 million in business annually, making profits of up to $250,000 a year and employing 100 people in the process. Early in 1929, in a kind of acknowledgement of his success, MacMillan made his first contribution to UBC, a $1,000 scholarship in economics that included a free study trip on a CTC ship to the Orient.

MacMillan had become a wealthy and powerful player in the BC business scene, with a growing reputation in the east, the UK, the US, Australia and Europe. Some of his success was due to good luck—happening to be in the right place at the right time—but mostly it was a result of his almost uncanny ability to predict the consequences of events, and to perceive and capitalize on business opportunities. Having spent the 1920s building a business, he would devote the next decade to protecting his "outfit," as he called it.

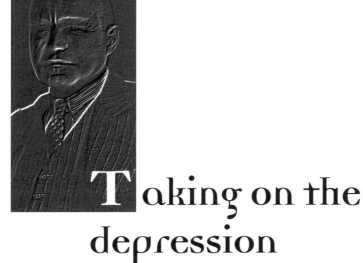

chapter 10

T aking on the depression

To an exporter such as MacMillan, the stock market collapse was not an event of momentous significance. Although he had ridden the crest of the boom through the late 1920s, several times during 1929 he commented critically on the speculative frenzy that had overtaken the business community. In many ways, as he indicated in a letter to Monty Meyer written three days before Black Thursday, the crash was welcome relief. "I have never seen business as bad as it is now," he wrote. "I think we are in for a few months of comparatively hard times in the lumber industry, which should produce some good effects."[1]

The H.R. MacMillan Export Company was in a better position than most businesses to face the coming depression. MacMillan had gained full control of the company from Meyer. His policy of reinvesting profits, which had risen strongly over the ten-year life of the firm, had fuelled steady expansion. One of the keys to the strength of the company was its internal coherence, largely a consequence of the working relationship between MacMillan and VanDusen. A unique kind of partnership had developed between the two very different personalities, giving a balance to the firm. VanDusen described the nature of their business relationship fifty years after it began.

> He was very imaginative. It was agreed in the early days that he would do all the speaking and the front work. I didn't like it, I didn't want to do it and he liked it. And he was always the fellow who had the last word, I recognized that right from the start. I put up my own views, but he had to have the last word. We have been close personal friends always. We worked most of the time. We always had far more fun working than we did playing. I don't think we ever disagreed on anything of major importance as far as the company was concerned. We always discussed it, and if I had an opinion and he said something else, well that was it. I didn't pursue it if I had a different view. I just accepted his. For instance, I was a little bit against the company getting too large, and I may have been wrong. I was a little opposed to expansion, but I went along. At no time did it ever get to a matter of principle though. We had a very good working relationship. We were a team and neither one of us would go and do something without talking to the other. . . . I never wanted to get into a position where I was, say, Chief Executive Officer, and he was the Chairman—where I would be responsible and he would be criticizing me for the things I did that turned out wrong. I would never do that, I never considered it. He often wanted me to be President, which I didn't want. So we did it some other way. He was made Chairman and I was Vice-Chairman. I think it was a very happy association and I think it worked.[2]

Much more significant to the export company than the collapse of the eastern US stock market was a looming threat to the core of the business, the UK market. In 1929 the Soviet Union, desperate for foreign exchange, launched a massive timber exporting drive that threatened to destabilize global timber markets. Initially, both Meyer and MacMillan looked on the Soviet trade as a conventional commercial operation and considered taking part in it. Meyer visited Russia in the summer of 1929 and discovered, to his mild annoyance, that MacMillan had been making inquiries through another British agent, Furness, Withy and Company, with a view to importing Russian timber into the US. They exchanged a number of letters on the subject, with Meyer pressing to act as agent on the Russian end of the venture. However, as the Soviet trading strategy became more clear, MacMillan backed off.[3]

It soon became apparent that the Soviet Union's communist government was not just seeking foreign currencies; it had a much broader agenda. In the case of the timber business, at least, the Russians were out to undermine the very foundations of the trade. The Soviet method was to ship an enormous volume of timber—800 million feet in 1930, about four times Canada's annual shipments to Britain—into the UK on a speculative basis. Instead of using the conventional CIF or FAS systems employed by the timber trade, the Soviets supplied Timber Distributors, a group of 100 British dealers organized to sell Russian timber, with sales agreements that included a fall clause. Under this arrangement, early buyers of Russian timber would be reimbursed if timber prices were to drop in the future, in response, for instance, to competitive action from exporters in British Columbia. No western-based timber company could compete with a state-run system utilizing what amounted to forced labour.

Through his contacts abroad and the experience gained in his travels, MacMillan developed a clear picture of Soviet intentions, as well as the advantages the Russians enjoyed over Canadian producers.

> The quality of the [Russian] wood is excellent—none better anywhere. The manufacture is much superior to Canadian manufacture in that every piece is sawn oversize so that when drying and shrinkage are finished the piece measures fully to its nominal size in all dimensions. The lumber is completely seasoned before shipment. . . . The freight rates on lumber from Russian exporting points . . . are about $5.50 Canadian funds per thousand . . . compared to $10 from Western Canada. Moreover the small vessels plying the short trip from Russia serve scores of small ports in the United Kingdom which cannot be reached by the larger vessels from Canada.
>
> This low freight and the Russian economic scheme combine to make possible a sale price which appears impossible for Canadians. Canadian prices are now based upon: no allowance for the standing timber, little or, in some cases, no interest or depreciation for plant and capital, and a level of wages which cannot fairly or wisely be reduced unless there is a further all-round reduction in the Canadian standard of living. If Russian authorities decide they must export timber to establish foreign credits, they will do so—they need make little or no provision for wages, which now constitute in the neighbourhood of 75% of the Canadian cost. . . . Should the Soviet organization succeed with present plans to rapidly increase lumber production from maritime Siberia they will seriously hurt British Columbia markets in Japan, China, Australia.[4]

At this stage, the nature of the Soviet state and its threat to free market economies was only dimly understood by the large majority of western business and political leaders. MacMillan was one of the first in Canada to grasp the consequences of Soviet actions for Canadians. Open competition with the Russians in the lumber export market would lead to drastically reduced wages for Canadian workers which, in the end, would further undermine the country's economic and social structures, already weakened by the depression. This realization prompted MacMillan's involvement in two distinct and time-consuming courses of action—taking on a new business, and taking the lumber export company into a new market.

In May 1930, while he was on board the SS *Europa*, MacMillan received a telegram from Stanley Burke, president of BC Packers, the largest fish processing company on the West Coast, inviting him to become a director of the company. BC Packers had been formed in 1928 through a merger of several partnerships and smaller private companies. Since its formation, the company had been controlled by Toronto-based directors, and it was losing money at an alarming rate. Due to prodding by J.H. Gundy, a principal in the Toronto accounting firm Wood Gundy, it was decided to place control of the company in western hands. Burke, a partner in the Vancouver investment house Pemberton & Son, was made president, and soon after MacMillan, food wholesaler Phil Malkin, BC Power chairman George Kidd, and Wood Gundy's Winnipeg partner Alan Williamson, joined the board.

The British Columbia fishing industry had begun in 1870 and grown rapidly over the next sixty years as export markets were developed, principally for salmon. By the early years of the twentieth century, there were scores of canneries operating at numerous locations on the BC coast. They were supplied with fish by a large fleet of fish boats, some owned independently, others by the canneries themselves. Over time, as the gear and the boats evolved, many of these canning plants became redundant. As fishermen were capable of travelling farther to deliver their catches, fewer canneries were needed. Modest technical improvements on the fishing grounds had enormous social and economic consequences for the canning sector of the industry. In part, it was these changes that led to the creation of BC Packers.

BC's annual salmon catch often fluctuates wildly from season to season. The year MacMillan joined the BC Packers board, 1930, spawning runs were enormous. Salmon, particularly sockeye, the prized canning species, flooded the canneries, which packed an unprecedented 2.2 million cases of fish. But that same year, the countries that bought BC salmon were plunging into economic depression. A potential bonanza turned into a financial disaster as the fishing companies were left with warehouses full of unsold salmon— fish they had bought from fishermen and paid to have canned. They went into the 1931 season with a huge inventory which had to be sold before they could think of marketing the next pack. That year they canned only 685,000 cases, spelling economic ruin for hundreds of fishermen and dozens of smaller canneries. For MacMillan, it was a dramatic introduction to the fishing industry.

That same year, MacMillan concluded a transaction at least as significant as his formation of the company with Meyer in 1919. In 1927 he had gone to Japan and China, in part to investigate the establishment of an East Asian office. He visited several large trading houses, including Jardine Matheson, the most prestigious and prosperous of all the western companies operating in the Orient, where he met the Keswick family, who controlled the company. It took three years of correspondence before the staid, secure managers of Jardine Matheson agreed to an alliance with the Johnny-come-lately from Vancouver who had so impressed them on his visit. In November 1930, the *North China Daily News* carried a large advertisement announcing that the H.R. MacMillan Export Company was open for business in Hong Kong, in association with Jardine Matheson. It was a move that established the export company, and MacMillan personally, as major figures in Pacific Rim trading circles.

At the same time, he was busily occupied with several other projects. In 1930, HRME faced an internal problem brought on by its small initial capitalization. Although the company was selling close to $20 million a year in timber, its capital still consisted of the original $20,000 put into it by MacMillan and Meyer. Until 1930, when federal tax laws were amended, financial surpluses could not be capitalized without paying substantial taxes. When the federal law was changed, permitting tax-free capitalization of surplus earnings, MacMillan decided to reincorporate as a federal company. At the same time, he and VanDusen redistributed the common shares among some of the company's key employees. When the transaction was complete in November, MacMillan held 3,173 shares; VanDusen 1,765; L.R. Scott 1,765; H.S. Stevenson, the manager of Canadian Transport, 1,200; Loren Brown 700; and Harold Wallace 262. All worked for the company, and anyone leaving HRME was required to sell his shares to the others.

Three young people from Vancouver joined the export company in 1930, none of them in positions that would necessarily have drawn them to MacMillan's attention. In time, however, they would become three of the most important people in his business life. In January, Ralph Shaw began as a junior clerk in the document department, assembling and processing the papers that made up the essence of lumber exporting. A few months later, a remarkably attractive twenty-eight-year-old woman was hired as a general office secretary. A native of the Kootenays in southeastern BC, she was almost single-handedly raising a large family of brothers and sister when she lost her job with a stock brokerage firm that went bankrupt in the crash of 1929. Within a year, Dorothy Dee had become MacMillan's private secretary, a role to which she devoted her complete attention for the rest of her life. In the fall, Bert Hoffmeister was hired as a junior salesman—in fact, a glorified office boy—at Canadian White Pine. Of the three, MacMillan probably noticed Hoffmeister first. The day before the new employee began work he broke his collarbone in a rugger game and appeared with a wooden cross taped to his back, forced to hold up his right hand with his left. He and Shaw became favourites of MacMillan and eventually assumed the top positions in his company.

During 1930, MacMillan also became a major participant in a campaign initiated by Martin Grainger to create a provincial park in the Cascade Mountains between Hope and Yale, BC. In the late 1920s, the eccentric mathematician, while serving as president of the Alberni Pacific Lumber Company, had made numerous trips along the old fur trails running through the mountainous area, and had built a cabin on the ranch of a local guide, Bert Thomas. MacMillan also became a friend of Thomas and hired him for his annual excursions into the area with his family, and for hunting and fishing trips with Don Bates, his Portland insurance agent.[5] Plans to build a highway through the area activated the trio. Grainger and MacMillan, whose dormant conservationist ethic was re-emerging, focussed their formidable energies on politicians, bureaucrats and journalists in a manner that ironically anticipated the later campaigns of environmentalists to save various parts of the province from logging by MacMillan's company. The men feared that once the area was opened up to vehicle traffic, its large wildlife population would be devastated by road hunters and the spectacular scenic beauty tarnished by commercial activity.

In June, MacMillan appealed to F.P. Burden, BC's minister of lands:

> I have been familiar with this region between the headwaters of the Skagit River and the headwaters of the Similkameen and Tulameen Rivers for the past 17 or 18 years. It is the best piece of high country available anywhere near Vancouver which is suitable for all forms of summer enjoyment, and when the road is finished will be within four or five hours by automobile from the city.
>
> Although it has been well known to the population since the Fur Brigade Trail was built in the 1840s no resources capable of development have been discovered. If a park were created no resources of any importance would be withheld from exploitation. A park would, however, enable quite a few people in the region to earn a steady living from the expenditures of travellers, but if a park is not created and the wild life protected the area will probably be devastated very quickly.[6]

By September, as part of its unemployment relief program, the government had begun work on the road from the Princeton end. MacMillan's fears had been realized, as he explained to Attorney General R.H. Pooley.

> Up until last year the greater part of this winter range was miles from a highway and could only be reached on horseback or on foot, which served as a protection for the deer, but now, over the newly constructed highway, anyone can drive into and through the winter range. The result of this accessibility was shown by last year's experience when hundreds of these deer were killed by all classes of people, who drove in cars, shot from the highway and from the car windows, and in a fair number of instances left the dead animals lying where shot.[7]

Another decade passed before a park was created in the area, just months before Grainger's death.

By 1931, it was becoming apparent to even the dimmest optimists that the depression was not going to go away at the command of pompous politicians and slick-talking bankers. The resource-based export industries of BC were in terrible shape. The fishing industry was prostrate under the burden of the previous year's overproduction. The lumber industry was beginning to feel the effects of Russian lumber dumping into the UK. A glut on world grain markets brought wheat shipments through the ports of Vancouver and Prince Rupert to a standstill, laying off hundreds of dock workers. Construction stopped. All of the suppliers and companies servicing these primary producers felt the pinch, and laid off even more workers.

Vancouver's benign climate was a magnet to the unemployed from across Canada, who poured into the city on every westbound freight train. Desperate men began to listen to socialists and communists who extolled their visions of a workers' paradise like the one they mistakenly believed existed in the Soviet Union. Demonstrations and strikes became common-place. The need for strong, intelligent leadership was obvious. Instead, in 1930, the country elected R.B. Bennett as prime minister, perhaps the most incompetent man ever to fill the position.

In the face of growing economic and social crisis, eastern financiers and bankers, who exercised significant control over western business, went so far as to appoint some western residents to their boards. MacMillan was one of the few so anointed when, in 1931, he joined the board of the Canadian Bank of Commerce—probably because of the influence of its chairman, Sir Joseph Flavelle. Once again MacMillan found himself becoming a focus of national attention.

In his first public speech after he was appointed a bank director, pre-sented to the Lumbermen's Club in Montreal, MacMillan dispensed the sort of banalities he usually eschewed. "If we will all get down to real solid hard work and quit thinking too much of depression," he pronounced, "we will soon return to prosperous times."[8] He went on to suggest that the threat of Russian timber exports was exaggerated, and that the root of the lumber industry's problems was low prices brought about by overproduction in a period of falling demand.

By the end of the year, his rosy outlook was replaced with a more hard-headed assessment of the worsening economic situation. In company with most of the Vancouver business community, he was becoming alarmed at the rapidly escalating provincial government deficit. He had been led to examine government accounts by reports that unemployment relief payments were higher than the wages being paid loggers.

In November, MacMillan and four other businessmen, including Austin Taylor and BC Power head George Kidd, met with Premier S.F. Tolmie and the government's executive council in Victoria. The ineffectual Tolmie government was floundering in the face of the growing economic crisis and, in a desperate attempt to cling to power, was increasing public expenditures by escalating the provincial debt. Even this attempt largely failed, and at one stage thousands of unemployed people converged on Victoria to stage a hunger march. Demonstrations of support for marchers passing through Vancouver on their way to Victoria led to violent clashes with police.

MacMillan and his colleagues proposed to Tolmie that he establish "a committee of non-partisan but representative citizens" to inspect the government books and to recommend how revenues ought to be spent to deal with the emergency while avoiding further deficits. The committee report, they suggested in a subsequent letter outlining their proposal, should be made public immediately.[9]

Leaving the premier to consider this proposal, MacMillan went to Europe in mid-December. He travelled by train to San Francisco, flew to Toronto, sailed from Montreal for Glasgow and arrived in London at the end of January. "The time I lose out of life in travel," he lamented in a letter written en route to L.R. Scott, "is a heavy drag."[10] He spent two weeks in Europe, visiting Norway, Germany and Holland, and after another few weeks in London, returned to Vancouver in mid-March. This trip did much to dispel any lingering sanguine thoughts of easy solutions to the depression and the threat to the BC lumber industry posed by the Soviet Union. In a long memo sent to various politicians and members of the Canadian Manufacturers Association, MacMillan detailed his reappraisal of the Russian timber export problem and his belief that the Soviet strategy was more than just a serious problem for the BC forest industry.

In a standing-room-only speech to a Vancouver Board of Trade luncheon in mid-March, titled "Impressions After a Visit to Europe," MacMillan provided a detailed and graphic description of the European political crisis resulting, he argued, from the inadequate peace settlements of World War One, which in time would produce a second global conflict. He was sobered by conditions elsewhere, but encouraged by British Columbia's relative position in the world, and more convinced than ever that government expenditures needed to be reduced.

> We are justly attached to our standard of living. We will seek to protect it. Let us remember that we are all here because and so long as we can make a living selling forest, mine and fish products in foreign lands. We cannot influence the sale price. That price at present goes chiefly to pay two charges, and two only—taxes and wages. The taxes come first. The higher the taxes the less there is left for wages. The wage earner has, of all people, the greatest interest in bringing about a decrease of all forms of public expenditure.

Two days later, MacMillan wrote to the BC leader of the opposition, Duff Pattullo, expanding on his theme.

> Speaking strictly non-politically, what bothers me is that the employment-giving,

revenue-producing, tax-paying industries in British Columbia have, by their own confession, and judging by the results, practically no more influence over the legislature at Victoria than they have over the legislature at Olympia, Washington. How is this ever going to be remedied? If it is not remedied it is going to bear down much worse on the farmer and on the wage earner than on the capitalist, because the capitalist will flee, but the farmer will lose his home market and the wage earner will lose his employment, and under present conditions neither one of them can leave the country.[11]

In Victoria, Tolmie was still contemplating the proposal for an independent examination of his government's books. To spur him on, in April MacMillan led another delegation to Victoria, representing twenty-two business organizations. The group again requested the formation of a nonpartisan committee. Tolmie was reluctant, but he was dependent on the business sector to keep him in office, so he gave in. Later that day, a five-member committee under the chairmanship of George Kidd was appointed and given complete access to all government documents. MacMillan took it upon himself to raise the $5,000 the committee needed for its expenses by soliciting $50 donations from Vancouver businesses. He also wrote a detailed analysis of expenditures within the Department of Lands and Forests, including suggestions on how they could be reduced, an exercise that provoked consternation and panic among the ranks of forestry bureaucrats across Canada.[12]

After a three-month examination of the government's books, Kidd delivered his report on July 12. It was a devastating indictment of the Tolmie government's program, and was regarded by many—in an era when totalitarian governments were appearing in Europe—as an attack on responsible government. The report recommended that the government's planned expenditures of $25 million be reduced by $6 million. The cut would be achieved by reducing the size of the legislature from 48 to 28 members; suspending work on and selling that perennial money-loser, the provincially owned Pacific Great Eastern Railway; ending public work projects; reducing the number of government employees and their wages; disbanding the provincial police force and contracting with the Royal Canadian Mounted Police; reducing the age for free education from fifteen to fourteen years; suspending the annual $250,000 grant to the University of British Columbia and suggesting that if the university could not function without this money it be closed and BC students provided with scholarships to study elsewhere; eliminating mothers' allowances; and a host of other means.

Tolmie's response was to sit on the report for as long as possible. By the time he released it in late August, Pattullo was making great political gains from its recommendations. When Tolmie proposed a union government to deal with the crisis, Pattullo turned him down, and the feckless premier and his government were voted out of office the following year.

Historians have tended to dismiss the Kidd Report as the product of a reactionary business mentality. That it advocated extreme measures is unquestionable; that its recommendations were unrealistic is debatable. Historians, for the most part, tend to be far more liberal-minded or left-leaning than business leaders, and are themselves the recipients of public funds, so their antagonism is as suspect as the report itself. Some of its recommendations were quite heavy-handed and if implemented would have caused much pain. On the other hand, much of the reaction against the report was because of its blunt and strident tone, which, in fact, was aimed at Tolmie's Tory government, which was being abandoned by the business community that had put it into office. In today's context, with government overspending once again an issue, the Kidd Report reads as a remarkably commonplace document.

MacMillan appears to have filled the role of ringmaster in this circus and he seems to have enjoyed himself. Over the next ten years, he continued to dabble at the fringes of politics, although his opinion of political parties was never very high.

> My view is that the Liberal Party and the Conservative Party both have so catered to the appetite of the citizenry for expenditures that they do not yet know how to play any other game and—foreseeing a political contest within twelve months—neither of them has enough faith in the common sense and patriotism of the electorate to adopt and announce a courageous plan of living provincially within our means.[13]

MacMillan was drawn into federal politics during the summer of 1942, when Ottawa cancelled $92,000 in shipping subsidies to two Canadian Transport subsidiaries, both operating in the Pacific. In defence of his objections to this decision, which he probably felt he must provide after his support of the Kidd Report, MacMillan argued:

> All our bulk British Columbia commodities in the export trade are competing with the United States. And the United States subsidizes steamship services to all important world markets. If the United States has the goods and the shipping services they are bound to benefit. If we have the goods but have not the shipping services we are bound to suffer.[14]

Although Ottawa-bashing was a practice MacMillan deplored, he also voiced the growing western Canadian lament that eastern interests were favoured over those west of Ontario, pointing out that Canadian shipping services on the Atlantic received twenty to thirty times more government assistance than Pacific cargo lines. Two months later, when the federal government said it would terminate the Canadian National Railway's steamship service between Vancouver and Montreal, MacMillan's response was to announce that Canadian Transport would provide regular service between the two cities. Since 85 percent of Canadian National's cargoes had consisted of HRME lumber, he had little choice.

In May 1932, the Canadian lumber trade was dealt a further blow when the Smoot-Hawley Act was passed in the United States, levying a $4 per thousand board foot tariff on lumber imported into the US. At that rate it was impossible for BC producers to sell any lumber in the US, and MacMillan's lucrative Atlantic market disappeared. When the effects of the tariff were added to the growing British appetite for cheap Russian lumber, the industry was headed for disaster.

MacMillan decided to take advantage of his new-found national status and exercise it at the forthcoming Imperial Conference in Ottawa. Until World War Two, these conferences were the main decision-making forums for members of the British Commonwealth. R.B. Bennett, an enthusiastic Commonwealth supporter, had launched a successful campaign to host the 1932 conference in Ottawa. For the lumber trade, the priority going into the conference was the growing volume of Russian timber flooding the UK market. Even the 10 percent tariff preference granted to Canadian lumber entering Britain was not enough to overcome the advantages the Soviets enjoyed, although a 20 percent increase in sales was experienced in 1932, the year the tariff preference came into effect.

MacMillan was somewhat predisposed to Bennett's Commonwealth vision. During the 1930 election campaign, and on numerous occasions after his election, the prime minister proposed a plan to increase Commonwealth trade through a system of preferential tariffs among members of the old Empire.[15] This was not incompatible with MacMillan's views on solving the Russian problem. "If it is expected or considered important that Canadian lumber should go to England in volume," he wrote in a memo, "Russian lumber, so long as produced at present, must be restrained in some other manner than by a 10% tariff preference, such as embargo complete or partial, such as a quota."[16]

Perhaps hoping to divert the depression-devastated Canadian electorate, Bennett staged an elaborate show for the hundreds of politicians, bureaucrats and lobbyists who flocked to Ottawa for the conference. MacMillan was among the throng. He rented a house in Ottawa's Sandy Hill district for the summer, and the entire family travelled east for the event. His assessment of the potential gains from the conference turned out to be far more accurate than Bennett's. The prime minister went into the conference convinced that member nations could sit down at the table as one big happy family and simply agree to favour each other's trade over that of non-members.

The British, of course, were not about to grant preferred access to Canadian lumber for nothing. In the first place the Timber Distributors group engaged in selling Russian timber in the UK constituted a powerful lobby. British manufacturing interests were also present, looking for the kind of preferred access to Canada that MacMillan and the lumber trade wanted in Britain. Behind a facade of pomp and splendour, the conference degenerated into a horse-trading free-for-all. MacMillan found himself in a unique and influential position. As he commented at the time, he was better known on the streets of London, where his wartime work for the Imperial Munitions Board was widely recognized, than in Toronto and Montreal, where he was regarded as a pesky western upstart. Through his association with Meyer and his many trips to the UK, he was well acquainted with many powerful and influential British politicians, government officials and businessmen. Behind the scenes, and with no official status, he was one of the most effective Canadian participants in the negotiations.

This did not endear him to eastern Canadian business interests, who were opposed to any kind of *quid pro quo* arrangement allowing increased Canadian lumber exports to Britain and a freer flow of British manufactured items into Canada. Whether it was because of these conflicting interests or something else, MacMillan was snubbed by Bennett throughout the conference. But while his pride may have been wounded, he came away with one of the few accomplishments of an otherwise unproductive conference: a British agreement to prevent the dumping of Russian timber into the UK and to set an annual quota on Russian imports. Observers and historians have tended to view these provisions as trivial, but within two years BC lumber sales to Britain had increased fourfold. If anyone in Canada gained from Bennett's fiasco, it was clearly MacMillan and the BC forest industry.

The gains were not automatic, however, and did not come easily. Through most of the following year, MacMillan kept up the pressure to have the provisions of the Ottawa agreement enforced, by maintaining his correspondence with H.H. Stevens, Canada's minister of trade and commerce, and marshalling the leaders of the BC industry. British dealers selling Russian timber fought back with pressure on their politicians and various subterfuges aimed at circumventing the agreement. The Bennett government was a reluctant defender of Canada's interests in this matter, reserving the use of its limited influence on Britain to aid eastern manufacturing interests.

MacMillan's presence in Ottawa over the summer had enhanced his national stature. He was invited to sit as one of two BC representatives on a committee of Canadian businessmen empowered to examine and make recommendations on federal government taxation and spending programs—a national version of the Kidd committee. He continued to beat the drum in support of Kidd's report, and in an October speech to the Life Underwriters' Association warned that British Columbians must economize on public expenditures or perish. Out of a provincial population of 700,000, he said, 86,000 people were on relief and an additional 75,000 Natives and Orientals paid virtually no taxes. The remaining workers and the industries that employed them were being taxed to death, he argued, to finance public programs established by politicians seeking office.[17]

chapter 11

Lumber wars

Early in 1933, MacMillan took on two new tasks, both of them raising his profile in the national business community. In January he was elected president of the Vancouver Board of Trade, with George Kidd succeeding him as vice-president. Then, in May, he took on the presidency of BC Packers, replacing Stanley Burke, who became a vice-president. His three years as a director had sparked his interest in the fishing industry in general and this company in particular. Even so, he and the other western directors appointed in 1930 had not succeeded in placing the company on a sound financial footing. It was still leaking red ink at a phenomenal rate—$1.6 million in 1932—and its largest creditor, the American Can Company, supplier of tins, was becoming restless over the continuing inability of the operation to pay its debts.

MacMillan's appointment as president was a final attempt to avoid placing the company in receivership.

His first move, which preceded his appointment, was to renegotiate credit arrangements with American Can and the banks. In March, he laid out the basics of financial reorganization in a letter to Ed Bell, the Vancouver representative of American Can.

> I think it is worth while to put this thing in permanent shape at the present time, particularly as the banks are inclined to desire to see something definite along these lines before they agree to the credit desired for this year's operation. . . . In the meantime, it perhaps might be sufficient for the banks if you could give them some assurance that your company will agree to some policy which will assist in the building up of the working capital of the Packers.[1]

Other letters to Bell, prior to the annual general meeting at which MacMillan was appointed president, outlined in detail the restructuring of the debt and the intent of the directors, for whom MacMillan appeared to be speaking. In retrospect, it is clear these measures saved the company from bankruptcy.

The first to recognize the astuteness of the move was Aemilius Jarvis, a colourful Toronto financier who had been associated with the company since the turn of the century. After congratulating the new president, he offered him some advice on the handling of unions.

> The chief deduction that I have always made has been the fear of the fishermen on the part of the managers, which I think is very much over-estimated, as the fisherman's interest is the same as ours; his season is just as short as ours and consequently the rank and file must dread a strike as much as we do, and I have always felt that if the Canners had a show-down once and for all, it would have a very salutary effect for many years. On this question I think Management often is too close to the picture to see clearly.[2]

In his response, MacMillan indicated why he had taken on the job of president of a near-bankrupt fishing company when his lumber exporting company was fighting for survival.

> I am going into a business which I do not know very much about, and I am doing so because I think that the only salvation that can come to this Western country will depend entirely upon the sound development of the primary industries, and if I can make use of this opportunity to do anything constructive over a period of time with the B.C. Packers, the results will be worth the effort.[3]

Under an agreement he made with BC Packers, MacMillan received a token salary of $5,000 a year. To compensate for time he spent away from the timber business, the H.R. MacMillan Export Company received BC Packers shares. Within a few years, this arrangement gave MacMillan a controlling interest in the fishing company.

MacMillan's capacity for taking on a diverse variety of tasks, without losing track of the details in any of them, was due to an unusual ability to concentrate on the job at hand, and then turn to something else. Once he had dealt with a matter as well as he could, he was able to put it out of his mind and move on. He was not known to lie awake at night worrying. At critical junctures in the company's history, he was apt to go off riding with Grainger in Stanley Park, apparently unconcerned about the precarious state of his business affairs.[4]

He did not waste his time on matters he thought were not worth his efforts. Over the years he had maintained a membership in the Canadian Forestry Association, and in the 1930s he agreed to become a director. This position gave him a closer view of an organization he had helped to found, and recalled his earlier interests as a forester. He did not like what he saw, as he indicated in a letter to C.D. Howe, dean of forestry at the University of Toronto, complaining that the association had neglected to accept or acknowledge his resignation.

> According to my view the Forestry Association never has been right. It has been partly a muffler in the hands of people who kept under control and observation any attempt to develop a sound opinion regarding forestry in Canada, and it has been partly an out-doors association, which had little or nothing to do with forestry, and which, as a matter of fact, put forestry in the wrong light.[5]

After fifteen years in the private sector of the industry, his views on forestry had begun to change. In a June speech to the Pacific Science Congress, he attracted much interest in Canada and abroad by predicting that, in time, the Pacific Northwest and British Columbia would follow the example of the Scandinavians and begin managing forest lands to grow commercial timber supplies. At the time this was a radical and unique opinion, appreciated far more in the UK, where forests had been managed for centuries, than in BC where it was widely believed the natural timber supply would never run out. In almost thirty years since he graduated from Yale, he had not observed in Canada any development of the type of forest management he had seen in his travels through Europe. Also he had lost his youthful faith in the ability of government to manage anything efficiently. He was not far from questioning the basic tenet of Canadian resource policy, public ownership of the land and its resources.

MacMillan's feelings about the suffering inflicted on his fellow citizens by the depression was somewhat at odds with his public pontifications on the virtues of hard work, public thrift and hope for the future. Although he still defended the Kidd Report and its drastic remedies, another more compassionate facet of his character emerged.

In October, with Vancouver heading into another dreary, wet and destitute winter, speaking in his capacity as president of the Vancouver Board of Trade, he issued a stirring appeal for private donations to the recently formed Community Chest.

> Over one-tenth of our population are on public relief, and public relief is not sufficient to provide a standard of living which makes life worth while. On private social service, dependent on voluntary gifts, rests the responsibility of providing those services which do something more than keep the body alive. Nursing services to the sick, fresh air camps for tired mothers and undernourished children, recreational activities for unemployed men, women and young people, preventive services, particularly in respect to physical health—these and many others must be continued, even enlarged, to meet the increased demands occasioned by unemployment.[6]

That same month he received a direct appeal from James Stewart, a young Scottish immigrant artist living in a relief camp in Cache Creek, who requested one or two dollars for art supplies. MacMillan sent him a cheque for $2.[7] It was one of his many charitable contributions, which totalled tens of millions of dollars by the time he died.

MacMillan's public presence during this period of political turmoil sparked suggestions that he run for public office. If he ever did entertain such thoughts it was only briefly; as chief forester he had been close enough to the political process to realize it did not suit him as a way of exercising what he saw as his public obligations. "You ask why some of us don't get into politics," he wrote one correspondent.

It has been mentioned to me several times, but I have always shrunk from it. I think I am assisting in keeping a fair number of people employed working where I am but that if I went into politics I would probably fail at it because I have not had the early training that appears to be necessary for success in politics and my ideas are so at variance with the ideas of ninety-odd percent of the people in the Legislature that I would be ineffective if I got there.[8]

The political situation in Victoria took a turn near the end of 1933, with the Tolmie Conservatives going down to a decisive defeat at the hands of Duff Pattullo's Liberals. MacMillan wrote a lengthy letter to Sir Joseph Flavelle, giving his impressions of the new government.

Since Premier Pattullo has been installed in office I called upon him and upon the Finance Minister and other Ministers. I find the Cabinet quite up to the average, I think, as compared with previous provincial cabinets. Individually, I think, they are much more able than the preceding cabinet. I think they intend to work hard, but I am somewhat disturbed regarding certain of their attitudes. I feel sure they have very little intention of reducing expenditures significantly. . . . A serious feature of this situation is that Mr. Pattullo's inclination seems to be to lay the blame for the distress, unemployment and lack of money upon the Dominion Government and upon the capitalists in Eastern Canada, thereby promoting discord and cleavage in this country, instead of gathering together the strength of the country—the whole of which is needed to meet our problems.[9]

The federal political situation looked even less promising in late 1933, particularly in the west. The Bennett Conservatives were resolutely digging themselves into an early grave with a combination of corruption and ineptitude. In the four western provinces the various socialist and reformist factions had coalesced into the Co-operative Commonwealth Federation under J.S. Woodsworth, a Social Gospel minister. In spite of his experience at the hands of R.B. Bennett the previous year, MacMillan was part of an official delegation welcoming the prime minister to Vancouver in early 1934, and introduced him at a banquet with the appropriate words of praise and respect. A year later he wrote the foreword for one of the prime minister's published speeches, giving him full credit for the achievements of the Imperial Conference. MacMillan had learned that politicians must be handled carefully and stroked continuously, particularly when they are in power.

He was also quite aware of the first intimations of an approaching international war.

There seems to be a growing belief that war is in the air somewhere. I was talking to a chap who had just come home a few days ago from Europe. He had been through Eastern Europe and was positive there would be a war within twelve months.

The Japanese I think protest too much that they are making tremendous arrangements for keeping peace. Their explanations are almost too perfect. I listened the other evening in a private house to the explanations made by Honourable Ikemasa Tokugawa, the present Japanese Minister to Canada, who returned from Japan this last week. He is an exceedingly clever man. I think though that behind all his protestations of the peaceful nature of the Japanese preparations there is visible the realization that war might easily occur.[10]

Astexo's sales agency, Seaboard Lumber Sales, had been mothballed with the adoption of the Smoot-Hawley tariff by the US. The Astexo mills were back to selling into Britain, largely through MacMillan, but as trade picked up in the wake of the Imperial Conference, they once again became restless, grumbling about his profits and generally feeling hard done by. They were particularly irritated to discover that MacMillan, through Meyer, had obtained a large order for railway ties at a substantial price, and was offering the order to the Astexo mills at a considerably lower rate.

In early January 1934, Astexo members attempted to restrict the amount of lumber MacMillan and other exporters could buy from non-Astexo mills. MacMillan related his version of events in a letter to his Toronto sounding board, Sir Joseph Flavelle.

> You may be interested in the character of the unusual developments in our business which kept me from attending the Bank meeting:
>
> I intended leaving for Toronto on the Thursday afternoon, but on Wednesday, while I was at lunch, I was asked to leave my lunch and attend a meeting of [Astexo]. At this meeting I was told by the chairman, our old friend, Mr. J.D. McCormack, that the export mills were dissatisfied with the way their product was being sold abroad; that they had determined to enforce a programme of orderly marketing and maintenance of higher price levels; that the chief handicap to the development of this program was that export merchants were allowing mills, which did not belong to the Association, to receive more than their fair share of business, and that therefore I was being notified that if I did not sign a contract—the essential points of which were to purchase 75% of my orders from the export Association and to place all orders with them at their fixed price within five days of my sale of the order—I would be put out of business.
>
> I was asked to give an immediate reply, but I said I wished to think the matter over for twenty-four hours.
>
> I met with them the following day and told them that I was willing to co-oper-

ate with them, provided they would establish a permanent committee of their association to deal with a committee of exporters. This is in the process of being worked out.

I feel that the mills have been so much encouraged by Empire Tariffs, which have greatly increased our lumber sales in Empire countries . . . that they are likely to attempt to move a little too fast in what they call "controlled marketing of British Columbia's export lumber."

I feel instinctively that some of their ideas are unsound, but I am inclined to work the situation out with them rather than to stand upon the principle of free buying and selling, which I have always believed to be a governing fact in any merchant trading business.

The air here seems to be full of desires on the part of the milk producers, the apple producers, the lumber producers and, to some degree, the salmon producers, to form comprehensive selling associations, which will fix prices and selling conditions, and have some power to penalize those individuals who do not care to join the associations.

In this present instance, about 25% of the lumber producers has seen fit never to join the selling association, seemingly feeling happier to individually manage their own businesses, and this present effort on the part of the Associated Timber Exporters of B.C., Limited, is to prevent the export merchants from buying material from such independent mills and thus starve the independent mills into the Association.

This program seems unsound to me, but the condition exists; the theory of controlled marketing has the support of a large proportion of the producers and if one is trading with these producers, one must study the programme and see what sound policy can be worked out.[11]

Ten days later, MacMillan wrote to tell Flavelle the outcome of his discussions with the Astexo group.

I have been continuing my negotiations with the associated mills and am entering into an agreement with them, which in general terms meets with their desires, and as a result of which they have made certain concessions, chiefly in the direction of restricting the number of exporters to whom they will sell.[12]

The one-year agreement signed by MacMillan and three other exporters stipulated that the exporters would buy 75 percent of their lumber from the Astexo mills, they would place their orders with the mills within five days of accepting them from buyers, and only a limited number of export companies would be permitted to buy from the Astexo mills. It was a short-term gain for Astexo, and both parties prepared themselves for further confrontations.

Throughout 1934, tensions between MacMillan and the Astexo princi-

MacMillan in his Metropolitan Building office, Vancouver, 1935.

pals increased. In terms of personalities, MacMillan's major antagonists were John Humbird, who had taken over the presidency of Victoria Lumber & Manufacturing from his father, and Henry Mackin at Canadian Western Lumber. They made quite a pair. Humbird was an enormous, overweight man who was an expert at performing magic tricks. Mackin was a short, slight figure with the personality and tenacity of a pit bull. They represented the two largest mills in BC, and eventually their dislike of MacMillan became almost pathological. Behind his back they accused him of taking advantage of his World War One position as timber commissioner to establish his own business. At times their actions to thwart his company took the form of a desire to best him personally.[13]

The UK lumber market was volatile, with the constant threat of Russian imports, as well as the dangers of a revived Baltic industry. The

situation was further complicated when a number of small British timber dealers, bypassing the established import agencies, attempted to create direct
links with some of the Astexo mills. Humbird, for one, went along with this
scheme for a while, almost destroying Astexo in the process.

In addition to Meyer, MacMillan had another excellent source of
inside information on the UK situation. His former employee and associate,
Loren Brown, became BC lumber commissioner in London during the Russian
crisis and was in an excellent position to keep MacMillan fully informed, as he
had been earlier when he worked in the Canusa office. One of Brown's less-
known functions was to act as MacMillan's intelligence agent in London,
reporting in a steady stream of personal, handwritten letters. When the
Astexo people eventually realized Brown was a shareholder in the H.R.
MacMillan Export Company and objected, MacMillan bought his 700 shares
for $8,750.[14]

By the fall, Astexo had decided to resurrect Seaboard Lumber Sales,
partially in anticipation of the elimination of the US lumber tariff, but also with
the intention of using it to export into the UK market in direct competition
with MacMillan and Canusa. Charles Grinnell, who had been operating his
own export company since the withdrawal of Seaboard from the American
east coast market, came back to manage the move into the UK. Having girded
themselves for further battle with MacMillan during the year their agreement
was in place, the Astexo directors were ready for another confrontation. In
January, Humbird and Mackin marched out of the association's office in the
Metropolitan Building, into the elevator and down three floors to MacMillan's
office, precipitating what has been called the most historic meeting in the
Canadian lumber business. VanDusen left an account of the meeting, in his
usual understated way.

> A group of the mills had a meeting and suggested that it was time for them to go
> into the export business with their own lumber and to do what we were doing.
> Well, they felt that they really should be offering us a little bit of the business any-
> way, we having developed it, they didn't want to cut us right off behind the ears,
> so they had a delegation that came in to see HR—I was in Toronto at the time.
>
> They told HR they were going into the CIF business and although we had
> done a very good job for them, there was no reason why they shouldn't do part
> of this business themselves. They offered HR 15 percent of the volume, which
> would have given us a nice living, and that was the deal, take it or leave it. HR said
> he'd let them know the following morning. He called me on the phone in Toronto
> and we talked about what we were going to do.

I told him I would rather raise the flag and go down with it flying than to take a deal like that. HR said he felt the same way and that we could make a go of it without the mills. At that time we were getting 75 percent of the export business, and the mills were offering us 15 percent.

So the next day HR called them up and said we were not going to accept it.[15]

The consequences of the meeting were enormous. It led directly to MacMillan creating Canada's largest integrated forest products company. And it resulted in Seaboard becoming one of the world's largest lumber exporting firms. But these developments were far in the future; the immediate outcome was less conclusive.

Astexo waited almost two months to make its move, then released a story to the *Vancouver Province* stating its intention to sell through Seaboard into the UK and US markets only, working through exporters to sell into other countries. Commentary in the media during the following weeks tended to welcome the move, with veiled suggestions in some publications that H.R. MacMillan, arch-capitalist exploiter of poor, hardworking sawmill owners, had finally received his just deserts. MacMillan himself had no time to concern himself with such innuendo. After giving his answer to Humbird and Mackin, and holding further discussions with VanDusen, he issued an order to the sales department at HRME: accept every offer to purchase that comes in. His plan was to tackle Seaboard head-on, cornering as much of the market as he could and filling the orders from mills outside the Astexo circle. He would try to freeze Seaboard out of the UK market.

It was a dangerous and risky tactic which, if it failed, would destroy the company and MacMillan's reputation along with it. Success in the lumber export business depends on an exporter's ability to deliver the lumber contracted for. Failure to do so cuts off future offers. At this time MacMillan controlled only one mill, Canadian White Pine on the Fraser River, which supplied about 10 percent of the volume HRME handled in a year. All of the large mills and most of the smaller ones on the BC coast were members of Astexo and were prohibited from selling to MacMillan, so he was dependent on two sources to fill the mounting orders: small mills in BC, of which only a few had ever cut to export specifications, and mills in Washington and Oregon.

With the Seaboard-affiliated mills in possession of most of the lumber and MacMillan holding most of the UK orders, the BC lumber industry was facing an immense crisis. The British were becoming concerned about the dispute; they could just as easily buy lumber from Russia or the Baltic and leave the squabbling colonials in BC to bicker amongst themselves.

There was more involved in MacMillan's refusal to accept the Seaboard offer than his desire to control the UK lumber trade. In March he sent a memorandum to Astexo president J.D. McCormack, outlining his position.

> I am somewhat governed by the fact that when the Seaboard Sales was put together some years ago to handle the Atlantic coast market, it seemed to be just as good as the idea now being adopted for the sale of lumber in the United Kingdom. Nevertheless, as the administration of the Seaboard Sales proceeded, the management, as the American market began to weaken, continuously forced all possible Canadian lumber into that market, thereby incurring the growing enmity of the American lumber producers.
>
> Although when the business was first developed by the Seaboard Sales in the Atlantic Coast market, the dealings were almost entirely with wholesalers, the policy was gradually changed to a point where wholesalers were eliminated to a great degree and sales were made direct to most of the wholesalers' good customers. This finally earned for the Seaboard Sales the hostility of the wholesalers.
>
> As the strength of the Seaboard Sales developed, the management discovered in their hands a great power in control over freight, which power they used almost entirely against the interests of the Conference American Intercoastal Lines by continuously chartering below Conference rates and making almost negligible the employment of American vessels, thereby bringing the American Shipowners' Association into the field as strong and aggressive enemies.
>
> The combination of these circumstances, so far as I have been able to learn during the past few years from my acquaintances in the shipping industry and the lumber industry in the United States, resulted in the imposition of the present prohibitive tariff on lumber into the United States. I believe this would not have happened if this power had not been gathered together in the hands of one person to be used as it was used.
>
> I am inclined to think that the same power placed in the same or similar hands in the United Kingdom market will, in the course of three or four years, bring about the same results, which I think is entirely too serious a situation to be contemplated by individuals deeply interested in the British Columbia lumber industry.[16]

In the midst of this tense situation, an incident occurred in MacMillan's family that was to affect the evolving dispute. Life in the MacMillan household during the hectic years of the depression revolved around the comings and goings of its sole male occupant. He was an attentive, dutiful son and husband, and an indulgent but strict father to his two attractive and popular daughters, Marion and Jean. They were not permitted phone calls in the evening, at least when MacMillan was home. They were allowed out one night a week and, if their father was home, he sat up waiting for them to come in. He actively discouraged visits from young men, although by the mid-1930s he was finding this a difficult rule to enforce. One evening in April 1935, a boyfriend of Jean's, Boswell "Bunny" Whitcroft, came to visit her. He worked in a junior position at Astexo. Although MacMillan was not particularly fond of him, and had never encouraged the courtship, Whitcroft was the bearer of some important news, which he passed on to Jean. A regular visitor to her father's office and the inheritor of his astute business sense, Jean immediately realized the implications of what Bunny was saying. He told her about a meeting held at Astexo earlier that day, after which, amid much jollity, various participants spoke of the imminent demise of the H.R. MacMillan Export Company. Their plan was to buy the output of all coastal BC mills, depriving MacMillan of a supply. Jean passed along this information to her father, who digested it and discussed it with VanDusen, then left for London the next day. It also seems likely MacMillan was not entirely in the dark about the Astexo plans, as his old friend Martin Grainger, through Alberni Pacific Lumber, was involved with Astexo and served for two years on the first Seaboard board of directors.

The critical information MacMillan received was that Humbird, Mackin and Carlton Stone, owner of Hillcrest Lumber at Cowichan Lake on Vancouver Island, were on their way to London to meet with British timber importers; they would impress upon them the Astexo group's ability to fill orders and reveal MacMillan's precarious position. Stone's son, Hector, later provided an account of their trip.

> The decision to go to London was made hastily. Frantic cables were coming in from all parts of Britain demanding clarification, asking to join Seaboard, launching accusations. Something had to be done.
>
> My dad packed his bag in a matter of minutes, caught the Victoria ferry to Vancouver and joined Mackin and Humbird in time to take the train. They were just beginning to catch their breaths when they boarded ship for London. According to Dad, they worked on strategy and prepared speeches and statements all the way across the Atlantic. What stories had MacMillan fed the industry? How would they counter them? What were MacMillan's remaining strengths and bargaining positions? How could these be matched and surpassed?
>
> Their one source of satisfaction was that they were beating MacMillan to the

punch. They would be able to meet with the British lumber industry first and tell the story accurately. They had five days on ship to rehearse and prepare.

When they disembarked at London, ahead of them on the gangplank was H.R. MacMillan, briefcase in hand and limousine waiting. Unknown to them they had shared the voyage with their chief adversary.[17]

The mission of the Seaboard group was to convince the British buyers they would save money by purchasing directly from Seaboard. And, they argued, since they had the lumber, MacMillan would not be able to fill the orders he had accepted. The latter suggestion was made by Mackin at a dinner of timber dealers, with practically the entire UK trade in attendance. There was an immediate flurry of concerned voices discussing this prospect. Meyer, who was present, got to his feet and in a quiet but firm voice said the H.R. MacMillan Export Company would deliver every foot of lumber ordered.[18] The dealers and their agents chose to take Meyer's word on this and stuck with MacMillan, leaving the Seaboard delegation fuming and reduced to making personal attacks on his character.

Returning to Vancouver with the market still firmly under his control, MacMillan was faced with yet another crisis. A US longshoremen's strike cut off the flow of lumber from the Washington and Oregon mills, upon which he was now heavily dependent. With that, his ability to accept more UK orders evaporated, and he was faced with a struggle merely to fill the stack of unfilled orders already on file. That task was one of Bert Hoffmeister's first important jobs after he was brought into the export company office from Canadian White Pine. MacMillan had maintained a long and friendly association with the smaller, independent mills, which until now had supplied no more than 20 percent of his lumber. He had a personal affinity for the smaller operators, as opposed to the often pompous and stuffy owners of the large mills who looked upon him as an upstart and a threat to their domination of the business. Many of the smaller mill men admired MacMillan. They tended to see him as one of their own, someone who, like them, had started with nothing. And they liked his directness; for example, the way he dealt with his first tie-supply crisis by moving portable mills into the Malahat. So when Hoffmeister came calling to ask them to take an order cut to precise specifications, involving delivery and documentation requirements they were unaccustomed to dealing with, they were prepared to give it a try.

The results of the October federal election in Canada were worrisome.

The erratic former cabinet minister, H.H. Stevens, had hared off on his own and started the Reconstruction Party, devoted to reducing the power of big business, only to be trounced at the polls. Not only was R.B. Bennett soundly defeated—in itself a welcome outcome—but in BC the socialist CCF party captured more votes than any other. MacMillan expressed his thoughts on the election to Sir Joseph Flavelle.

> I think we are all pleased here that the Liberals are in with such a clear majority. There may be very little difference in the tariff policy between the Liberal and Conservative parties, but if there is any difference it will undoubtedly be in the direction of a somewhat lower tariff under the Liberals, which will assist this community, which lives so largely on export goods.
>
> Furthermore, the Liberals might be able to reopen the United States and Japanese markets under conditions that might be helpful to us and they might succeed better than the Conservatives in holding our present favourable position in the United Kingdom.
>
> The chief views one hears here are: Sympathy for Mr. Bennett personally. Satisfaction at the disappearance of the Reconstruction Party. Fear that the CCF, who polled the most voters of any party in British Columbia, will carry the next provincial election.[19]

The fall of 1935 was one of the lowest points of MacMillan's life since his ordeal with tuberculosis. The civilized world was under threat in Europe and in the Pacific. His company, to which he had devoted sixteen years of his life, and for which he had forsaken his passion for forestry, was in a precarious position. Then, on November 30, his mother died at age seventy-seven.

MacMillan never discussed or committed to paper his feelings about Joanna or how her death affected him. Later, he often acknowledged the difficulties she had endured after his father's early death, and the sacrifices she had made to raise and educate him. She had an enormous influence on the shaping of his character, beginning with the letters she wrote him during their long separations in his childhood and youth. He fully believed Joanna had saved his life by coming to care for him at Ste-Agathe. But after that point, their roles had reversed. From the time Joanna accompanied MacMillan and his young wife to Victoria, he cared and provided for her, and to a great extent directed the course of her life. It is almost as if he was making up for the circumstances of his own childhood by assuming responsibility for the lives of his immediate family, often down to the most minute details.

The only real difference of opinion between MacMillan and his mother appears to have been his refusal to permit her to live in her own residence. But even this was not a great issue between them; she lived quite contentedly in her rooms on the second floor of the MacMillan household, immersed in church missionary work and a close circle of friends. Much later, in his own retirement, MacMillan devoted considerable time and effort to exploring and commemorating his maternal ancestry. For the moment, there were other, more pressing matters to concern him.

Life becomes art

While others in the company dealt with the direct threat posed by Seaboard, MacMillan devoted his energies to expanding his operations. One addition was a plywood plant built next to the Canadian White Pine mill in 1935. BC Plywoods Ltd. was under the direction of Blake Ballentine, who had developed a highly successful plywood door plant. When Mackin's company, Canadian Western, cut off the plant's only source of supply, MacMillan and Ballentine built their own plywood mill. Within a few years it was the largest producer of plywood in Canada.

But MacMillan's major thrust was toward acquiring sawmills. Although he later stated on numerous occasions that he went into milling only in response to the threat from Seaboard, his earlier moves in this direction indicate otherwise. Now, however, the need for sawmilling capacity was urgent; acquiring them was a question of survival. MacMillan's first step was to buy Dominion Mills, a non-operating sawmill, also adjacent to the Canadian White Pine Company mill. The Dominion mill was an early victim of the depression, and in 1933 MacMillan and a dozen others—including, curiously, Humbird and Mackin—formed a syndicate and bought it. Two years later, CWP bought out the syndicate's $100,000 investment for $150,000, a carefully planned transaction that gave the group a healthy 50 percent profit that was not subject to taxes.

The main object of MacMillan's attention continued to be the Alberni Pacific Lumber mill at Port Alberni on Vancouver Island, which Grainger and Aird Flavelle were still managing for its absentee British owners, Denny, Mott & Dickson. Although the mill had begun to show a profit by the mid-1930s, it owned a limited timber supply that did not augur well for future revenues. But up the nearby Ash River valley, there was an 18,000-acre tract of magnificent Douglas fir owned by John D. Rockefeller, Jr., which could keep the mill running for another twenty years. When Rockefeller announced he would sell the timber, interest was aroused in many quarters—particularly in the offices of Bloedel, Stewart & Welch.

The woods end of BS&W was under the control of Sid Smith, the woods manager, and it was he who was MacMillan's major competitor for the timber. Smith was one of the most experienced and respected people in the coastal forest industry. He was a forceful person, with physical attributes like MacMillan's. He usually got what he wanted, and what he mostly wanted was to acquire forest land. Smith wanted the Ash River for BS&W's new Somass mill in Port Alberni. It was a dicey situation for MacMillan. Without more timber, the APL mill was almost useless to him. But the combined price of the two was very high—almost $5 million. Through a convoluted set of long-distance negotiations and another trip to London, MacMillan arranged financing with a number of British firms. As he juggled the parallel negotiations to purchase the mill and timber, he received a disquieting bit of information.

Most of the players in this drama occupied offices in the Metropolitan Building and were members of the Terminal City Club, located on its ground floor. They lived and worked in close proximity to each other, and were aware of each other's business dealings, often in detail. One day, while he was eating lunch in the club, the MacMillan company's accountant, Harold Wallace, over-

heard someone remark that Sid Smith was on his way to New York. He quickly informed MacMillan, who immediately boarded a plane to New York and
met with the Rockefeller people. As he was about to leave, a secretary came
in to announce that Sid Smith was in the outer office. MacMillan, who did not
want to encounter Smith, hid in the office washroom while the disappointed
woods manager was told he was too late.[1]

What MacMillan obtained in New York on June 11 was a conditional
sales agreement on the Ash River timber, subject to his purchase of the
Alberni Pacific mill. Rockefeller gave him until the end of June to conclude the
mill purchase. From New York, MacMillan hastened to London, arranging to
meet Grainger there to continue negotiations with Denny, Mott & Dickson in
person. They returned home in early July, making headlines in London newspapers because it took them only six days to travel via the *Queen Mary* to New
York, and by air to Seattle and Vancouver. As the paper noted, it was the quick-
est passenger trip ever made from London to Vancouver. They needed to
hurry, for the deal was in trouble. Although MacMillan and Grainger worked
out most of the details of the Alberni Pacific purchase in London, final agreement had not been reached. MacMillan convinced the Rockefellers of his ability to consummate the deal as agreed, so he was not held to his June 30 dead-
line. After his return to Vancouver, further complications developed. In an
August memo, Grainger summed up the situation in his colourful language:

> All of which indicates that deals totalling $4.9 million arranged between principals
> should not be ruined by revamping by solicitors, or chiselling by ambitious inter-
> mediaries, or personal frictions, or wounded amour propre. If the edifice has tot-
> tered for any of these reasons, saner people must clear away the fog of non-essen-
> tials and again expose the naked essentials for consideration by balanced minds.[2]

It took the saner minds more than a month to straighten out the difficulties, but eventually they prevailed and the twin sales were announced to
the press. During the summer's negotiations, Rockefeller visited Vancouver
Island to take a look at the timber he had never seen and was about to sell.
MacMillan had planned a full tour of island logging operations for his distinguished visitor, who was accompanied by his wife. Before they arrived, however, he came down with a severe case of phlebitis and was confined to his
bed. He refused to meet the Rockefellers in his convalescent condition, so
Jean was recruited to guide the visitors on their tour.

The Alberni mill and timber purchases took some of the supply pressure off the export company. It was one of the biggest mills on the coast, and when its output was combined with that of the Canadian White Pine and Dominion Mills, as well as the five portable sawmills still producing ties near Duncan, it made the MacMillan company the largest private timber producer in Canada. The acquisition of the Ash River timber had a profound influence on MacMillan. He was now the owner of his own forest, and before long he began to resurrect the knowledge and training he had received as a professional forester, and to apply it in his own inimitable fashion.

Typically, even the harrowing work of contending with the Russian timber threat, the Seaboard struggle and the Alberni purchases, did not fully occupy MacMillan's restless mind. From the mid-1930s until the war began, he devoted considerable time and effort to collecting and breeding exotic birds at his Qualicum Beach home. After perusing catalogues and talking to other bird fanciers, he placed his orders with British dealers and had the birds delivered on a returning lumber ship. One of his 1937 orders, for instance, was for a pair each of Diamond doves, Cinnamon Wing budgerigars, Bengalese finches and African Silver Bells. He was sufficiently involved with this pastime to accept the honorary vice-presidency of the Vancouver Poultry and Pet Stock Association.[3]

Nor was he too busy to renew an old family acquaintance. While in Britain in 1935 to fend off the Seaboard threat, he looked up one of his many cousins. Now known as Mazo de la Roche, she had become an enormously successful writer, having published several of her Jalna novels, and was living in England with her cousin and constant companion, Caroline Clement. After hearing MacMillan's account of his activities during the previous twenty-five years, she decided to write a novel based on his life. She was wildly enthusiastic about the project and enlisted his assistance. After he returned to Vancouver, she flooded him with letters and they began a lengthy correspondence. Initially MacMillan was ambivalent about becoming the subject, even in disguised form, of a novel.

> I have also felt somewhat deficient about getting myself into a book (even by an outstanding Canadian author), which, when it began to appear in my neighbourhood might give me cause for embarrassment and incline me to hide from the public eye. One is not accustomed to having oneself uncovered and one feels, no

Greyshakes, MacMillan's Qualicum Beach home on Vancouver Island.

matter how one goes about it, the result can only be something like an expressionless figure in a stained-glass window.[4]

But he did agree to help, which consisted of answering dozens of her questions about his early life and providing detailed descriptions of the places he had been and the things he had done. She sent him many long, effusive letters, addressing him by his childhood name, "Reggie." His lengthy responses, which were interrupted by a two-month stint in the hospital in 1937, contained a detailed and thoughtful look back upon his life as a youth, a student and a young forester. He also provided a long description of his experiences at the Saranac and Ste-Agathe sanatoriums.

De la Roche completed the book in about a year and sent him a draft copy in 1938. "I have looked it over and, as you requested, I have not put it into circulation," he responded.

> I do not know what to think about the book. Your capacity for imaginative effort is so well known and well established in your previous books and so well maintained in this book, that it leaves me nothing to say. When I approach comment on a book which so far as I am concerned makes me feel half shadow and half substance, I do not know what to think of it. For some strange reason I feel some

discomfort in attempting to read a book which uncovers some of my "danger" spots and I suppose causes me to feel self-conscious or sensitive. I do not know whether or not this is something that is peculiar to me. I think I would find it quite impossible to read any intimate biographical publication relating to me—if I were important enough to justify its production.

It also occurs to me that you may have had fun writing this book searching through your memories and re-constructing from their shadowy forms or scenes and building into living characters the ghosts of departed members of the family. I think I missed a great opportunity in not seeing you when you were doing this work. It would have been interesting to have assisted in piecing together what we could dredge up from the depths of our respective memories. Needless to say, I feel highly complimented by your having chosen me for any part of one of your outstanding Canadian series. It will be a matter of great interest to read the reviews and criticisms that will be appearing before very long.[5]

He declined her offer to dedicate the book to him, feeling it would identify him clearly as the protagonist. The gesture was in vain, as he explained after *Growth of a Man* was published in October 1938.

The book reviews began to appear a few days ago. I am sending you copies of those which appeared in our local papers. A shipment of books has reached Vancouver, but I do not know how many. There were none left when I sent out a few days ago, however, I understand that more have now arrived. It will be interesting to see how they sell.

I think the "cat seems to have got out of the bag already", as some people who have read the reviews have immediately spotted me. I suppose this is of no consequence. I am very pleased to know of the flattering review which appeared in the Boston Herald. I hope the book justifies your expectations and adds at least one more leaf to your laurels.[6]

Growth of a Man is an interesting but, in the end, not very accurate portrayal of MacMillan's early life. While it is probably a faithful rendering of life in rural Ontario toward the end of the nineteenth century, it overstates the harshness of the treatment MacMillan received at the hands of his maternal grandparents, his aunts and uncles. MacMillan's own writing indicates that, apart from the drudgery of farm life, which he hated, he was quite well treated during the time he spent growing up with his mother's family. Mazo's account of MacMillan's student days and his stay in the sanatoriums, while not highly faithful to the facts, captures the spirit in which MacMillan experienced these aspects of his life. The novel ends with his move to British Columbia, and anticipates his business success in heroic but fairly realistic terms. Her portrait of him is true in the sense that the ideals she ascribes to the protagonist, Shaw Manifold, as he sets out on his lifelong career are those that MacMillan himself later tried to fulfill. "He saw himself as sending ships out to the Seven Seas laden with sweet-smelling lumber, lumber not butchered from the forests, leaving a wasted land behind, but chosen with care, with a new growth to follow."[7]

Although MacMillan never admitted to having read it, in the end he was immensely pleased by the book. And its writing and publication meant that his relationship with Mazo and Caroline was re-established. The two women lived among artists and writers, and may have been lesbians as well— all of which must have seemed exotic and almost bizarre to MacMillan in his staid existence. Although they later visited him in Vancouver, and for some time considered moving to the West Coast, Edna and his daughters did not approve. But whether or not the two women were lesbians, and whether or not MacMillan knew it, he corresponded with and visited Mazo until she died in 1961, and continued his friendship with Caroline after that.

In spite of the critical situation facing him in the forest sector, MacMillan found himself drawn deeper into the affairs of BC Packers during the 1930s. He never established an office of his own at the company, but he was very much a hands-on president. In 1935 he appointed Secretary-Treasurer John Buchanan as general manager. This turned out to be one of his most fortunate decisions, as he and Buchanan established an effective working relationship that eventually turned into a lifelong friendship. He had also acquired a valuable source of advice and experience in the person of Stanley McLean, the president of Canada Packers. An older man, but self-made like MacMillan, McLean lived in Toronto and served as a distant and somewhat detached sounding board for MacMillan's ideas and proposals for BC Packers. In time they became close friends and fishing companions, and McLean took a position on the BCP board, where he contributed his knowledge directly.

MacMillan's relationship with McLean is a good example of how he sought out and cultivated powerful, influential people. He understood that they could assist him in his endeavours, although apparently he never pursued the alliance if no personal rapport developed. These personal relationships lay at the heart of practically every one of his successful business ventures. At the same time, MacMillan had much to offer in such associations, and his friendship was highly valued.

MacMillan took his usual direct approach to learning the fishing business. One habit he developed early on was to visit grocery stores, where he examined displays and had long talks with proprietors. An engaging conversationalist who knew how to listen, he soon acquired a network of dedicated retailers who provided him with the best possible information on marketing fish products. He inundated Buchanan with long memos and letters detailing his findings, making suggestions and issuing orders. Part of the success of their relationship was Buchanan's ability to deflect the frequent tirades MacMillan directed at the people working close to him. On occasion, when MacMillan telephoned and launched into one of his browbeating sessions, Buchanan simply hung up. Eventually, MacMillan would call back, and conclude the conversation in a much calmer manner.

In 1936, the year the company showed its first profit, almost $68,000, MacMillan began to make annual tours of the company's operations, which consisted of a few dozen canneries, reduction plants, stores and fishing stations at various locations between the Fraser River and the Alaskan border. These were usually one- to two-week excursions on one of the company's work boats. Each tour could be counted on to produce many pages of written commentary and instructions to Buchanan.

MacMillan was very concerned about the appearance of the plants, feeling that it was important for a firm in the food business to practise "good housekeeping." His suggestions after one trip included painting cannery roofs, cleaning up discarded equipment and planting flowers along the sidewalks— all to be done as economically as possible.

> Amongst the directors we could get a good supply of bulbs this fall which could be sent out at no cost to the plants at Alert Bay, Kildonan, Ecoole or other places where they would use them. A continuous encouragement of the maintenance of flowers and gardens at little or no cost to the company would in my opinion be a very wise policy.[8]

These inspection tours, during which he spent long hours meeting and visiting with residents of the coastal communities, made him acutely aware of the important role companies like BC Packers and H.R. MacMillan Export played on the coast. He felt they should take a leading role in promoting the development of the communities and the interests of local residents, whether they worked for the companies or not.

MacMillan developed a habit on these tours that would later drive senior managers in both companies into fits of exasperation. Wherever he went he got to know people at all levels of the operation—not because of their rank or position, but on a more personal basis. He was just as apt to talk to a whistle punk in a logging camp or a janitor in a cannery, as the manager. In this way he obtained a ground-level view of the operations. On one trip to the canneries he seems to have sought out foremen, and duly reported to Buchanan those whose breath smelled of whisky in the morning.

In 1937 he was the main speaker at the twenty-fifth anniversary celebrations of the BC Forest Service. After a brief review of changes in the industry and the Forest Service, he went on to discuss the future of forestry in the province. Noting that the chief commercial species, Douglas fir, was disappearing fast, he argued that conditions in coastal British Columbia made it one of the two best sites in the world, along with the Baltic, for growing timber as a commercial enterprise. Chiding the Forest Service for the effort it expended on administration, he called on the government agency to come up with a long-range policy that would be in the best interests of all the people of BC, not just the bureaucrats, loggers and sawmill men.

> The people of B.C. have been allowed to assume all is well with the forests. They assume the forests are permanent. They feel secure in their dependence on this source of wealth. They spend lavishly on education, the care of the feeble-minded and the sick, bringing water or transportation to comparatively unpopulated areas. Having done this we must give them the faith to devote a reasonable share to the perpetuation of the resource that is their chief source of livelihood. In my opinion they won't forgive those entrusted with this responsibility if it is neglected. We must have a policy and put it to them clearly, forcibly, quickly and constantly.[9]

MacMillan was not encouraged by the state of forestry in BC or, in fact, any place in Canada. Having lost his youthful, zealous belief in the "cause" of publicly administered forestry, he had withdrawn from its primary organization, the Canadian Forestry Association. In BC, virtually every other professional forester worked for the Forest Service, an uninspired and neglected agency lacking the leadership needed to provide direction for government, industry and the public. And, with the availability of large volumes of cheap, publicly owned timber, there was little incentive for private timberland owners in the province to even consider managing their forests for the future. In an article he wrote for the *London Daily Telegraph* and *Morning Post*, MacMillan was even more critical of what he perceived as a growing failure of Canadian forest policy.

> "How long can it last?" it may be asked. "What of the future?" Canadians have listened to such tales of Canada's limitless resources that they are prone to avoid the answer rather than seek it. The national attitude is that of Micawber—something will turn up. . . . Meantime it is generally known among the well-informed that the forest is being overcut at a devastating rate in every forest province in Canada; that Canada, an essentially forest country, lags far behind India, the United States, Norway, Sweden, Finland and France in forest policy; and that forest schools and forest departments in Canada are half-starved and failing to lead or influence a Canadian people, who are still bent on exploitation rather than conservation of their great natural resources.[10]

Although MacMillan had been elected president of the Vancouver Board of Trade in 1933, the twin crises of the Seaboard threat and BC Packers' financial distress prevented him from playing much of a public role after the release of the Kidd Report. In 1938 he presented, and probably played a major role in writing, a brief to the Rowell–Sirois Royal Commission on Dominion-Provincial Relations. This commission had been established to examine the relationships between the provinces and Canada in light of the disastrous financial state of most provincial treasuries. MacMillan stunned the commission and outraged provincial governments across the country with a Vancouver Board of Trade brief bearing the support of thirty-one of the largest business organizations in the province. Among other measures, it called for a plebiscite on the question of abolishing the country's nine provincial legislatures. Until a public vote could be held on the question, he went on to argue, the number of provincial legislators should be drastically reduced and all provincial and municipal borrowing should be controlled by Ottawa. These measures were required, he said, because provincial debt was out of control. Even though the BC government was collecting the largest revenues in its history, he pointed out, it was still running a deficit.

Premier Duff Pattullo's reaction verged on apoplexy. In a series of speeches and statements, he denounced the proposal as absurd and impractical. One of the premier's chief supporters on this issue, Senator J.W. deB. Farris, produced a telegram from the BC Lumber Manufacturers' Association, repudiating the Board of Trade brief.[11]

The telegram reflected the ongoing battle between MacMillan and the Seaboard Lumber group. Following their 1935 standoff, the two firms shared between them, more or less equally, about 80 percent of the province's lumber export business. Their old antagonisms had been incorporated into their respective operations. When MacMillan began to acquire and operate his own sawmills, the membership of the BCLMA, the primary sawmilling association, was dominated by Seaboard companies. Consequently, the MacMillan mills, along with several independents, formed a second association, the Western Lumber Manufacturers Association, with VanDusen as president.

In 1937, during a trip to the West Indies, MacMillan discovered that the provincially appointed BC Lumber Commissioner for the Caribbean, C.D. Schultz, was in fact an employee of the BCLMA. He was using his semi-official position to gain confidential information about MacMillan's business in the Caribbean and to pass it along to the Seaboard group. Further investigation revealed that newly installed lumber commissioners in several foreign postings were all BCLMA employees, paid for by a government grant to the association. In theory, these commissioners represented the entire BC industry; in fact, as MacMillan found out in the case of Schultz, they were working on behalf of BCLMA members and to the detriment of its competitors, such as MacMillan. In other words, what they were doing was not all that different from the role Loren Brown had performed for MacMillan in London.

MacMillan immediately launched a campaign in Victoria to have the public subsidy to the BCLMA terminated. In spite of repeated assurances from cabinet ministers over the next few years, he continued to find evidence of provincial payments to the mill association. In this context, the BCLMA position on the Board of Trade brief takes on a different hue.

Eventually, world events overshadowed the intense and essentially personal competition between the MacMillan and Seaboard groups. War was declared on September 3, 1939, and the demands it created for the lumber industry did not long permit the pursuit of business feuds. On September 19, the two factions sent a joint telegram to the UK Trades Federation, which began by stating: "We wish to assure your organization of our fullest joint co-operation in any plan evolved by the timber industry of the United Kingdom to meet the present emergency."

Out of the confusion following the declaration of war, it quickly became apparent to BC's lumber exporters that a shortage of shipping was the greatest obstacle to fulfilling the large wartime lumber orders. MacMillan had received a huge order directly from the British Admiralty, and Seaboard was also getting a flood of orders. There were simply not enough ships available to carry the required volumes of timber. Having worked out a plan in Vancouver, MacMillan and Seaboard's vice-president, Bruce Farris, took it to Ottawa. They proposed to alleviate the shipping shortage by transporting BC lumber across Canada by rail and shipping it from the east coast, cutting the maritime shipping requirements by more than half. The Canadian government was sympathetic, but it took MacMillan several months of intense negotiations to get the two Canadian railways to lower their shipping rates to a level the cash-strapped British would accept. The following March, the first lumber headed east by rail, a joint shipment by MacMillan and Seaboard. By that time, since the mills were working at full capacity, the yards in Vancouver were clogged with lumber; it was even stored in great, floating rafts in the Vancouver harbour.

At the outbreak of war, MacMillan was probably better informed about global conditions than anyone in BC or, for that matter, most people in Canada. Some time before, when his daughter Jean graduated from Stanford University, he had given her a trip around the world for two, as a graduation present. When he got around to asking her who she would like to take with her, the answer was simple: "You." He agreed, and in March 1939, a few weeks after his oldest daughter Marion's wedding to John Lecky, the two of them sailed for Japan on the *Empress of Canada*.

They spent several weeks in the Orient, assessing the growing Japanese nationalist fervour that had led to restrictions on lumber imports from BC and other disruptive trade measures. On a visit to the Keswick family, they were brought into direct contact with the Sino-Japanese war near

Shanghai, where they were stopped and searched before passing through Japanese lines. And they were confronted face-to-face by the growing Japanese antagonism to the West when they took a Japanese ship, the *Osama Maru*, to Hong Kong.

For Jean, the trip was an exciting, romantic adventure. Wherever they went she was the centre of attention among young Canadian diplomats working abroad, to her father's consternation. Even more perplexing to MacMillan were the large bouquets of flowers that greeted her in every city. They were from her future husband, Gordon Southam, who obtained an updated itinerary from Dorothy Dee, MacMillan's secretary, and carried on his courtship as Jean and her father continued around the world.

They flew on one of the first commercial flights from Hong Kong to Bangkok. It was a harrowing twelve-hour flight through fog and storm on an aircraft that was ancient even by pre-war standards. MacMillan, seasoned traveller that he was, wrote a four-page letter of complaint to the head of Imperial Airways.[12] In India, while on a tiger hunt with the Maharajah of Gwalior, Jean contracted what was later found to be amoebic dysentery. By the time they reached Alexandria she had a fever the doctors could not diagnose. Nevertheless, they carried on to France and London, and eventually

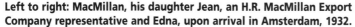

Left to right: MacMillan, his daughter Jean, an H.R. MacMillan Export Company representative and Edna, upon arrival in Amsterdam, 1932.

New York, where they met Edna at the World's Fair. After four months of travelling, they returned to Vancouver.

For MacMillan, World War Two began when he arrived home. Within a few days he delivered a lengthy account of his observations to a packed luncheon meeting of the Vancouver Board of Trade. On the European situation, he made it clear war was coming and forcefully argued that Canada's role was to support Britain. Pointing out that 89 percent of BC's lumber exports were going to the UK, he argued it was not the time to follow the American lead and remain aloof from the coming conflict. But his audience was most interested in his description of conditions in the Pacific. Although BC enjoyed a good deal of trade with Japan, there was scant knowledge in Vancouver or anywhere in Canada of events and conditions in the Far East. MacMillan stated his suspicions that by taking advantage of distractions in Europe, and of isolationist sentiment in the US, Japan hoped to gain control over the Pacific.

> Japan is a neighbour with whom we have long traditions of friendship, but there have been many changes since she developed an intense nationalism, got her finance well organized with good government machinery and tremendous industrial expansion. Japan is trying to emulate the United Kingdom as the world's workshop, and is doing this without raising wages or production costs.
>
> Politically, Japan scorns western civilization and any left wing movement is feared by those who seem to be threatening a swing to military dictatorship. . . . Japanese warfare [in China] is one of unparalleled atrocity, of oppression, indiscriminate slaughter and utter disregard of all civilized laws. In China the Middle Ages have returned—it is a case of woe to the conquered.[13]

While events in Europe did not particularly disturb MacMillan, Japan's activities were another matter. He was intrigued by the Japanese and their culture and history. He had been doing business with them successfully for more than fifteen years, and he enjoyed social events with visiting business and political representatives. Consequently, he was shocked and appalled at Japan's conduct in the Asian conflict.

There had been waves of anti-Japanese sentiment in BC since the turn of the century, part of a longstanding, deep-seated anti-Asian racism. By 1939, large portions of the population—most of them of European heritage—believed there were hundreds of illegal Japanese immigrants in the province, that the community contained Japanese spies and military officers, and that a Japanese fifth column was being prepared. MacMillan did not subscribe to these near-hysterical views, and there is no evidence he had any connection with or sympathy for anti-Japanese agitators. He did, however, have strong feelings on the question. A month after his return to Vancouver, he took up the issue of the Japanese in Canada. Throughout August he exchanged several unpublished letters on the subject with *Saturday Night* magazine editor B.K. Sandwell, who had published an editorial deploring anti-Japanese feelings in Canada.

> I agree with the principle of justice and liberty which led you to write the editorial; nevertheless, I think that there must be some policy we in Canada could adopt which would be an improvement upon the methods and practices now in existence with respect to residents in Canada of Japanese racial origins. . . . The Japanese have many virtues: industry, thrift, technical knowledge of and aptitude for many occupations. Certain types of Canadians object to Japanese more for their virtues than for their vices. . . .
>
> My view respecting the Japanese problem arises from my opinion that the Japanese are not becoming Canadians. Even if they remain racially exclusive over a long period of time, the danger to the country would not be so great, provided they became Canadians, but they remain racially exclusive and do not become Canadians and continue to increase in numbers, the result of which will create a national problem concerning which the people in Canada should now become informed, and when Canadians become informed they should not depend too much upon doctrinaire opinions produced by international sentimentalists, but should be prepared to learn that the Japanese is not a civilized person who has adopted or is ready to understand and accept the fundamentals of our form of government and beliefs in justice, equity and liberty.
>
> The opinion of many people who have had close contact with Japanese in our schools and universities and who are well acquainted with second and third generation Japanese is that all Japanese—first, second and third generation—are still Japanese in every way and their Canadianism consists chiefly or solely in their ability to inhabit the fabric which Canadians have erected in this country. They contribute nothing to supporting this national edifice. They are ardent Japanese nationalists and they "gang" up here to achieve their ends. Even if born in this country, they support and patronize with national fervour all Japanese institutions here from ships to shops, and when a problem arises affecting their interests they treat it as a racial problem which they settle by taking it to the Japanese Consul here, who secures the help of the Japanese Minister in Ottawa. Many of them also send their savings back to Japan. They exist as Japanese cells in the Canadian body politic. . . .
>
> Japan is committing a series of premeditated war-like acts against Britain in China; we are members of the British family, therefore Japan is committing war-like acts against us. Japanese residents of British Columbia are sending their nationals back to Japan to take part in this wholly Japanese war and, so far as one

can judge, Japanese in Canada are—so far as their resources allow—sending material aid to Japan. This is regarded by them as being not only natural but a national duty.

If these things are found to be facts, is it not reasonable therefore for Canadians to re-examine any policies that might lead to building up a Japanese colony in Canada?[14]

In early September, MacMillan sent copies of his correspondence with Sandwell to federal transport minister C.D. Howe and, in a more practical vein, wrote to Sherwood Lett, a director of the Vancouver Club:

I believe that we cannot accept at face value any expressions by the Japanese of friendship for the British (which includes Canadians). Everything that can be studied about us and used to our disadvantage is being done and if for no other reason than this, I believe a change should be made in conditions of employment at the Vancouver Club. It would appear fitting that preference be given to our own nationals for positions available, particularly to those who have military service. There appears to be no justification for such a representative institution as the Vancouver Club giving preference in employment to those who are neither our nationals nor our friends.[15]

His was one of a growing chorus of voices that three years later led to the expulsion of all persons of Japanese ancestry from the West Coast for the duration of the war.

After spending the fall arranging for rail delivery of lumber across Canada, MacMillan headed back to London in early January 1940 to negotiate a major lumber order with Britain's wartime Timber Control Board. He took Ralph Shaw, the young document department employee, with him; it was Shaw's first trip beyond Calgary, and they were to meet Humbird and Grinnell in London. It was already agreed that Seaboard would take 60 percent of the order and MacMillan 40 percent, and that the two groups would negotiate together with the Timber Control Board.

MacMillan spent several days in London, showing Shaw the sights and introducing him to the major figures in the British timber business. Humbird and Grinnell were delayed in New York, waiting for berths on a US ship. While they waited, MacMillan and Shaw continued discussions with key figures in the timber industry, and learned that the Admiralty was about to place heavy restrictions on the use of ships for transporting lumber. If this decision had been implemented before a lumber order was placed, it would have been disastrous for the BC lumber industry, forcing the closure of many mills and throwing thousands of people out of work.

When the Timber Control Board indicated it wanted 300 million feet of lumber, and the Seaboard contingent still had not arrived, MacMillan went ahead and negotiated an arrangement, stipulating that the agreed-upon 60 percent would go through Seaboard. As he told Shaw, if they waited for Humbird and Grinnell to arrive it would be too late; the Timber Control Board would be unable to secure ships and would not place the order.

When Humbird arrived three weeks later and discovered the deal had already been concluded, he was outraged, mostly because he had looked forward to the ritual and prestige of negotiating it himself. In a temper, the overweight Humbird threw himself into the rear seat of the local Seaboard agent's prized Lancaster automobile, and broke the axle. Later, back in Vancouver, when various distorted versions of the story were circulating suggesting MacMillan had double-crossed Humbird and attempted to negotiate the whole deal for himself, MacMillan responded, "You can't be put in jail for breaking your word." He had, however, earned the lifelong enmity of Humbird.[16]

At about the same time, he was approached by a young newspaper reporter and asked, "Do you know what they are saying about you, Mr. MacMillan?"

"No. What are they saying about me?"

"They say you are a buccaneer."

"That's right," MacMillan responded, smiling broadly. "I sink them all."

Leaving Shaw to return to Vancouver on his own, MacMillan took a tour through Italy and France, both of which were teetering on the brink of all-out war. If he had ever had any doubts about the nature of the war, they were dispelled by the time he left Europe. This, clearly, was not going to be a conflict like others, a contest between competing military forces, but total war between nations in which every citizen was in some way mobilized and subject to attack. Nor did he have any illusions about the evil of the forces confronting the UK and its allies. Through Meyer he had a window into the fate of European Jewry, and he came back to Canada grimly determined to do everything within his power to advance the Allied cause.

While he was away, MacMillan had been elected president of the Canadian Chamber of Commerce. To acknowledge his new position, the

Vancouver Board of Trade invited him to speak at a dinner in April. His speech, "Thoughts After a Month in England," was delivered to a capacity crowd and made a deep impression on those who heard it. Thousands of copies of the printed text were distributed to schools and other public institutions throughout Canada and the UK. He gave a long and poignant description of conditions in Britain and the mobilization of her people. He described how the global trading situation affected the course of the war, and how the economics of war would affect Canadians. He called on the Canadian government to clarify its war policies, and on citizens to discuss and criticize these policies so as to improve them. "We can contribute our share toward shortening this war by putting our whole strength into it," he said, "a policy which should appeal to all Canadians as being the sole means of preserving our real independence and self-government."[17]

In the following weeks he made dozens of similar speeches across Canada and, occasionally, in the US. At this time he was the most vocal and articulate spokesman for the British cause in North America.

In late May 1940, Transport Minister Howe asked MacMillan to go to Ottawa. On June 5, he met with Howe, who described the imminent creation of the Department of Munitions and Supply to oversee the production of war materials in Canada. He asked MacMillan to become steel controller. MacMillan demurred, and suggested Howe find someone acquainted with the steel business. The following day Howe proposed instead that MacMillan take charge of an airplane factory planned for Vancouver, which MacMillan said he was willing to do if he had complete control of it. Howe wanted a few days to think it over.

MacMillan left for a fishing trip with Stanley McLean in the Gaspé while Howe sorted things out in Ottawa. A few days later, he was again summoned to the capital. This time Howe asked him to become timber controller for Canada. MacMillan accepted, and he was off to war.[18]

Dollar-a-year man

In the weeks immediately preceding MacMillan's appointment as timber controller in 1940, the war in Europe took a dramatic turn. The so-called Phony War, during which both German and Allied nations marshalled their military forces, came to an abrupt end. Germany successfully occupied Denmark and Norway, then marched through France. British troops, which had landed in France the previous year to assist the French, were forced to beat a hasty retreat at Dunkirk in June 1940, leaving the enemy in control of continental Europe. Until this juncture, Canada played a token role in the gathering conflict. After following an isolationist foreign policy for more than a decade, the country was now stuck with the consequences. The armed forces were almost nonexistent. One infantry division had been mustered and dispatched to

England where it was preparing for action. It was accompanied by a single RCAF squadron, assigned to provide air cover. The navy, whose fighting ships consisted of four destroyers, was largely occupied with convoying men and equipment across the Atlantic. The most significant Canadian contribution to the Allied cause was the British Commonwealth Air Training Plan. Under this agreement, Canada provided facilities and personnel to train British airmen for the predicted air war. Although the plan had been agreed to the previous fall, like most Canadian military efforts, it was slow to begin.

With the collapse of France, the only opposition to Hitler in the summer of 1940 was the United Kingdom, backed by its Commonwealth members. Of these, the strongest and most strategically located ally was Canada. MacMillan described the situation and the importance of Canadian lumber in his first report as Timber Controller:

> Canada is now almost the sole supply of wood products for Great Britain. The whole of Europe is cut off by war and the United States is almost entirely cut off by the scarcity of American dollars. Other parts of the world supply no soft woods and only limited quantities of hardwoods. Great Britain is the world's largest importer of wood. Her pre-war requirements were supplied about 75% from the Baltic countries, France and the Danube valley. Even during the Great War, most of these supplies were maintained. Now for the first time since Napoleon's continental policy Great Britain is cut off from European wood supplies and is dependent wholly upon Canada.[1]

The spread of this realization across the country, aided by the exhortations of a few strong Commonwealth supporters such as MacMillan, encouraged a growing clamour for increased Canadian involvement in the war. In response, Prime Minister Mackenzie King called a snap election in March and won an easy victory. He ignored Opposition calls for a union government and appointed a new, all-Liberal cabinet, with C.D. Howe in charge of the new Department of Munitions and Supply, J.L. Ralston as minister of defence and J.L. Ilsley as finance minister. Howe, an engineer from northern Ontario, was a straightforward type who shunned the convoluted rituals of conventional politics and unhesitatingly rode roughshod over the standard procedures of parliamentary government. Ralston, a World War One veteran, was a calm, collected professional politician with a distinguished air, who had recently been a hardheaded minister of finance. Ilsley, a tall, academic-looking Nova Scotian with an enormously strong character and a simple, direct approach to complex problems, furnished a balance to Howe's rough and tough methods. These three men, plus King, formed the core of the cabinet war committee.

Among the committee's first decisions was one to mobilize the country in support of a large Canadian expeditionary force in Europe. To facilitate this enormous task, a large contingent of able individuals was recruited from the private sector, all to work under Howe's Department of Munitions and Supply. None of them was paid by government; the firms they owned or worked for paid their salaries and, in most cases, expenses as well. Later, in the US, these "recruits" were called dollar-a-year men; in Canada they never even got the dollar. MacMillan was the first Canadian controller appointed under this scheme.

Even before the discussions with Howe that led to his appointment, MacMillan had expressed concern about Canada's war effort and the government's ability to meet the emergency. In late May, he explained his misgivings to Stanley McLean.

> We have seen tremendous disappointments occasioned by lack of foresight and lack of organization, and it is within the realm of possibility that our form of government may be lost through such failures in the past on the part of our political leaders and higher state officials; and as a result of the sudden discovery by the people in a dramatic fashion of the governmental errors, is the rising tide of uninformed public opinion which is likely to force the government into more errors.
>
> One of the next possible errors is that the public will force parliament and parliament will force the government to make immense expenditures which will be embarked upon so rapidly that value cannot be got and co-ordination of production will be absent, with the next possible result that the country will not secure from the expenditures the volume of military strength they are seeking, nor will they get it at the appointed time . . . everything that happens every day and every announcement that is made from Ottawa carries deeper into my mind the conviction that the small group of over-worked men have had thrust upon them an immense task—far greater in intricacies and volume, far more exacting in the element of speed, more foreign than any past Canadian experience, and of more grave import than any task that has heretofore fallen to any group of Canadians.
>
> It would not be surprising therefore if bedeviled and driven as they now are they fell into the common error of burying themselves in the work, with the consequence that they are not left sufficient time to think, plan, and organize, and with the further dangerous consequence that the men at the top will become too tired in the beginning of the job to deliver the best that is in them, and will consider that they have not the time to devote to building an organization which would put at their services a sufficiently large number of the best men of Canada, selected for their experience of the job in hand, and that they will fail to keep their organization ahead of the increasing volume of work and will fail to keep

plenty of free brain power at the top. I do not desire to make this too critical, and I have hesitated about writing it even to you.[2]

Holding these gloomy opinions, MacMillan was anxious to tackle the responsibilities of timber controller. Howe's instructions had been scant and vague. "He said there were some matters respecting aeroplane spruce which required attention, but as to further duties he had no definite ideas at the moment," MacMillan confided to his diary. "He expected that I would be able to handle it from Vancouver."[3] After his final meeting with Howe, he stayed in Ottawa for another day in a vain attempt to gather more information on the country's wartime timber needs. "I asked Mr. [Ralph] Bell, in charge of aeroplane production for information as to the Department's probable wood requirements, but no one in Ottawa had any knowledge of this."[4]

Realizing he was wasting his time, MacMillan returned to Vancouver. He met with L.R. Scott and VanDusen at the export company office, obtaining their views on the wartime timber situation and preparing for his absence from company activities. The following day he met with the heads of the various timber organizations in Vancouver, and that evening he flew to Victoria

MacMillan and his senior executives, early 1940s. Left to right: H.H. Wallace, MacMillan, W.J. VanDusen, L.R. Scott, E.B. Ballentine.

for dinner with the minister of lands, A. Wells Gray, and Chief Forester E.C. Manning. Gray consented to Manning serving as MacMillan's assistant in charge of BC's airplane spruce production—at provincial expense. MacMillan's first instruction to Manning was to purchase the cutting rights on two large tracts of spruce in the Queen Charlotte Islands from their owners in Japan. The following day, back in Vancouver, he called together everyone with an interest in spruce production, including Brian Gattie, the Vancouver representative of the British timber board.

> Explained to them situation so far as I understood it and said I desired to get everything on commercial lines to make timber available without delay at fair prices to both consumer and buyer, and to secure the advice, best intentions and help of everyone in the industry. I explained that 23 years ago the Imperial government had paid for spruce cutting-rights $6 for #1 grade and $2 for #2 grade and that now, after 23 years' carrying charges, with spruce scarcer now than it was in the Great War, it might, in cases of expropriation or by the bringing of pressure on people to make them sell, be necessary to pay somewhat higher prices than in 1917 and 1918.[5]

Within a week he had put in place a basic organization for handling the spruce problem. By then an even larger problem loomed—the logistics of purchasing and delivering the enormous volume of lumber required to build about eighty aircraft hangers, which were needed immediately by the air training program all across Canada. The program created enormous difficulties. Workers had to be found for the accelerated logging and sawmilling operations, orders had to be placed, specifications had to be modified and delivery had to be arranged in a timely manner—all without driving up lumber prices and wages.

The amateurish nature of the Canadian military apparatus at the beginning of the war is described in a letter from MacMillan to C.G. "Chubby" Power, Minister of Air Defence, written in the midst of this hectic period.

> An application had been made to me for financial assistance to provide aeroplanes for training pilots in the vicinity of Vancouver. I did not understand what your policy might be and therefore I took the liberty of telephoning you. After your exposition of your policy and your position with respect to training pilots, the directors of this company have decided that the H.R. MacMillan Export Company will contribute $8,000 for the purchase of a training plane.
>
> We regret that facilities are such that this contribution will not result in there being one additional plane, but believe that it is advisable at the present moment

to give every possible stimulus to morale, and therefore make the contribution for that purpose and also, if it is in line with your policy, to increase by one the number of training planes maintained in the area of British Columbia.[6]

It did not take MacMillan long to run head-on into the type of bureaucratic proceduralism that had played such a large part in his decision to leave government service in 1919. At the end of his first week on the job, he communicated his frustrations to Howe.

> I enclose copies of telegrams which I have addressed to the Deputy Minister rather than to you because I thought such procedure would result in only the necessary matters being brought before you, the others being passed along through ordinary official channels without taking your time. I have not received acknowledgement of or reply to any of these telegrams. . . . I telephoned the Deputy Minister this morning to ask how I should go about getting an answer to some of these things or to secure other information, and I was told on the telephone that he was too busy to talk to me today. Perhaps I have been following the wrong procedure, a continuance of which is going to place this appointment in a very embarrassing position with the trades concerned here and with others from whom one desires complete confidence and expedition in the performance of work, and therefore I am troubling you with this letter.[7]

He also communicated some of his confusion to Stanley McLean.

> I have not yet been able to form any opinion respecting the responsibilities of the office of Timber Controller. Because of my position in the lumber business, I did not feel that there was any valid reason for refusing the position. The Order-in-Council appointing me has not yet been finally passed, consequently I have no idea of the duties and authorities, but I am led to believe that the authorities are all inclusive. It was not possible in Ottawa to learn anything significant about the general duties. Possibly the position was created and someone appointed so that the government might have the power and means of dealing with any trouble affecting timber for war purposes. It is possible that in cases leading to a breakdown in the smooth commercial operation of the export lumber business to which people have become accustomed during the past few years, this office might be called upon by public clamour to deal with the ensuing disappointments and chaos.[8]

A few days later Howe called MacMillan and asked that he take over responsibility for all of the department's timber concerns. Now that he had a clear idea of his responsibilities, his authority and the nature of the task ahead of him, MacMillan realized he could not fulfill his duties from Vancouver. In early July, he moved to Ottawa, setting up an office in the

Parliament Buildings and moving into a suite in the Chateau Laurier. He arranged for Dorothy Dee to join him a few days later.

By the time of their move to Ottawa, the relationship between MacMillan and Dorothy Dee had evolved into a more personal and intimate one than that of employer and secretary. Over the course of the previous decade the tall, striking blonde had become a fiercely competent manager of MacMillan's affairs and an indispensable employee. Almost from the day she began as his secretary, she devoted herself completely to his interests. Perhaps because of the strength of her personality, which was a match for his, the two formed a close emotional partnership that lasted as long as they both

Dorothy Dee, at the time she joined the H.R. MacMillan Export Company, c. 1930.

lived. Edna was and always would be MacMillan's wife, but it was not in her nature to be his partner. She was a quiet, retiring woman who occasionally travelled with him on his journeys abroad, but was neither capable of nor interested in operating on the same intense mental plane as her husband.

Whatever the intimate nature of the relationship between MacMillan and Miss Dee, as she was known, it was discreet. Unlike many of his contemporaries, such as Montague Meyer, who often appeared at public events with attractive young companions, MacMillan was careful to observe the proprieties. He and Miss Dee did not appear inappropriately in public, or otherwise provide fuel for the Shaughnessy gossip mills. His family relationships were too important to him to be jeopardized, and it was not in his nature to cause pain or embarrassment to his wife and daughters.

Those who knew MacMillan well occasionally described him passing through their lives like a hurricane or a tornado, or some other elemental force of nature. His presence created turbulence and upheaval in others. While he himself lived calmly, in some ways shyly, in the vortex of his own life, few others could live with the same intensity. His daughters could, each in her own way—Marion quietly and with reserve, Jean flamboyantly and with gusto—sometimes at great cost to them personally. Even after they married and had families of their own, their father continued to occupy a central position in their lives. Besides Marion and Jean, the only person capable of functioning within MacMillan's highly energized and emotionally intense ambit was Dorothy Dee.

Life in Ottawa during the summer of 1940 was tumultuous. Howe had recruited into the Department of Munitions and Supply many of the country's most powerful and capable businessmen. The Chateau Laurier, where several of them stayed, took on the atmosphere of a private boarding school. It was an enterprise fuelled on adrenalin manufactured out of the belief that if they failed to mobilize the country, they and everything they valued would be destroyed.

By the time MacMillan arrived in Ottawa, several other controllers had been added to the Wartime Industries Control Board. Initially, the board was chaired by Hugh Scully, the commissioner of customs, who also served as steel controller. George Cottrelle, the chairman of Union Gas, became oil controller. George Bateman, a mining engineer, was made metals controller.

Before the end of the year a machine tools controller, T. Arnold, and a power controller, Herbert Symington, were added. They held unprecedented powers over their industrial sectors and, apart from regular meetings, conferred with each other constantly, from early in the morning until late at night. Their job was to oversee the production of the various critical materials that fell under their jurisdiction, allocate the distribution of these materials, and guard against an inflationary rise in prices that would undermine the whole effort.

Many other prominent businessmen worked in the department. Henry Borden, a corporate lawyer from Toronto, occupied various positions and served Howe primarily as an advisor. W.C. "Billy" Woodward, head of the Vancouver retail firm, was one of two executive assistants to Howe, and until he was appointed lieutenant governor of BC in 1941, commuted between Vancouver and Ottawa with MacMillan. One of the most controversial and effective of Howe's dollar-a-year appointments was E.P. "Eddie" Taylor, who started out as director general of the Munitions Production Branch. It was a heady atmosphere. The department controlled virtually the entire Canadian war effort, apart from the military services. These were all powerful individuals, and when they were thrown into close proximity with cabinet ministers, politicians and senior civil servants, the atmosphere was electric in the dining rooms at the House of Commons and in the Chateau Laurier.

None of these businessmen had any patience for the snail-like procedures of the civil service, and their interface with the federal bureaucracy was, initially, a battle ground of the first order. Howe was cut from the same cloth as many of his industrial conscripts. He was brilliant, decisive and impatient with the normal pace of government activity. Having accepted the job, he proceeded to perform it in spite of the confusion over direction that prevailed in the first months. He tended to operate the department out of his vest pocket, issuing contracts for materials long before formal requests were received from the military or other end users. He never bothered with tenders or other procedures usually employed in government–business transactions. That there were few violations of public trust is probably due to Howe's ability to convince contractors that they had a civic duty to act properly, backed by the contractors' knowledge that he would destroy them if they stepped out of line. Of all the politicians in the King government, Howe was probably the only one capable of marshalling the energies of the Canadian business community and focussing them on the vital task ahead.

Even before he moved to Ottawa, MacMillan had taken the bit in his teeth and begun the complex process of reorganizing the forest industry to meet war needs. It probably gave him a special pleasure to dispatch a brief communication to his rival, Charles Grinnell, at Seaboard.

> This confirms that in response to a special request from the Honourable C.D. Howe on behalf of Sir William Glasgow, Australian High Commissioner in Canada, you will ship via the *AORANGI* direct to Sydney about August 4th between 10M feet and 20M feet of aeroplane spruce on the order placed by Astexo with the Sitka Spruce Lumber Company Limited.[9]

For the duration of the war, the battle with Seaboard was on hold.

It took much time and a firm hand to impress upon most of the country that a war was under way and the future of Canada at stake. When MacMillan learned that loggers hired to get out the spruce logs were unable to obtain passage to the camps, he wrote to Gordon Farrell, a director of Union Steamships Limited. The steamship company's manager, he told Farrell, believed his company was under no obligation to give loggers any priority. With uncharacteristic tact, MacMillan went on:

> I feel that there must be some error in this report, but I am assured that it has been impossible on the vessels of the Union Steamships to find sufficient accommodation for loggers and that those travelling to and from the logging camps are being given the impression that the tourist traffic has first call upon that accommodation. The maintenance of logging crews at full capacity is important to Canada's present aeroplane production programme. It is considered that if the programme is to be continued in its present form, it probably will be necessary to increase the number of men working on the Queen Charlotte Islands; therefore, as the Union Steamships at the present time is the only transportation company operating to the Queen Charlotte Islands, I should like to have assurance from them that they will take care of the traffic offering if they are given fair warning of the time and movement of men to and from the Islands.[10]

Even during the early, hectic stages, MacMillan found time to offer suggestions on matters well outside his jurisdiction. On July 4, he wrote to the BC minister of trade and industry:

> It has occurred to me that the virtual obliteration of France is going to open up a possibility for the production of certain French specialties, such as brandy, certain wines, and other alcoholic drinks, which can be produced in parts of British Columbia now that competition from France has probably been removed for a considerable number of years into the future.
>
> The first step would be to try to bring into this country from France … some

of the French personnel who have been engaged in these industries in France.

British Columbia is one of the few places in Canada where grapes and other fruits can be grown from which the French have supplied large parts of the world with alcoholic drinks. It is quite possible that British Columbia could not compete in world trade, but conditions might permit British Columbia to supply an important market in Canada.[11]

By early July, MacMillan had transformed the fiercely competitive BC lumber industry into a co-operative war effort. The entire industry was geared up to provide, in order of priority, British needs, aeroplane spruce, Canadian government needs and, lastly, existing domestic and foreign markets. The speed with which this was accomplished was a testament not only to MacMillan's knowledge of the industry, but also to the respect he commanded, even from his toughest competitors. The rapid mobilization had placed the industry in an extremely vulnerable position at this stage of the war. All the mills and camps, on the strength of MacMillan's assurances, had gone into full production. Yet in the chaos of these early months, the lumber markets could collapse overnight, breaking the mills and throwing thousands out of work. On July 6, 1940, MacMillan upped the stakes even further by banning all log exports to non-British destinations, primarily the US and Japan.

By August, the timber controller's office was swamped with work. Isolated in Ottawa from direct contact with the lumber business, and dependent on people not directly employed by him, MacMillan vented his frustrations on one person—Manning, his assistant in Vancouver. By late July he was dispatching caustic cables to the chief forester, and prevailing on VanDusen to intervene. Manning responded in kind.

> *You* have surrounded yourself with some men of outstanding ability in Ottawa, but I wonder if you don't need a junior executive to keep track of the continuity of things and to read our wires *carefully*. I am using the personnel that others can spare me, not selected for me on the basis of experience for this particular job. At that I am of course getting the loyal help of everyone working for me. . . . If you or any of your staff get too critical I am going to go right back at you if any useful purpose is to be served. You are a little inclined to expect the *impossible* but I shall do my utmost to deliver you the maximum *possible*. I shall do my best to keep on top of things.[12]

A slightly chastened MacMillan wired back immediately.

> HAVE ENJOYED AND BENEFITTED BY READING YOUR PERSONAL LETTER OF AUGUST TENTH STOP IT WILL HELP US IF YOU KEEP POUNDING AT THIS END AS WELL AS THAT

END STOP I KNOW YOU ARE DOING FINE AND I CAN ASSURE YOU I WILL CONTINUE TO EXPECT THE IMPOSSIBLE AND I REALLY BELIEVE I WILL GET IT.[13]

He followed up with a conciliatory letter which included a look ahead to the next few weeks.

> I am apprehensive respecting the next four weeks because I know the volume of work is going to be increased five, ten or twenty fold as a great number of projects, which are now in the blueprint stage, pass into the building stage, the start of which will be this week.[14]

Although his relations with Manning improved somewhat after this exchange, MacMillan continued to be irritated by Manning. In October he again worked himself into a state of agitation over his assistant's activities, or lack thereof, in securing an order of lumber for construction of an army camp in Nanaimo. After sending several telegrams, MacMillan mailed the kind of letter every employee of his companies dreaded, a written version of the verbal tirades for which he was well known.

> You may wonder why I am telegraphing you about the Nanaimo job—perhaps it is because I have not expressed myself clearly that you still seem to have missed the point.
> This is a war. The Army wants this job done. The Army has the idea that we are not giving them lumber fast enough. I telegraphed asking certain questions on which I wanted immediate definite answers . . . I still don't know why you don't answer the questions. It is now four days since the Director General of Purchases, after conversing with the D.O.C. in Victoria, telephoned me that work was being held up for lack of lumber. That was an official complaint. I sent you immediately official instructions, and I am still without any knowledge of what should be done or how you are coping with the situation. To tell me that you are sending 100,000 feet of lumber means nothing. . . . If you were not in British Columbia, I would get the answers myself from army officials in that area, and I would have the answers in three hours, but working through you, I have not got them in four days, which makes me think that I have failed to make myself clear; therefore, I am telegraphing you again today asking for the answers. This is what makes these government files a foot thick.[15]

For MacMillan, even at this early stage, the war was of paramount importance. His impatience with anyone who did not feel this way, and act accordingly, became notorious. He cared little for personal feelings and refused to coddle others. There was a job to be done, and anyone who was not getting on with it would feel his wrath. He tried to lead, but those slow to follow found themselves driven or, in rare instances, shoved aside.

Shortly after MacMillan's arrival in Ottawa, his daughter Jean came to stay. Ostensibly, she was there to keep her father company, but the presence of Gordon Southam, who had joined the navy and was stationed in the city, was a major attraction. He received MacMillan's approval as a future son-in-law. A few weeks after her arrival, Jean was stricken with a severe attack of the dysentery she had contracted the previous year in India, and spent most of the summer in the hospital. Although the state of her health was an added concern to MacMillan, he was relieved that she was safely confined, and he was blissfully unaware that she was sneaking out through a side gate for long summer drives through the Ottawa valley with Southam.

In late August he was elated when Marion Lecky gave birth to his first grandchild, Johnny. The pressure of events in Ottawa kept him there for almost a month before he was able to manage a quick trip to Vancouver for the primary purpose, he confided to a few friends, of seeing his new grandson.

By mid-fall the Canadian lumber industry was operating at its highest level in history. More than a thousand loggers were working on the Queen Charlotte Islands, getting out spruce logs for aircraft construction. Despite disruptions to shipping and the collapse of the continental European market, timber exports were double those of the previous year due to the increased shipments to Britain. With the timber sector more or less under control, MacMillan began to devote more of his time to wider policy questions. He entered into a steady correspondence with Howe and Graham Towers, the governor of the Bank of Canada, on matters of fiscal policy. In September he wrote to the munitions minister about conditions in Vancouver.

> We are entering upon a period of activity when both numbers of employed and rates of wages will increase. The great increase in wealth in Canada during this period will be chiefly in the form of extra hundreds of millions of dollars paid in wages to the employed.
>
> Unless a large share of these extra wages are returned to the State through war savings, they will contribute to and be a chief cause of rising price levels, rising volume of imported luxuries, rising operating costs, interference with budgets and with export trade and creation of social discontent between those who enjoy high wages and those who cannot get them.
>
> While in Vancouver recently, I learned that in spite of a drastic drop in tourist trade, the race tracks are more crowded than for years past, the jewellery stores are doing a larger business than for a long time, and retail trade in general is at a

high level, but that although the income of the people is at the highest level ever, the war savings are only about 90 cents per month.

Briefly, the people do not understand the national necessity of saving to enable the country to sustain the war effort now begun. It is essential therefore that the War Savings Campaign be geared up, be organized more minutely and be given greater prominence by the leadership of the country.[16]

MacMillan's other major fiscal concern was currency. The Commonwealth countries, alone at this stage in their war against Germany, had an enormous need for foreign currencies, particularly American dollars, in order to purchase war materials. He began urging restrictions on the domestic use of certain exportable lumber products in order to increase exports to the US. Although they were still emphatically neutral, the Americans were beginning the cumbersome process of gearing up for war, and the demand for timber in the US was increasing dramatically. The revenue from increased lumber sales in the US, MacMillan argued, could be used to purchase aircraft engines, armaments and other items urgently needed in Britain. Further, he suggested lumber imports should be prohibited, as should the use of steel where lumber could be substituted.

He also revived his campaign against the Japanese, who he believed to be in an undeclared war against Britain, hence the Commonwealth. After asking John Buchanan to estimate the number of Japanese citizens working for BC Packers who would be returning to Japan that fall, he argued persistently with Towers that they should be subject to the same kind of restrictions imposed on other Canadians.

I am informed that about 10% of the British Columbia Packers' Japanese fishermen from the Skeena River, B.C., are planning to go to Japan for the winter, and some of them are taking their families with them, with the object of retiring in Japan. The manager of the British Columbia Packers' operation at Steveston, B.C., on the Fraser River, estimates that about 75 of the Japanese at that operation are planning to go to Japan this winter.

It will appear strange to the white residents in the neighbourhood around Steveston if Japanese are permitted to go to Japan and Canadians are not permitted to go outside Canada unless they produce a doctor's certificate together with a very strong story. Practically none of the white residents of the Skeena River and Steveston would be able to secure permits to go to the United States. . . .

My contacts with the fishing community indicate that the Japanese who go to Japan are able to build up foreign exchange reserves for use in Japan in ways not available to white Canadians. Some of the Japanese are engaged in carrying fish in their boats from Canadian points to the United States, where they are able to secure payment in American funds to a degree which cannot be thoroughly checked by a Canadian official; others export products in small quantities via Japanese vessels for which payment is made in Japan. I cannot prove these statements, but it has been a current practice and is still current gossip amongst the Japanese with whom one comes in contact, that funds are accumulated in this manner by persons desiring to use the proceeds in Japan.[17]

Ten days later he wrote to an official in the finance department, pointing out that although much of the Okanagan apple crop had not been picked that year due to poor markets, distributors of Japanese mandarin oranges were preparing to ship several hundred cases into western Canada. This, he maintained, was an economic loss and a waste of the American funds used to purchase the oranges in Japan.[18]

In typical MacMillan fashion, once he had committed himself to a task—in this case, winning the war—his mind ranged widely over the problems and opportunities involved. Had he not been so well informed and so well connected, he would probably have been dismissed as a crank by the frantically busy people to whom he offered his advice. Instead, he was taken seriously and his stature in Ottawa rose steadily.

Power struggles in ottawa

Busy as he was in Ottawa during the fall of 1940, MacMillan maintained a close watch on activities at H.R. MacMillan Export and BC Packers. For the most part, he did so in a steady exchange of letters and memos with Buchanan and VanDusen. Both companies were enjoying a brisk business, with mounting orders for lumber and salmon coming from the UK.

By mid-fall the press was raising the question of a conflict of interest among all the controllers except Scully. Although there were no complaints of improprieties, the enormous power wielded by controllers over the industries in which they owned companies or worked was becoming an issue. MacMillan's case was perhaps the most obvious. He was the principal owner of the country's largest forest products company. As timber controller, he had unlimited authority to move the indus-try in one direction or another. While there was no suggestion that he ever made decisions favouring his company over others, he did enjoy certain advantages from his position in Ottawa. Chief among these was advance knowledge of government initiatives and other confidential information. Some of this was classified information and, as required by the oath he took on assuming the office, he kept it to himself. But other, less sensitive information he did pass along in his regular communications to VanDusen and Buchanan. In October, he wrote to VanDusen:

> Confidentially I may tell you that the munition factories built and being built in Canada will require 500,000 employees. You can see what that is going to do. This confirms my previous suggestions that you take those measures possible to develop string men for your technicians and keep your key men with you, and to select the type of men who are likely to value long term employment in a peace-time industry rather than temporary attractive employment in a war-time industry. Confidentially we are on the eve of a much more rigid control of trade, particu-larly imports; there will soon be licences. The intent is to restrict licences to those people who can say that the imports are for war purposes, the first of which is likely to apply against machine tools, which may be interpreted to mean every kind of machinery, such as wood-working machinery, sanders, anything used in a power plant, sawmill machinery, cannery, logging camp, or plywood fac-tory machinery. You might confidentially pass this along to J.M.B.[1]

A few days later, the question of conflicting interests was given promi-nence in a *Saturday Night* magazine opinion piece, titled "The Problem of Controlling the Controllers."

> The real problem in the administrative set-up is that of the Controllers in the Munitions and Supply department. Here we have men with the greatest possible power and in a position to use that power without any safeguards to protect members of the industry whom the Controllers are appointed to judge and sen-tence without appeal.
>
> Here we have the Controllers in the position of controlling the industry with powers in some aspects so absolute that they are staggering. Yet there is nothing to prevent anyone appointed a Controller from continuing in a dominant position with the industry he is controlling. The power of death over even an unimportant part of industry should not, as a principle, be entrusted to a member of that industry unless there are sufficient safeguards to protect members of that indus-try who may not be members of a dominant or even influential group. . . . Here we have six gentlemen who have power as individuals without restraint, which for

their own sakes as well as for the good of the industry and the country as a whole they should not have.[2]

Although this article did not document any specific complaints against any particular controller, it did question the procedures of Howe's department, which soon was plunged into a raging controversy over other issues.

By early November, the King cabinet was faced with a difficult problem. The Canadian treasury was about to run out of gold and American dollars because of vast expenditures on the mobilization program. Finance Minister Ilsley brought to cabinet his concerns about Howe's management of the Munitions and Supply finances. When Howe responded with vague answers about the amount of money already committed in materials contracts, and was unable to rank the expenditures in order of priority, the rest of the cabinet sided with Ilsley. On November 9, the War Committee of cabinet received a report from the Economic Advisory Committee recommending the creation of a wartime planning or requirements committee.[3] Howe was able to defer consideration of this proposal. By the following day, however, word had leaked to the press that the government was considering the creation of a federal economic planning body, headed by someone from the business sector, to oversee the country's war effort. MacMillan was named as the proposed head of this body.

A few days later, during a speech in the House of Commons, Howe announced that a Wartime Requirements Board had been appointed. MacMillan would head this body, while Loren Brown would take over as Timber Controller. Other WRB members included representatives of the three military services, Graham Towers from the Bank of Canada, Deputy Finance Minister Clifford Clark, R.A.C. Henry from Munitions and Supply, and Carl Goldenberg as secretary. An order - in - council passed two days later granted the WRB the power to secure from any source information respecting war-related production, and to use this information to formulate plans to ensure wartime needs took priority over all others. The Board was to report to the War Committee of cabinet through the Minister of Munitions and Supply.[4]

Whether Howe selected MacMillan or had MacMillan forced upon him by cabinet is a matter of some conjecture. Howe's biographers portray MacMillan as the villain in the crisis that developed over the next few months.[5] But there is considerable evidence suggesting the two never were seriously at odds and that the controversy was mostly a product of Opposition politicians and some of their supporters in the press.

The creation of the WRB provided a focus for the widespread discontent with war preparations which had been growing across the country throughout the fall. In answer to critics who argued the Canadian war effort was insufficient, journalist Bruce Hutchison published a column in the *Vancouver Sun* dealing with MacMillan's new appointment.

> The question is not size. The question is whether this enormous, sprawling, many-sided effort is going to crystallize and jell into something worth while. The question is whether all the loose ends can be gathered together and welded into an effective and deadly weapon. There have been grave doubts about it. That is why they have appointed Mr. H.R. MacMillan's super-planning committee. Mr. MacMillan's job, in pulling the loose ends together, is perhaps the biggest job in the nation, outside that of the Prime Minister and the governor of the central bank.[6]

There were great expectations in many quarters for the WRB. The Opposition shared these expectations but had its own agenda, which was to secure cabinet seats in a union government. When Prime Minister King flatly rejected this proposal, Opposition Leader R.B. Hanson called for stronger powers for the WRB, including authority to override Howe. Several newspapers, chief among them the *Financial Post*, took up the argument to limit Howe's powers and put more control in the hands of a strong business leader. Hanson referred repeatedly to making MacMillan another Joseph Flavelle, who had been granted enormous authority over war production in World War One. There is no indication MacMillan himself had any desire for more authority than he was granted by the Privy Council, and it is highly unlikely he would have accepted the appointment if he was not satisfied with the board's advisory role.

Later there were assertions that MacMillan believed Munitions and Supply to be "out of control," but such claims are not substantiated by statements he made during this period. For example, on November 2, he sent a long letter to Howe on the question of British timber purchases going through Munitions and Supply. Until that point, on MacMillan's recommendation, they had gone directly to Canadian mills from the British Timber Controller. This procedure created several problems, one of them painfully close to

MacMillan. Since the outbreak of war, Britain had made large purchases from BC's two plywood mills, utilizing almost all their output. Under pressure from the British Timber Controller, both these mills, one of which was owned by the H.R. MacMillan Export Company, had invested large amounts of money to expand their operations. Then, in late 1940, the British announced they were making their 1941 plywood purchases in the US. Having cut off their domestic markets to supply the British during 1940, the mills found themselves in a critical situation. MacMillan recommended to Howe that Munitions and Supply begin making all Canadian timber purchases on behalf of the British. He would not have made such a recommendation if he lacked confidence in the operations of Howe's department.

By this time, Howe was dealing with a more important problem. Since the summer, a breakdown in communications between the British and Canadian officials had hopelessly confused material requirement priorities. It was becoming increasingly apparent Canada was not gearing up to produce what Britain needed. Howe decided to go to London to straighten things out, taking Eddie Taylor, Bill Harrison and Bill Woodward with him when he sailed from New York on the *Western Prince* in early December. The ship was torpedoed and sunk as it approached the British Isles. The minister and his entourage spent a harrowing eight hours in a lifeboat before they were rescued and carried on to London, where Howe stayed for almost two months working out details of Canadian war production.

In his absence, some newspapers, with the active assistance of the Opposition, launched a fierce attack on Howe. Their basic argument was that Canada's war production was in chaos, Howe was responsible, and the entire operation should be placed under the control and direction of a powerful figure from the business world, with MacMillan most often mentioned as the prime candidate.

In his first speech as WRB chairman, to the Canadian Club in Toronto on December 9, MacMillan addressed the controversy obliquely but made it quite clear where he thought the ultimate authority should lie. "Management in war production implies the timely use of assets and opportunity; objectives are too important to be left to chance," he declared. "War against totalitarian States makes it necessary for us to adopt a certain measure of totalitarian methods, but these must always remain under the control of Parliament."[7]

The new year began with a January 1 editorial in *Maclean's* magazine containing most of the standard criticisms of Howe.

> . . . inside observers have begun telling that he is over-optimistic, that he tries to
> do too much. They claim he lacks knowledge of basic conditions and limitations,
> or ignores them; that he launches upon projects without adequate planning; that
> his promises too often exceed his capacity for fulfilment. Further, complaint is
> heard that Mr. Howe keeps too much of his own counsel, that he is contemptuous
> of advice, that he is too stubborn in his snap judgements. . . . [The Wartime
> Requirements Board], filling an obvious need, was not the work of Mr. Howe.
> What the "inside" voices tell is that it was forced upon Mr. Howe; put through the
> Cabinet, without Mr. Howe's support, by Defence Minister Ralston, its job to co-
> ordinate war effort, to give it plan and purpose. . . . [It] is pretty limited in its pow-
> ers. It has no executive authority. It reports to Mr. Howe; tells him what it thinks
> should be done; but with Mr. Howe, or with the War Committee of the Cabinet,
> rests the final decision. Mr. Howe, back from England, may not feel disposed to
> take the War Requirements Committee's advice; may feel that he knows better. In
> that event, not at all unlikely in the light of the past, there may be interesting
> developments. Indeed, it may be worth while to keep an eye on MacMillan.[8]

Meanwhile, MacMillan was hard at work trying to give substance to
the WRB and disposing of the last of his Timber Control responsibilities. In
early January he received a warm letter from Manning, who was planning to
visit Ottawa later that month. The BC chief forester ended on a personal note.
"When this trouble is all over I am counting on you for that day or two of hunting," he wrote. "Let's hope we won't be too old."[9] MacMillan responded: "If we
get too old before the war is over, we can surely sit in a duck-blind during the
warm spots in the best afternoons, so don't get too discouraged. It is a pleasure to have our present experience, and I am sure you won't suffer by it."[10]

He was sadly mistaken. It was their last communication. A few days
later, Manning was killed in a plane crash. Ironically, the BC government chose
to honour Manning by naming the Cascades park, for which MacMillan and
Grainger had campaigned, after him. MacMillan commented on this twenty-
five years later.

> I thoroughly agree with you that the Park now known as "Manning Park" should
> have been named for Grainger. It was Grainger who discovered it, studied it and
> made it. I am quite sure that Manning never saw the area. It happened that
> Manning was killed in an aeroplane accident just before the Park came up to be
> named, and a New Westminster politician named it—probably instead of provid-
> ing a pension for Manning's family. That was the way things went in the past even
> more so than now.[11]

In the midst of this busy period, MacMillan found time to write a brief article as part of a series *Saturday Night* magazine was running on Ottawa's wartime organization. In it he included a recommendation to establish a Priorities Office which had been described in a letter he had received a few days earlier from Deputy Finance Minister Clark's office. In his *Saturday Night* article, MacMillan wrote:

> If a shortage should develop of commodities, facilities or services, that the supply is not adequate to supply both war and civilian needs or even all war needs, it will be necessary to allocate the available supply to the most important claimants in the order of their war and national urgency and importance. This will be cared for by a Priorities Office to be set up under the direction of The Wartime Requirements Board.[12]

Much of the ensuing controversy hinged on this last sentence.

During Howe's absence, MacMillan, at the request of cabinet, prepared a report on the state of war production. Howe was not aware of this report until he returned from abroad, although there is no indication it was done behind his back. One of the major problems Howe had gone to England to resolve involved aircraft production. The British minister of aircraft production was the erratic Lord Beaverbrook, appointed the previous year by Prime Minister Churchill. One early victim of Beaverbrook's capriciousness was the Canadian aircraft industry, which had received large orders to produce Anson training planes for the Commonwealth Air Training Plan. To co-ordinate Anson production, Howe established a Crown corporation, Federal Aircraft, and placed it in charge of Ray Larson, an independent-minded man who owned a printing company. Larson and Ralph Bell, Director General of Aircraft Production, were soon at each other's throats. Their feud, combined with the problems created by Beaverbrook, had brought Canadian aircraft production to a virtual standstill by early 1941. At the request of the War Committee of cabinet, MacMillan's survey of production took a thorough look at the aircraft situation.

During Howe's absence, his political enemies within cabinet and in the Opposition generated a hostile atmosphere for his return. He arrived back in Ottawa on January 26, and within twenty-four hours the press was full of stories about chaos in the administration of Munitions and Supply and a rift between Howe and MacMillan. The origins and substance of this controversy have been a subject of dispute ever since.

The day before Howe arrived home, MacMillan's article, in which he discussed the formation of a priorities office, appeared in *Saturday Night*. On the day of his arrival, Howe received the first copy of the report MacMillan had prepared for cabinet. The next day, a story under Bruce Hutchison's byline appeared in the *Vancouver Sun*, written to suggest it was based on an interview with MacMillan and accompanied by a photograph of him with a caption claiming he had provided the information. It said a "super-planning committee" would be set up to oversee the production of all war materials, and that MacMillan was expected to head it. Subsequent articles by Hutchison and other journalists characterized the committee as an attempt by the King government to control Howe, and speculated about a rift between MacMillan and Howe. The *Financial Post*, a strong supporter of the Opposition, predicted a major cabinet shuffle, claiming in front-page headlines: "control and direction of war effort at stake."[13]

Eventually, word of MacMillan's report leaked out, although copies of it were not available to reporters. Several highly speculative stories claimed it was critical of Howe, and called for a drastic reorganization of Munitions and Supply, including the dismantling of Federal Aircraft. Hutchison, who would later become one of Canada's most respected journalists, wrote several front-page stories for the *Vancouver Sun* on the controversy, describing in detail the escalating dispute between Howe and MacMillan.

In retrospect, it is clear there was little substance to most of these claims. Hutchison's initial story, which touched off the whole affair, was fabricated from MacMillan's *Saturday Night* article in which he had mentioned the possibility of establishing a Priorities Office, as suggested by the Department of Finance. Hutchison claimed, in a curious *mea culpa* published thirty-five years later on the occasion of MacMillan's death, that he had obtained the information for the story from another member of the Wartime Requirements Board. After being confronted by an irate MacMillan, Hutchison wrote, he sent a letter to Howe absolving MacMillan of any blame in misinforming the press, but Howe was not satisfied with this explanation.[14]

Throughout the ensuing tempest, MacMillan made no statements to the press, except to deny rumours. The most virulent attacks on Howe were published in the *Financial Post*, whose publisher was supporting the creation of a joint Liberal–Conservative wartime government and the appointment of a business leader to take over war production from Howe. When other papers began to lose interest in the story, the *Post* took the unprecedented step of buying space in them to publish its own, highly speculative articles.

In early February, eastern papers began running stories suggesting MacMillan was going to resign because Howe would not accept the recommendations in his report. When he left for Vancouver to attend his daughter Jean's wedding, speculation increased, in spite of articles in the *Vancouver Province* saying that he was not going to resign and, in fact, was not even in disagreement with Howe. Before leaving Ottawa he issued a brief written statement to the press. "I shall be prepared to remain in Ottawa if there is a job in which I can be sufficiently effective to justify my continued neglect of my interests, which are all in British Columbia."[15] According to Howe's biography, MacMillan at this time made several statements critical of Howe and the administration of his department, including an indiscreet conversation with a group at Toronto's York Club. However, all of the comments quoted by the biographers are third-person accounts from people with political interests in the situation.[16]

At the Cabinet War Committee meeting on February 18, Howe raised the issue of the Wartime Requirements Board. It was not performing the functions for which it had been designed, he said, and its usefulness was being adversely affected by speculation in the press. He suggested that if the board was going to continue functioning, it might be transferred to another department or required to report through the secretary of the Cabinet War Committee. Defence Minister Ralston recommended the board be dissolved and its functions transferred to a war production council made up of representatives from the Department of Munitions and Defence.[17] A decision about the future of the board was deferred, but the committee did appoint a priorities officer, as suggested by MacMillan on the advice of the Department of Finance. When called by the press for his reaction, MacMillan expressed delight at the appointment of R.C. Berkinshaw, general manager of Goodyear Tire, as director general of Priorities.

Just as the controversy began to die a natural death in the press, it got new life in the House of Commons. Opposition leader Hanson used events surrounding the furore to launch his strongest attack on the King government since the war began. He focussed on two issues: the state of the Canadian airplane industry, and the supposed conflict between Howe and MacMillan.

The following day, while MacMillan escorted his youngest daughter down the aisle in Vancouver, Howe rose in the House of Commons to defend himself. After reading extensively from MacMillan's report to deny claims of its critical content, he tabled all but the confidential sections of it. One by one he refuted the rumours in the press, then directly addressed his relationship with MacMillan.

> What has been the effect of this on Mr. MacMillan? Let us talk about Mr. MacMillan, since his name has been mentioned a good deal. Before he left Ottawa he remarked to me that he was being sabotaged, and I agreed with him. The story was put out that he had quarrelled with me when as a matter of fact there has been no quarrel. At any time when we were both in Ottawa Mr. MacMillan has been in my office every day since my return, and we have discussed many problems. He has made many recommendations all of which I think have been accepted in whole or in part. . . . Does anyone suggest that a senior executive can quarrel with me and still continue to work with the department? I set the policy of the department, and if any executive does not follow it, he has no option but to resign. . . . I think the job of sabotaging Mr. MacMillan has been fairly successfully done, but I hope he is able and big enough, and I think he is, to live it down and to continue to carry on the very useful work he is doing for this government and for Canada.

Howe went on to attack the press, zeroing in on the *Financial Post* as the most reckless in its speculations on the rift between himself and MacMillan. "I make this charge, and I do it after careful consideration with a full knowledge of the facts of which I speak: the number one saboteur in Canada since the beginning of this war is the Financial Post of Toronto."[18]

In the ensuing uproar, in which some papers backed Howe while others rebuked him, the issue of his "dispute" with MacMillan was dropped. MacMillan returned to Ottawa in early March and, during his first day back at the WRB, met with Howe. He declined comment when asked by the press about the meeting. The atmosphere surrounding the Wartime Requirements Board had changed during the preceding weeks, however. The military departments—army, navy and air force—were now unwilling to refer projects to the board and wanted it weakened or even dissolved. There was widespread disagreement among civilian officials and politicians about the board's future. On his departure for Vancouver, MacMillan had been asked by Howe to give some thought to redefining its functions, and they agreed to meet and discuss the matter on his return. After a week back in Ottawa, meeting with other board members and discussing the situation with a number of his friends and

MacMillan at his Wartime Merchant Shipping Company office in Montreal, June 6, 1941.

associates, MacMillan outlined his views in a letter to Howe. Noting the appointment of the Priorities officer, and re-examining the mandate of the board, he concluded:

> The members of the Board feel it misleading and unwise to continue to carry, in the eyes of the Government and the public, responsibilities as stated in the Order-in-Council, the assumption of which have been beyond the Board's control.[19]

By this time Howe was preoccupied with another major problem. He had returned from the UK two months earlier with a new task for Canada—the production of cargo ships to replace those vessels lost to German submarines. This was probably the most critical task facing his department at this stage of the war, as the transport of all the war materials produced in North America depended on the availability of ships. British shipyards had fallen behind and were now unable to keep up with losses. Canada's east coast shipyards were devoted to repairing the damage suffered by the existing merchant fleet in the North Atlantic, while those on the West Coast were practically idle. A March 8 memo from MacMillan to Howe outlined delays and problems encountered in the Halifax and St. John's shipyards, which were symptomatic of the state of the entire shipbuilding industry. Three days later, in a debate on the war appropriations bill, Howe revealed that Canada was about to embark on a major shipbuilding program, but offered little explanation.

The following week, MacMillan drafted a letter of resignation as chairman of the WRB.

> The Order-in-Council creating the Wartime Requirements Board provides that it shall have power to secure from any source information respecting existing or projected war needs involving the use of materials etc., and refers in particular to obtaining information from each of the fighting services and war purchasing agencies. The object was to enable the Board to plan for Canada's war needs. The fighting services and war purchasing agencies have not made available the necessary information concerning their projects as required by the Order-in-Council, and it is obviously impracticable for the Board to force disclosure.
>
> My letter as Chairman to you March 11 set forth this position. Since writing it, at your request I have discussed the position with the Defence Ministers but see no likelihood of any change.
>
> Under the circumstances the Wartime Requirements Board cannot function as required by the Order-in-Council.
>
> My personal position as Chairman of the Board is invidious—I appear before the public as holding an office of responsibility and importance and at the same time am rendered unable to perform effective service, therefore, I hereby regretfully tender my resignation to which I desire you will give effect at your earliest convenience.[20]

It appears, however, that MacMillan did not submit this letter to Howe, at least officially. Two weeks later Howe announced his appointment as head of a new federal shipbuilding agency, Wartime Merchant Shipping, which was assigned the task of overseeing all cargo ship construction in Canada. On April 3, MacMillan sent a brief, clearly pre-arranged note to Howe.

> I find that in order to carry out my duties as president of Wartime Merchant Shipping Limited, it will be necessary for me to move to Montreal, where I shall be in closer touch with the shipbuilding and engineering industries; therefore, I desire to resign as chairman of the Wartime Requirements Board.[21]

He was succeeded by R.A. Henry, an economist in the munitions department, and the WRB subsided into insignificance.

The controversy that swirled around MacMillan and Howe for the first three months of 1941 constituted a significant, if short-lived, political crisis for the King government. Had it not been defused when it was, it could easily have mushroomed into a debate that would have weakened Canada's war production capability seriously. The press of the day pinned the blame for the situation on Howe; historians have tried to make MacMillan the villain of the piece. It appears neither was to blame, that there was no significant conflict between them, and that the affair was largely a tempest concocted by some newspapers and a few Opposition politicians. Apart from hearsay accounts, there does not exist any written version of MacMillan's supposed criticisms of Howe's management of Munitions and Supply.[22] Hutchison's claim, in his column at the time of MacMillan's death, that Howe had fired the WRB chairman is contradicted by the correspondence concerning MacMillan's resignation from the board. Nor is it likely Howe would have appointed someone he could not trust to the vitally important shipbuilding job. The two remained friends long after the war ended, and MacMillan staged a large dinner in Howe's honour in Vancouver. Several years later, at the outbreak of the Korean War, Howe again asked MacMillan to take on a high-level job. What did occur was the near-destruction of the effectiveness of two of Canada's most powerful and proficient wartime administrators.

chapter 15

Canada's "buzz saw"

Within a few days of the announcement of his new job in 1941, MacMillan moved out of his parliamentary office and the Chateau Laurier, and headed off to Montreal, taking Miss Dee with him. Office space was scarce in wartime Montreal, so he acquired a warehouse at 420 LaGauchetière Street West and hired carpenters to build offices using plywood from his company's plant in Vancouver. To assemble the office staff, he got in touch with the heads of several large Quebec corporations, who provided 100 of their most capable people. In the midst of the move, he was the subject of a laudatory article in *Time* magazine, titled "Canadian Buzz Saw," which seemed to capture the tone of his new undertaking.

Canada's war effort last week showed belated signs of shifting into high gear. Into the driver's seat of the important Shipbuilding and Shipping Program moved Harvey Reginald MacMillan, a hard-driving, hardheaded lumberman who believes in getting things done. No business-as-usual fuddyduddy, MacMillan is a reminder that Canada also produced Lord Beaverbrook.[1]

The task that lay ahead of him was to mobilize and co-ordinate the expansion of Canada's existing shipyards. It was an enormous undertaking. At this stage the Canadian shipbuilding industry was at a virtual standstill. The business of building ships is largely one of bringing together the various components—steel plate, engines, propellers, navigation equipment and so on—and assembling them. Few of these components were made in Canada, so MacMillan had to find manufacturers of other machinery and persuade them to produce ship parts. Thousands of workers had to be pried out of other wartime occupations and taught the shipbuilding trades. It was the type of job at which MacMillan excelled. He had an easy rapport with practical, down-to-earth working people, and they were quick to respond to his patriotic appeals.

Wartime Merchant Shipping was concerned only with building cargo ships, of essentially one design. The British had developed a reliable, versatile ship known as the North Sands type. MacMillan described it in a memo to Howe's assistant.

> This type was chosen because the components were suited to mass production by engineering shops and foundries not heretofore accustomed to serving shipbuilding, and the ship itself for construction in all types of shipyards, including new shipyards yet to be built and equipped hurriedly. The engines and boilers were of a rugged construction simplified for operation by persons of limited experience and easy to repair or maintain in all ports of the world. The ship was suitable for carrying all types of cargo: general merchandise, crated motor vehicles, bulk grains, coal, ore, lumber, or munitions of all types ranging up to the heaviest of guns and tanks.[2]

It was the same ship as the Liberty ships being built in US yards. A slightly modified version was later built in Canada and called the Victory ship.

Within a few days of his move to Montreal, MacMillan called together the heads of every shipyard in Canada; in one day they hammered out the basics of the program. The British had agreed on a fixed price per ship, out of which the builders had to finance their own expansion. Planning the enterprise as a means of expanding and enriching the Canadian economy, MacMillan predicted to Howe that Canadian shipyards could produce the ships at a lower price than American yards. He also told him work could begin within a week if the program was authorized.[3] Howe quickly obtained the required approvals, and four days later MacMillan announced the first contracts for the construction of 100 ships, costing $175 million. Burrard Drydock and North Vancouver Ship Repairs received the bulk of the contracts, with orders for ten vessels each going to Yarrows and Victoria Machinery Depot in Victoria, and Canadian National Drydock in Prince Rupert. The builders began laying keels immediately. The BC business community was ecstatic; it was the largest single injection of cash into the local economy in history.[4]

Eastern yards, still occupied with repairing damaged ships coming in off the Atlantic convoys, were a bit slower to get new ships under way. Within days of the West Coast announcement, contracts for another eighteen North Sands-type vessels were let at yards in Lauzon and Sorel, Quebec. A short time later, three new shipyards in Vancouver, Montreal and Pictou, Nova Scotia, were opened and received contracts. Just as things began to roll, the British government asked Canada to build several naval escort ships, and this task was passed on to MacMillan, who incorporated the construction of minesweepers at various Great Lakes shipyards into Wartime Merchant Shipping.

The speed with which this operation got going was a reflection of the desperate state of the war for the Allied forces. The Germans were beginning their aerial probes of the British Isles and their U-boats were inflicting heavy damage on Atlantic shipping. In March alone, 119 British ships were sunk, about one-quarter of the UK's annual building capacity.

MacMillan lost no time recruiting some of his trusted Vancouver colleagues into his new setup. H.S. Stevenson, the manager of Canadian Transport, was called to Montreal to work as his assistant, and his longtime companion in similar ventures, Austin Taylor, was appointed a director and made vice-president in charge of western operations. Howard Mitchell, a Vancouver public relations consultant who had worked for MacMillan at the Wartime Requirements Board, was assigned the vaguely defined task of "special duties."

By July, work was under way on two dozen ships on both coasts, and MacMillan was kept busy speeding up production and ironing out bottlenecks. He was back and forth across the country, constantly shuttling between Montreal, Toronto, Ottawa, Vancouver, Halifax and St. John's. Apart from Miss

HR MacMillan

MacMillan in Montreal, 1941.

Dee, he had few friends or acquaintances in Montreal, which was an added inducement to travel. Jean, whose husband was now stationed in Halifax in preparation for duty overseas, was living in the Lord Nelson Hotel, and whenever MacMillan was in Nova Scotia he visited her there. Similarly, in Toronto he got in the habit of visiting Mazo and Caroline, who had become permanent fixtures in his life. They provided him with a glimpse into the exotic world of artists and writers. They fascinated him, and he visited them whenever he was able to find time.

That summer he broke off relations with one of his oldest and closest colleagues. Percy Sills severed all connections with the H.R. MacMillan Export Company following a dispute with MacMillan that neither of them ever explained. Sills moved to Victoria, where he established a small lumber company. More than two years later, in one of his lengthy epistles to VanDusen, MacMillan wrote: "I notice that there is a Mrs. Sills in the UK Department. If she is the same family I think it is a mistake to bring such people in. It is always prudent to keep as free as possible from certain types of connections."[5] Whatever the cause of the breach, it was sufficient to draw out the most vindictive side of MacMillan's nature. A close colleague and friend, defended loyally and advanced steadily, would suddenly be cast aside—it had happened earlier with Montague Meyer, and it would occur again.

In early October the first WMS ship, the *Fort Ville Marie*, was launched by Howe's wife Alice at the Canadian Vickers yard in Montreal. A few days later the first West Coast ship, the *Fort St. James*, slid down the ways at Burrard Drydock. That same month, MacMillan's oldest and closest friend, Martin Grainger, died. The eccentric Englishman, who had been responsible for bringing MacMillan to BC, then followed him into the job of chief forester and later into the lumber business, had spent the last of his active years roaming the Skyline Trail in the country that is now Manning Park. He introduced MacMillan to that country, inspiring in MacMillan a lifelong love of travel in the back country. In normal times, Grainger would have been sorely missed. Now, preoccupied with his efforts to speed up ship production, MacMillan had little time to dwell on past friendships.

Under the Hyde Park agreement between Canada and the US, signed earlier in 1941, the American government paid for the production of many Canadian war materials—including the ships WMS was building—in US dollars, then turned them over to the British. This, and the rapidly accelerating preparations the Americans were making to enter the war, led to increased co-operation between the two countries. In November, a twelve-man US–Canada Defence Production Committee was established to co-ordinate munitions production throughout North America. MacMillan was appointed to this board which, along with the growing ties to the US shipbuilding industry, sent him to New York and Washington a good deal of the time. The integration of US and Canadian shipbuilding, along with other joint wartime projects, was a precursor to the post-war integration of the countries' economies.

Then, on December 7, the entire war changed dramatically. The Japanese attacked the British in the Pacific and simultaneously bombed the US base at Pearl Harbor. That same day, WMS's first ship, the *Fort Ville Marie*, went to sea. For Canadians, the war had entered its grimmest phase. The losses of air force and naval personnel and equipment in Europe were mounting almost daily, and the first glimmerings of victory were months, even years away. For western Canadians, the distant war had suddenly landed on their doorstep. Everyone was caught up in the fight, some to the point of hysteria at the prospect of a Japanese invasion of BC. The brightest hope was that, after more than two years of fighting alone, Commonwealth forces were joined by the most powerful nation on earth, the United States.

Once his work in Montreal settled into a routine—to the extent that was possible during a war—MacMillan turned some of his attention to his BC interests. The markets for Canadian lumber and fish had stabilized as a result of the war, with most of the lumber and all of the fish going to the UK. In early 1942, the financial press carried several articles describing the excellent performance of BC Packers, giving full credit to MacMillan for rescuing the company over the previous decade.

The strong wartime market for plywood encouraged his export company to build a second plywood plant, this one in Port Alberni adjacent to the Alberni Pacific sawmill. It was a state-of-the-art plant, built under Blake Ballentine's direction. While it was being planned and constructed, MacMillan had thought it should cost about $700,000. When the final accounting came in and the plant cost $1 million, it caused him to reflect on some longer-term questions in one of his regular letters to VanDusen.

> There is a natural and considerable difference in the point of view between the shareholders like you and me, who are in the 50's, and must try to prepare financially for the inevitable at the worst time in history for such preparations, and those in the late 30's and early 40's, who have not that problem on their minds but who want to make their mark to the best possible advantage by using other people's capital, which they can hope to enjoy for 20 to 30 years, and who probably can take more pride in plant layout and perfection than a person can whose first thought when he sees it might naturally be how are they going to value this thing, and how are succession duties going to be paid on it. . . . I am really writing this not as a criticism—and I am very anxious that it should not appear so—but to point out the difference in interest between the younger group, to whom the management of this company is now passing, and that of the older group, who have some difficult financial problems to face, all of which are being predicated upon the ability to liquidate a part of the capital and the ability to provide fresh capital from what is left after taxation, which I expect we will be able to do, but it is just a little bit too near a finesse to suit me.[6]

In the early months of 1942, he began to think about the post-war lumber market and the value of the untouched tracts of BC coastal timber. Western hemlock, which had been shunned in the past, was beginning to find a market in the UK. It could be substituted for Douglas fir, particularly in many wartime applications, leaving the fir to be sold after the war when price controls were dropped. Hemlock could also be used for pulp, which MacMillan and others expected would be in great demand after the war. He also became aware of American timber dealers and lumbermen buying great tracts of hemlock with a view to persuading the timber controller to lift the ban on log exports, enabling them to ship BC hemlock logs to their mills in Washington and Oregon. Of particular interest at this time was the wheeling and dealing of Ossian Anderson, a US lumberman who was aggressively buying timber on the coast. In May, Harold Foley, the head of the Powell River Company, stopped in Ottawa to visit MacMillan en route to his family home in Florida, and passed on his knowledge of Anderson's activities. MacMillan immediately relayed his findings to VanDusen.

> Harold Foley was in to see me last night. He is greatly worried about the hemlock situation, is now seeing the officials in Montreal and will be in Ottawa next week.

I do not know whether or not one can take his statements at face value, but if what he says about their position is true, it is very serious indeed. . .

Foley says that Ossian Anderson is negotiating with the CPR for all the timber land available in the E&N grant, particularly between Cowichan Lake and the south west boundary line of the grant, and is apparently getting money in California and from his pulp companies. Foley says he knows Anderson well, that he has been broke three or four times, is full of imagination, and is a great salesman. Foley says his company has been asleep and most of the hemlock supplies south of Queen Charlotte Sound, available in any size logging tracts on any accessible ground, have been alienated; that they are now trying to buy what they can, but have come too late into the field, and are now buying Kitimat, which Pacific Mills [at Ocean Falls] have not bought up to date because they have always thought it was not necessary to buy it to have it held for them.[7]

In subsequent letters, MacMillan and VanDusen tracked the timber trading situation closely. What they were observing was the beginning of another influx of American capital into the BC forest industry. At various periods in the past, when US timber supplies were scarce, American lumbermen had moved into BC in search of logs or, failing that, to buy operating companies with substantial timber reserves. Now it appeared they were back, and as MacMillan realized, if their presence continued and grew, it would alter the industry enormously. He wrote to VanDusen.

It begins to appear that Anderson is going to clean up the whole country, and with the over-building of sawmills and pulp mills on the Puget Sound, the accumulation of capital in Puget Sound, the immense political pressure they can exert through Washington on Ottawa, I would say that this is the darkest cloud there has yet been on our horizon. It is easy to see . . . that we are headed into a new era.[8]

He went on to talk about changes he felt were coming in another area—labour relations—in a way that was almost heretical for a capitalist of his stature at this time.

I have a feeling that we are quickly heading into a new era also in labour relations and cost of production. We are too far out of line with some American practices, rates of pay, and overtime; therefore it would appear that because of active and well-informed leadership of the workers and the tendency to eventually make all things even on both sides of the Border, we will have to be brought up to their levels.[9]

He pursued the theme in another letter the following day. "I think negotiating with employees will be a permanent industry henceforth," he wrote, "concerning which the laying of a sound foundation should be the best preparation. It is very possible that the good companies will be cursed by the evil practices of other companies."[10] A few months later he returned to the topic:

> I come into contact with people who are resisting the formation of labour Unions. The attitude of these employers as exhibited in their conversations seems to be that a Union is almost as bad as an invasion—they consider it at least an invasion of their privileges.
>
> I am beginning to think that it might be better to try to work out an arrangement with some Union and get peace and confidence in a given plant rather than to be gradually isolated as being an enemy of Unions and inferentially an enemy of the worker and thereby have no hope ever of having peace and constructive relationships. It seems to me that the whole of Canada is almost certain to be unionized before this war is over and that all governments actively or passively will assist in this process.[11]

Although MacMillan increasingly found his mind drawn to interesting developments in his own businesses, the more mundane affairs of shipbuilding required his attention. Throughout much of 1942 he was preoccupied with two major problems.

The first was the availability of housing for shipyard workers in the smaller centres. MacMillan ended up in an acrimonious dispute with two officials of Wartime Housing Limited, J.M. Pigott and a Mr. Goggin, who were responsible for providing accommodation for the workers. The dispute dragged on for weeks in typical bureaucratic fashion, finally provoking a caustic letter to MacMillan from Howe's always-diplomatic deputy, G.K. Shiels.

> I do wish that, having regard to
> (a) the shortage of paper,
> (b) the shortage of hours available for the transaction urgent business, and
> (c) the maintenance of the utmost co-operation between the various branches of this Department,
> you would make it a point to sit down with Messrs. Pigott and Goggin or either of them the next time you are in the same city with them and iron out once and for all these arguments as to who is responsible for delays, etc., in connection with these various projects. We would then be able to get on with the important aspects of our job and by pulling together instead of at cross-purposes could undoubtedly make better progress.[12]

It was about as close as MacMillan ever came during the war to receiving an official rebuke.

The second problem, steel shortages, was much more substantial. Aside from the general wartime shortage of steel, there was a trickier problem requiring a great deal of diplomacy and tact to resolve. As Wartime Merchant Shipping was increasingly successful at reducing the time required to build its ships, and in lowering their costs, the US government began to buy Canadian-built ships for its own use, as well as paying for those provided to the British. Many American officials chose to buy ships at Canadian yards, rather than in the US. Because the North American economy was becoming increasingly co-ordinated, it made little strategic difference where the ships were built, but it was an affront to American shipbuilders, who were loath to part with orders and have their steel allocations redirected to Canada, which desperately wanted the business because of the US currency it provided. The resolution of this problem required MacMillan's attention during much of late 1942 and early 1943 in the US. He worked closely with Eddie Taylor, who had been dispatched to Washington to co-ordinate American and Canadian munitions production.

It was typical of MacMillan that he could still find time to worry about the financial affairs of a small farming operation he had begun on Kirkland Island at the mouth of the Fraser River. Although the main purpose of the property was to provide a hunting retreat close to the city, MacMillan had decided to put to use the knowledge of farming he had acquired as a child and during his time at the Ontario Agricultural College. Throughout the war, he grew sugar beets and potatoes on the property. Occasionally he bothered VanDusen with long letters involving rather insignificant sums of money—curious behaviour for someone centrally involved in the workings of the North American steel economy.

There were, as well, larger private business concerns that attracted his attention. In the fall of 1942, Taylor discussed with MacMillan his plans for forming an investment company, which eventually became the giant Argus Corporation, and invited MacMillan to become involved as an investor and director.

> I should like very much if you would associate yourself with me in this venture. If Canada is to go ahead after this war (and we must see that she does) industrial shares of well-managed companies should appreciate very much in value over

present quotations. I don't think the bargain sale which is now in progress will last more than another six or nine months, hence my anxiety to get this thing under way.[13]

MacMillan declined the offer, pleading a need to begin acquiring funds for succession duties. However, it marked the beginning of a friendship and a business association that would flourish in the post-war years.

By the end of 1942, MacMillan found time to do something for his friends and acquaintances living in wartime Britain. Aware that they were getting by under a system of strict food rationing, he began regular shipments of canned salmon to a list of people that, by the end of the war, was several pages long. He was particularly concerned about the fate of Tom Meyer, Monty's youngest son, who was a prisoner of war and, because of his Jewish background, being badly mistreated in Italy. After much difficulty, MacMillan was able to arrange for weekly food parcels to be shipped to Meyer from the US. In November he wrote a lengthy circular letter which he mailed to dozens of people he knew in the UK, apologizing for his lack of communications during the past two years and offering a detailed description of conditions in Canada, along with predictions about the future of the country.

> The drawing of the population into the production of munitions and war necessities—such as minerals and certain agricultural products—and thus into very much higher monthly family earnings than have ever before existed in Canada, will permanently change the distribution of our population. Canada will be a much more urban country after the war. . . . As is elsewhere in the world, the general trend of thought is toward a more controlled capitalism, toward a greater opportunity for all working people, and toward a higher standard of living for the average of the people, which might be loosely described as a "movement to the left." As might be expected, the great increase in the number of persons continuously employed on payrolls has led to a great pro-Union movement. Unfortunately, in Canada, as in the United States, there are two or three opposing Trade Unions attempting to secure membership in a great many of the same categories of employment. Also the Trade Unions' memberships suffer by not having developed as large a proportion of patriotic and far-sighted leaders, as has been the case in the more mature history of Trade Unions in the United Kingdom. These developments will probably come with time, but in the meantime there may be some headaches. . . .
>
> Canadians do not know what is going on in this country physically as well as

socially. When the war is over they, as well as you, will have an opportunity of seeing this country in both respects through new eyes. . . . I think that we are all discovering that Canada in many important respects is very different from most other countries. It is harder to govern and is really not the kind of country that either its Western portion or its Eastern portion have assumed.[14]

Just before Christmas, MacMillan returned to Vancouver for the launching of the one hundredth 10,000-ton ship built by Wartime Merchant Shipping. The *Fort Frobisher* entered the water at the North Vancouver Ship Repairs yard on December 21, almost a year after the first ship went to sea. It was an enormous achievement. Nearly 80,000 people had been mobilized to build the components and assemble them into operating ships. About 94 percent of the work had been done in Canada. But it was not a time to sit back and rest, as MacMillan told the shipyard workers in a stirring speech which earned enthusiastic applause.

> You may not fully realize the critical need for ships. It is common knowledge that the United Nations now, after three years and three months of war, have available only about one half the tonnage of merchant ships possessed by the Allies after three years and three months of the first Great War.
>
> Furthermore, the need for merchant ships is greater in this war. We have no bridgehead in Europe. We cannot use the Mediterranean. We are fighting and supplying armies and navies across more oceans. Mechanization has trebled and quadrupled the tonnage required per soldier.
>
> Menaces to merchant shipping have multiplied. Submarines are bigger, stronger, faster. Bombers operate far out to sea. Torpedoes and bombs are larger and more deadly. The submarine is not licked. The building of merchant ships is not yet sufficiently greater than losses. We cannot yet rest on our achievement. . . . In spite of all optimistic statements and budgets, the time is not yet in sight when there will be enough ships. . . . The eyes of Canada are on us. We must make good.[15]

By the end of 1942, as well, the rush to buy timber had become a frenzy. Realizing that it was indeed the beginning of a permanent shift in the industry, the H.R. MacMillan Export Company began to search aggressively for timber. From early in the new year, MacMillan and VanDusen's regular correspondence was filled with discussions of various tracts and the possibilities of obtaining them.

In my judgement there are only a few things on which one can have relatively strong convictions respecting business future: one of which is the prolonged demand for competitively produced lumber—a demand which will arise either from war or from the creation here and elsewhere of employment by Government credit or from a social security programme, which will be pushed strongly to prevent a destructive slide of voters to the Left, which seems to be the thought uppermost in the minds of statesmen. Therefore I have no doubts whatever about the wisdom of acquiring competitive timber. I also have great confidence that we can swing it financially.[16]

Most of the timber being bought and sold was on the east coast of Vancouver Island, part of the original Esquimalt & Nanaimo Railway land grant, and competition was intense. HRME's first major purchase was the Thomsen & Clark Timber Company which, in addition to its extensive holdings, carried on its books an accumulated loss of $3.5 million. Through a convoluted series of deals devised by a Price Waterhouse accountant, Thomsen & Clark was purchased for $100,000, then renamed MacMillan Industries Ltd. and made the parent company of H.R. MacMillan Export. By the time the paperwork was finished and the income tax department satisfied itself, tax advantages worth $5.3 million had been gained, as well as ownership of an enormous stand of top-grade timber.

In the same area were several small tracts and one large parcel of timber owned by Japanese Canadians. Following Japan's entry into the war, all residents of Japanese descent—many of them Canadian citizens—had been expelled from coastal BC. Their property was seized and sold at public auction by the Custodian of Enemy Alien Property. It was argued at the time, and confirmed after the war, that some individuals and companies profited greatly at the expense of those who were interned and dispossessed.

MacMillan's long association with the Japanese market, and his acquaintance with people in the lumber business, gave him a slight advantage in securing this timber. Instead of buying through the Custodian, he had VanDusen negotiate directly with some of the owners. Eventually he acquired many of the smaller parcels as well as the largest, containing about 15 million board feet, owned by a man named Kagetsu at Fanny Bay on Vancouver Island.

MacMillan also had an interest, through BC Packers, in the hundreds of fishing boats seized from Japanese Canadians, who before the war were a dominant force in the fishing industry. In this instance the fear was that the recently seized boats would not be available for the approaching fishing season. MacMillan was publicly accused by a virulently anti-Japanese Liberal MP, Tom

Reid, of attempting to use his influence in Ottawa to let Japanese Canadians keep their boats in order to profit from their expertise. In one of his rare public disputes with a politician, MacMillan vehemently denied this accusation.

MacMillan's views on the expulsion are difficult to determine. There is no indication that he felt strongly about the matter one way or another. Before the war he had made comments on the inability of the Japanese to become Canadians in a cultural sense, but there is no trace in his remarks of the kind of anti-Asian frenzy that ignored even the advice of the police and military authorities, and led to the expulsion. According to the documentation, MacMillan dealt fairly with Kagetsu in the Fanny Bay timber purchase. And there is no indication that BC Packers took advantage of Japanese Canadian fishermen. MacMillan was as close to and informed about the situation as anyone. He was vitally interested in two of the main areas of Japanese Canadian economic activity, fishing and lumbering, and he was a close friend and associate of the man in charge of the commission to oversee the expulsion and look out for the welfare of the Japanese Canadians—the ubiquitous Austin Taylor. Like most British Columbians, MacMillan was neither actively supportive of nor opposed to the expulsion. Engaged fully in a global conflict he believed had to be won at almost any cost, he either perceived there to be no injustice or judged it to be of little consequence under the circumstances.

Through the spring and summer of 1943, the competition for timber intensified, and VanDusen was spending much of his time engaged in the byzantine world of timber trading. For part of this time, he and MacMillan worked closely with Harold Foley, who was also searching for timber. The two companies, H.R. MacMillan Export and the Powell River Company, were not in direct competition with each other as the latter required pulp wood and the former needed saw logs. At one point, seventeen years before it finally came about, Foley proposed a formal association or even a merger of the two companies which, considering their compatible needs for wood, might have been advantageous. VanDusen wrote to MacMillan:

> Harold has mentioned on one or two occasions that he is looking forward to an association broader than this present deal that we are working on. I have not pursued this at all as I thought it as well to let it develop naturally. He is quite "hopped up" on the result of two years research that he has put on the recovery

of pulpwood from logged over lands. He has considerable hopes of drawing gradually an increasing quantity of his raw material from salvage operations.[17]

By mid-summer, MacMillan's attention focussed more and more on
British Columbia. During a visit to his Qualicum Beach house, he purchased a
large nearby farm, and spent his spare moments planning changes in its operations. Although the fishing industry was experiencing a good season and BC
Packers was maintaining its profitable course under John Buchanan's management, trouble was brewing in the forest industry. A shortage of loggers the
previous year had led to the selective service commission practically forcing
many out-of-work or otherwise-employed loggers back to work in the woods
to meet the wartime need for lumber. Wages in the forest sector were still
rigidly controlled, and the loggers were keen to work where they could earn
more money. The loggers' union, the International Woodworkers of America,
had grown in membership and a strike was brewing. MacMillan's perspective,
from eastern Canada, was markedly different from that of many owners in
Vancouver, who were intent on destroying the union.

> Being a diehard and staying in business to service your capital are frequently incom-
> patible these days. What the diehards don't recognize is that many of the people
> who counsel moderation and compromise with the present labour leaders don't like
> the outlook any better than the diehards, but are trying to be more practical. . . .
>
> No matter what you think about Ottawa, it is still the Government and I don't
> like to see the leading logging operators—who operate under the handicap of
> being on the fringe of the country, but still governed by the Government in the
> centre of the country—sticking to a line of argument and a course of action which
> gradually convinces those responsible for Government that the loggers are reac-
> tionaries and are willing to carry their viewpoint to the length of cutting off war
> production.
>
> The more I see of it the more I realize as Government becomes more active in
> all the affairs of our daily lives, the greater becomes our handicap in British
> Columbia, where we are so far from the seat of Government that we cannot maintain
> the constant contacts and opportunities for explanation and representation that on
> the one hand would enable us to influence the viewpoint of the Minister, and on the
> other hand would enable him and his officials to influence our viewpoint.[18]

He was tired of living in the east, away from the action in BC that increasingly
captured his interest. But he was not complaining about his position; he still
believed his job to be critically important and had no desire to escape it so
long as he was needed and useful.

HR MacMillan

In November 1943, King George VI appointed MacMillan a Commander of the Order of the British Empire. He was pleased with the honour, but he did not consider it significant enough to cancel an important meeting about ship-building, and he asked officials in Britain to mail his citation.

By December, it was becoming clear that the war in the Atlantic, while far from over, was being won. Shipping losses were falling off and cancelled orders began to trickle into the Wartime Merchant Shipping offices. As affairs began to wind down, MacMillan realized it was a propitious time to take his departure. Maintaining a shrinking operation did not require his presence, and events in British Columbia were laying greater claims to his attention and interest. After conferring with Howe, he submitted his resignation as president on December 16. The war was over for MacMillan.

Debating forest policy

In spite of a gruelling three and a half years in Ottawa and Montreal, MacMillan's wartime experience left him highly energized for the work ahead of him. He returned to Vancouver in 1943 with an enthusiasm that was astonishing for a man heading into his sixtieth year. While many of his contemporaries settled into late middle age with their minds firmly fixed on comfortable retirement, MacMillan came home brimming with ideas. One of his first acts, upon which he declined comment to the press, was to resign from the board of BC Power Corporation. Privately he made it clear he had no time for the staid utility company, tightly controlled by its dominant British directors. He was clearing the decks of past encumbrances so he could devote his time to endeavours he believed would help shape the post-war West Coast economy.

Before he set out to change the world around him, he applied some of the knowledge he had gained during the war to the operation of his own company. His return to an active role in managing the MacMillan Export Company left a lasting impression on most employees, including Charles Chambers, who later became a vice-president of the company.

> On his return he was highly critical of the way HRMCo. did business and everybody was put to work making up organization charts and setting out clearly areas of responsibility. Consternation reigned and a great deal of criticism was levelled at Harold Wallace for his somewhat unorthodox methods of getting a lot of work done in a very little time. Understandably, Harold resented this somewhat sudden attack first because everything got done and secondly these were the way things were done before the war and there certainly had been no criticism at that time. In addition he had been getting severe migraine headaches and had been "putting out" under these adverse health conditions. So the upshot of it was that he took early retirement and HR brought in Geoff Eccott who had worked for him in the East and was just the ticket for the job ahead.[1]

Outside the company, MacMillan's major concern was a royal commission on BC forest policy. Although the commission had not been officially announced until the last day of 1943, it had been clear for more than a year that major changes in the administration of BC's forests were under consideration. In 1942, the province's new chief forester, C.D. Orchard, wrote a confidential memo proposing sweeping changes in the provincial Forest Act. Premier John Hart had circulated copies of this memo to a select group of people in the forest industry, including MacMillan. Although Hart made it clear to Orchard he intended to implement the suggested changes, he realized that politically, he could never get away with making the changes without first preparing voters. With this in mind, he appointed Chief Justice Gordon Sloan to conduct a one-man royal commission, which, as far as Hart and Orchard were concerned, would recommend their predetermined conclusions.[2]

At that time, administration of the province's forests still rested on the same 1912 Forest Act which had created the Forest Service and brought MacMillan to BC as chief forester. Except for a large tract of land on the east coast of Vancouver Island granted to the Esquimalt & Nanaimo Railway, and similar but smaller areas in the interior of the province, almost all the forest land in BC—about 93 percent of it—was owned by the provincial government. Legislation still in force prohibited its conversion to private ownership. At the same time, there was no legal way to regulate or restrict the annual timber harvest, which had been increasing dramatically. Nor did existing legislation provide for restoring and maintaining the productive capability of logged-off lands with the objective of sustaining a permanent forest industry. As MacMillan had commented a few years earlier, "there is little or no rational forest management in Canada," and that certainly included British Columbia.[3]

From its beginnings in the 1860s, the forest industry obtained timber from Crown-owned lands through a system of leases and licences which terminated when logging was completed, leaving title to the land with the province. For most of this time, the predominant commercial interest was in Douglas fir, the other species having little or no value in the market. The best fir stands were on E&N lands, while the other major source of supply was on the McBride leases which had been staked early in the century. By the end of the war, most of the timber available in these tracts had passed from the hands of speculators into the control of large companies like MacMillan's. By the early 1940s, it had become clear that this timber supply would not last forever under prevailing conditions. The only mechanism available for obtaining Crown timber was a short-term timber sale, which was not used by the government to convey access to the large tracts of timber the big companies wanted under their control before they committed funds to build new mills. This situation, which existed in Washington and Oregon as well as BC, produced the intense competition for timber that MacMillan had observed while in Montreal.

Professional foresters like MacMillan and Orchard understood that the industry could not expand indefinitely. If logging rates were not regulated, as they had not been in eastern North America, the timber would run out one day. The difference was that there was no new timber frontier to move to when the Pacific Northwest was depleted. And given the geography of the region, there was scant possibility of an agricultural economy establishing itself on the cut-over lands, as happened elsewhere. Foresters also knew, at least in theory, that it was possible to sustain a permanent forest industry by growing timber. It had been done in other parts of the world—in some cases, such as Japan, for many centuries. The question was how to establish the conditions in British Columbia that would make forest farming a viable proposition. Although MacMillan and Orchard were generally in agreement over the need for cut regulation, they remained at odds for the rest of their lives over the way to initiate forest management in BC.

MacMillan renewed his interest in forestry some time before the launching of the Sloan Commission. During the war, freed from day-to-day business pressures, his mind occasionally returned to the ideas which had captured his imagination during the early years of his career. He was not impressed with what he saw, as he indicated in a 1941 letter to Fred Mulholland, then serving as president of the Canadian Society of Forest Engineers.

> Public interest in forest management in Canada appears to be at a lower level than that which existed in 1906 to 1912. The reason for this may be that the Canadian public has reached a state of complacency because it has been fed so much unjustified and optimistic press material during the past twenty-odd years respecting the forest resources of the country, and the expenditures on the forest services of the country, that it feels that the situation is in good hands.
>
> The public has voted large sums of money to Federal and Provincial forest organizations, and most of this money over a period of twenty years seems to have gone into business administration and office personnel. Very little has gone into the character of work that would result in the public demanding such silvicultural practices as are necessary to maintain the forest industries of this country for succeeding generations.
>
> There must be something wrong with the forest schools of Canada. I see very little leadership emanating from them. We have partial foresters, but through apathy, or ignorance, or complacency, or lack of having developed pure foresters, as opposed to engineers with a knowledge of botany, we have not produced sufficient real foresters to work in the vanguard of public opinion and to make themselves the "spark plugs" from which there would be continuously a radiation of such leadership as would slowly but surely develop in Canada the realization that sound silvicultural practices must be adopted if we are to avoid the fate of some of the States of the United States, and some other well-known parts of the world, as a result of forest exhaustion.[4]

Following the acquisition of the Rockefeller lands in the Alberni Valley almost a decade earlier, MacMillan's company had made some of the first, tentative moves toward the creation of a sustainable forest operation. His interest in managing forests increased with the purchase of additional land on Vancouver Island during the war. Under questioning at the Sloan Commission, MacMillan explained that it was company policy to retain forest land after it had been logged in order to grow another crop, instead of letting it revert to the Crown in lieu of taxes, as was the common practice at the time.

MacMillan was not alone in his thinking about growing future forest crops. Prentice Bloedel, who was taking control from his father of the Bloedel, Stewart & Welch holdings, and Bob Filberg at Canadian Western's woods affiliate, Comox Logging, were planning along similar lines and beginning to make provision for a successive timber crop on logged-off lands. While Bloedel, an introspective man with an avid interest in scientific methods, leaned toward replanting logged lands, MacMillan began experimenting with logging practices that would provide natural regeneration of the forests. His approach, called "patch logging," was to reduce the size of clear-cuts to encourage the natural reseeding of logged areas. MacMillan was skeptical about the silvicultural value of tree planting, conceding only that it was useful "to impress the public."[5]

To oversee his forest management program, as well as to assist in preparing his presentation to the Sloan Commission, MacMillan hired an old acquaintance from his days in government service. John Gilmour was a brilliant forester with a distinguished career that was only slightly impaired by his love of Scotch whisky. He was probably the most experienced and accomplished industrial forester in Canada, if not North America. After serving in the BC Forest Service under MacMillan as district forester for Cranbrook, he spent many years managing large industrial forests in Newfoundland and Quebec. As the MacMillan Export Company's chief forester, he was first to perform that function for any company in BC. In the ensuing debate, he would become the chief spokesman for those opposed to Orchard's scheme, including MacMillan.

A preview of the policy struggle that would occupy MacMillan's attention for the next dozen years occurred early in 1944, at the annual meeting of the Western Lumber Manufacturers Association. Both he and Orchard spoke, articulating fundamentally different positions, but never directly disagreeing with or attacking each other in public. The two foresters agreed on the basic objective—the maintenance of the province's forest-based economy in perpetuity. In Orchard's view, this could be achieved by state regulation of private forest companies. What he neglected to reveal at this meeting—and kept to himself until he gave his submission to Sloan's commission—was that he favoured continued public ownership of forest land which would be leased to private firms in Forest Management Licences.

MacMillan, with almost forty years of experience under his belt, equally divided between the public and private sectors, had long since lost

MacMillan and Chief Forester C.D. Orchard at the opening of Pinewoods Lodge, Manning Park, BC, June 3, 1950.

his faith in the ability of public agencies to manage any type of business enterprise, particularly forestry. He saw that the crux of the matter was ownership of the basic resource, forest land. Having read Orchard's memo, and being aware of his intentions, he addressed the issue of forest ownership directly.

> It doesn't appear possible that the [forest] business could have been developed so well on a different basis because free enterprise, for all its individualism, has always been subject to checks and balances and forced by public opinion to protect the rights of society. Yet there are those who remember old grievances who refuse to believe that private industry has reformed and who believe that there should be a new system under the domination of the state.
>
> We cannot ignore the danger. It is our duty, as common sense business, to defend the system that has developed in our industry and made it prosper. Given the opportunity to apply the same resourcefulness and enterprise that we used in meeting other problems, we can successfully deal with the problem that faces us now with the exhaustion of our Douglas fir. . . .[6]

Although it was over this fundamental question of land ownership that Orchard and MacMillan disagreed, the basis of their dissension lay as much in their personalities as in their opinions on private versus state ownership. At heart, Orchard was a bureaucrat, having come to his position via the Forest Service's operations division. He had very little experience with, or direct knowledge of, the province's forests or its forest industry, and had spent virtually no time in the woods or around logging and milling operations. He was reserved and withdrawn, a characteristic his colleagues interpreted as arrogance, and generally was MacMillan's opposite in character. The one trait the two men shared was a stubborn determination to have their own way.

MacMillan, in his submission to Sloan at the end of August, based his arguments on a restatement of the conservationist viewpoint that had evolved forty years earlier. The challenge, he explained, was to maintain the productivity of the province's forest lands. It was an argument he had articulated while discussing the commission in a letter to his old Yale roommate, Thornton Munger.

> My opinion is that this whole question is very simple. Plant up our derelict land over a period of years, bring about a new forest on our logged-over lands, and protect it all from fire. If this is done during the next two decades, there will be excellent growth and at the same time we can be training a staff of foresters to administer the new crop.[7]

He placed little stock in Orchard's grand plans for administrative reorganization, believing it would be best dealt with after effective protection and reforestation measures had been put in place. Consequently, his sixty-page brief was a concise, detailed plan for restoring the province's forests to a healthy state. Like almost everyone else outside the bureaucracy, MacMillan was not particularly concerned with the question of land tenure. While fully aware of Orchard's ideas on that issue, he did not consider them something the Coalition government would take seriously. Instead, his recommendations dealt with tax laws that could be used to encourage sustainable forestry over the long periods needed to grow timber.

Sloan concluded his public hearings without the differences between Orchard and his fellow professional foresters becoming a matter of public debate. MacMillan's brief attracted little attention, devoted as it was to technical, administrative matters. He did end up at the centre of a controversy as a result of the commission's hearings, but it had nothing to do with the proposals he presented.

When the Sloan inquiry first opened in Victoria, the MacMillan company immediately came under attack. A year earlier it had purchased a run-down sawmill and a large tract of timber on southern Vancouver Island from the Shawnigan Lake Lumber Company and, after a brief attempt to make the mill profitable, closed it and began shipping its logs to the Canadian White Pine mill in Vancouver. The local board of trade appeared at the commission and argued that shipping logs off Vancouver Island was bad for the local economy. MacMillan was called before the commission and grilled by its counsel. He described how, over the previous few years, most of the logging companies from which his mills bought timber had either been purchased by competitors or run out of timber. He explained how control of the coastal industry was being concentrated in fewer, larger companies, most of which were located in Vancouver and other large urban centres. "It is a question of each individual seeking to survive," he said. "As it is, there are nine cats after one mouse, and somebody has to close down, or have to peter along for three or four years, partly operating and partly closed."

Over the course of the Sloan Commission's hearings, competition for timber became intense, fuelled by a strong wartime lumber market and rising expectations of a postwar boom in UK and European markets. While MacMillan was in the east, his company had made several major purchases of former E&N timber on Vancouver Island, including the one at Shawnigan Lake, a 35,000-acre stand at Northwest Bay near Nanaimo, and another large tract at Iron River, farther north.

A variety of other actors appeared on the scene at this time, among them a group of wealthy European refugees, some with long experience in the timber business. They were referred to collectively as "the Czechs." Especially prominent were two groups that arrived just before the war and began aggressive timber-acquisition campaigns. One, involving Poldi Bentley and John Prentice, Czechoslovakian textile manufacturers, set up a veneer plant which quickly became the world's largest supplier of birch plywood to the British aircraft industry. In 1943, after buying several tracts of timber on the mainland, they obtained one of the best stands of fir left on Vancouver Island, at Nimpkish River. Another Czech family of Jewish refugees, the Koerner brothers, arrived in 1939 and, in spite of a deep-seated anti-Semitism in BC, established a sawmilling company that processed the long-neglected Western hemlock. They called it Alaska pine and were highly successful selling it to their pre-war contacts in Europe and the UK. In 1942, they obtained title to 20,000 acres on northern Vancouver Island.

At this point one of the largest and oldest forest companies on the island came up for sale. The American owners of Victoria Lumber & Manufacturing Company, faced with new taxes on their timber and the need for capital to cover impending inheritance taxes, decided it was a propitious moment to sell. Company president John Humbird had no choice but to accede to their wishes, although he did extract one concession: the company could not be sold to H.R. MacMillan. He had never forgiven MacMillan for out-manoeuvring him in London during the war, and he was not prepared to see his life's work in the hands of his rival.

At the midway point in the war, VL&MC was the fifth largest lumber producer in the province, outranked only by MacMillan's Alberni and Vancouver mills, Bloedel, Stewart & Welch's Alberni plant, and Canadian Western's Fraser Mills. VL&MC's greatest asset, even after five decades of heavy cutting, was a large tract of high-grade Douglas fir on southeastern Vancouver Island. To the timber-hungry MacMillan company, it was a prize worth scheming for.

MacMillan made his initial move when the mill came up for sale in 1942. Knowing Humbird refused to sell to him, he arranged for E.P. Taylor to make an offer. Humbird, of course, was suspicious of Taylor's new-found interest in the BC lumber business and tried to pry from him the identity of any associates who might be involved in the deal. Taylor adamantly refused to part with this information. Negotiations dragged on through 1943, while the property was thoroughly appraised and MacMillan lined up the financing. By early 1944 he had funds in place for Taylor to make an offer to Humbird. In a highly secret series of negotiations, the Canadian Bank of Commerce granted a note on the Canadian White Pine Company for $3 million, placing the money in a Royal Bank of Canada trust, Roytor, in Toronto. Taylor then arranged to meet Humbird in Vancouver and made him an offer—$3.2 million in cash and $6 million in debentures—well above what he and MacMillan knew Humbird expected. In the transaction, VL&MC was sold to a new company, Victoria Lumber Company, in which Taylor and half a dozen other front men held one share each. Roytor, acting on behalf of MacMillan, held 600,000 shares.[8]

The sale was announced in April 1944, with no mention of MacMillan's involvement. Two months later, what had transpired became clear to anyone with the slightest bit of knowledge about the industry. The H.R. MacMillan Export Company had been given a contract to manage and act as sales agent for Victoria Lumber, and within months the fiction of Taylor's ownership was revealed. Coming in the middle of the Sloan Commission hearings, and hot on the heels of the charges of monopoly surrounding the Shawnigan Lake Lumber sale, the news was a bombshell. It provided a dramatic conclusion to the Sloan Commission's public hearings. The stepped-up competition for timber and, in particular, MacMillan's purchase of VL&MC, prompted Sloan to address the issue at his final public session, just before he settled in to write his report.

> The position [MacMillan takes] is that this is our timber, and we propose to do what we like with it. There is no law in the country to prevent me from buying a mill, closing it down, and disposing of my timber as I see fit. That is the privilege the buyer has. Now should the present state of the law be continued from the point of view of an economic policy is a question including many factors. Should a small community suffer because of modern trends concentrating industries in large cities—that is something to be considered.[9]

Within a few weeks, the issue was being hotly debated along the entire coast, particularly on Vancouver Island. In the local press, at public meetings and in resolutions by local governments and boards of trade, MacMillan was either being criticized as a monopolist stripping local economies of their raw materials to feed his mills in Vancouver, or applauded as a public benefactor for introducing far-sighted forest management practices on the land he owned.

As he had promised Humbird almost thirty years earlier, MacMillan did own the Chemainus mill the next time he walked through the office door. Over the years, that door assumed legendary status within MacMillan's company. Long after he died, in 1985, the mill and office building were torn down to make way for a new mill, and a worker not familiar with its historic significance tossed the old door in a garbage pile. It was rescued only a few hours before the pile was burned, and it was installed in the new office building in time for a ceremony marking the 100th anniversary of MacMillan's birth.

Some of the criticism of MacMillan's purchase of the Chemainus mill was prompted by older animosities. In one swift move he had dealt Seaboard a devastating blow, removing one of its largest lumber suppliers and forcing the resignation of Humbird as its president. The balance was restored almost immediately, however, when one of the Czech groups, the Bentley company, joined Seaboard. As for Humbird, he did not appear to be particularly upset about the deal. A few months later he told one of the backroom negotiators of the sale that he was satisfied with MacMillan's management of the company.[10]

The Associated Boards of Trade of Vancouver Island lodged a strong protest over "the way the island's lumber industry is getting into the hands of one or two operators." The *Vancouver Sun*, long a devoted fan of MacMillan's, pontificated nervously from atop its editorial fence.

> Our news columns tell of a battle of titans in the lumber business of British Columbia. Billions of feet of our choice Douglas fir, cedar and spruce, units of great value in the natural resources of the province which heretofore have been crown-granted or sold away by license from public ownership, are being used as pawns in a fight for supremacy.[11]

The strongest attack on MacMillan came from Tom Reid, the Liberal MP from New Westminster who made a political career out of vilifying others, especially MacMillan and Japanese Canadians. In a lengthy House of Commons speech, he claimed MacMillan had used his time in Ottawa and Montreal to further his business interests, and he demanded a federal government investigation of MacMillan's companies under the War Measures

H R MacMillan

Act.[12] In one of his rare responses to hostile politicians, MacMillan dismissed the accusations, adding an observation that was taken up by the press and public, effectively squelching Reid. "Mr. Reid has attacked me on numerous occasions in the House and knows from experience that, so far as he is concerned, it is an occupation totally without risk," he told the *Province*. "The political baiting of business men is about as dangerous as shooting at a draught horse that is pulling a load, and equally useful."[13]

This response was an opening salvo in an often acrimonious debate between MacMillan and certain segments of BC's political left wing that continued for many years. It was a curious debate at times. Although he was occasionally attacked by communists and socialists in the labour movement, he had no quarrel with unions or workers. In fact, he returned to Vancouver convinced that unions were a permanent fixture in the industry and—contrary to the views of many forest industry managers and owners—that their development as responsible representatives of employee interests should be encouraged.

> The old hire-and-fire days in industry are gone. Wise and constructive policy regards the labour factor in production as being far too important to leave to unscientific methods. The chief controllable factor in production cost is the labour cost. Plenty of experience now exists to prove that the enterprises deliberately working on labour relations in an enlightened and progressive way, without developing paternalism, are producing an employee attitude that not only creates a sounder situation within those industries but affects their public relations favourably.
>
> We are going to need the co-operation of all our management groups to make this organization one of the leaders in this province in the development of good employer-employee relations. Insofar as it is practicable, our policies will be standardized and will be clearly stated to you and conveyed to employees so that they will know where we stand.[14]

Labour leaders with collectivist political agendas found it difficult to generate much anti-MacMillan sentiment, largely because he was so popular with workers. A frequent visitor to his mills and logging operations, he was known as an owner who talked freely with employees, no matter what their job. In later years this practice annoyed his senior managers and executives, but it made him very popular in the workplace.

One example of MacMillan's habit occurred when a major fire broke out on Alberni Pacific land. The entire crew was mobilized to fight the blaze in a desperate round-the-clock operation. A young cat skinner was dispatched in the middle of the night with a locomotive and a flatcar to bring a bulldozer to camp and build a fire guard in front of the advancing flames. Arriving in camp, which was a chaos of frantic activity, he looked around for someone to help him unload a pile of railway ties over which the cat could be driven off the flatcar. A tall figure stepped forward and the two of them hastily unloaded the ties. As the young man drove the cat off, its lights swept over his helper—H.R. MacMillan.[15]

Over the years, hundreds of employees had conversations with MacMillan. He remembered them, occasionally looked them up when he was at their operation and, most importantly, took seriously what they said. He had similar relations with the owners and operators of small milling and logging companies along the coast. In the face of this, attempts to portray him as the enemy of the working class for the most part fell flat. This did not stop ideologues of the left from making him a focus of their anti-capitalist calls to action; and he, in return, did not hesitate to attack head-on the policies and proposals of the left. After repeated assertions by leading CCF politicians that, when elected, they would expropriate private property, MacMillan's in particular, he made a number of speeches directly confronting the socialist agenda.

At this time the CCF was rapidly gaining popular support and many, including MacMillan, believed they had a good chance of defeating the Liberal–Conservative coalition. With the growth of a strong export market following the war, profits soared, and were attacked by the left as evidence of capitalistic greed. BC's CCF leaders, such as Colin Cameron, Harold Winch and Dorothy Steeves, used MacMillan as their chief target. At one point Winch explained why MacMillan was the focus of their anger. "It should be understood that Mr. MacMillan presents a symbol to our party, a symbol of the capitalist system which we attack," he told the *Vancouver Sun*. "That is why he is singled out, certainly not because we have personal criticism of Mr. MacMillan himself."[16]

MacMillan, for his part, did not see the CCF as a symbol, but as a direct threat to the future of the province. He feared people might vote the party into power out of ignorance of its intentions. "If the people of this province are socialists, why then, you can't stop it, but it is very probable that only a few of them are socialists," he told a convention of truck loggers in 1949. "It is very probable that a great many unthinking people look at this party full of promises and . . . support it in error."[17] His answer to the threat was the same as he had proposed during the debate on the King government's wartime mobilization—public debate and informed democratic decision-making. The task of keeping the socialists out of power was not made any easier by his belief that politicians in the ranks of the coalition were a fairly sorry lot. In the end, MacMillan had little time for any politicians and gave no money or support to any party, choosing instead to take on the socialists himself.

Having been made a symbolic target by the political left, he became the recipient of scores of letters, some asking for or demanding help, others berating him for his success, often in vituperative language. He was quite capable of writing long, reasoned responses to even the most vitriolic correspondents.

Especially after his experience in the war, MacMillan's faith in democratic processes was deep-seated and profound—his earlier support for abolition of the provincial legislatures notwithstanding. He believed it was extremely important that information be available to the public, and that people participate in public discussions of important issues. It was a faith that extended even into his own business practices. He elaborated on this idea in one of his many exchanges with E.P. Taylor.

> I believe that public companies should give the shareholders all the information that they can possibly be given without jeopardizing the interests of the company; otherwise, our system of limited-liability companies will sooner or later break down. Firstly, the shareholders will become apathetic; secondly, they will lose confidence in the management, and thirdly the management—lacking guidance from the shareholders—might be inclined to run the company from a personal standpoint rather than a public standpoint.[18]

Taylor must have chuckled when he received this advice, knowing that at this stage the H.R. MacMillan Export Company had no public shareholders, and that the small group that owned the company was firmly under the control of MacMillan. Here, as on other occasions, he was a regular practitioner of the adage: Do as I say, not as I do.

The preparation of his brief for the Sloan Commission revived MacMillan's interest in conservationism. In fact, he had written a strongly worded piece on salmon conservation in that year's BC Packers annual report, blaming a decline in salmon stocks for a fall in profits. MacMillan attributed the decline to lack of proper regulation on the part of the federal government, particularly in protecting streams and spawning grounds. He called for increased fisheries staff and the creation of a fisheries department at the University of BC.

At about the same time, he took out a membership in the Society of American Foresters, the US professional association, which supported conservationist forest policies. MacMillan's entire career was very much an outgrowth of his deep involvement in the conservation movement which had flourished forty years earlier, so he welcomed the reborn interest in forest policy as something he had given up for lost.

That fall, at a Hotel Vancouver dinner for 400 export company employees celebrating his company's twenty-fifth anniversary, MacMillan included in his speech a lengthy discourse on the dangers of public ownership and state intervention. The pending report of Chief Justice Sloan and the growing popularity of the CCF were on his mind, and he was concerned that the public's flirtation with socialist ideology would lend credence to Orchard's proposals. In the same speech, however, he spoke optimistically about the company's future.

> We tonight are at the end of only the first 25 years. The assets have been acquired and coordinated for the next 25 years. The start for the next 25 years is hundreds of times stronger than for the first 25 years. The depletion or extension of these assets, their use and management is in your hands. This company is not a closed dynasty—it is an open association of workers in which the road is open to the top for those who qualify as the best managers, the best custodians of the traditions and responsibilities. As in Napoleon's army, there is a field marshal's baton in every worker's knapsack.[19]

His concluding statement about top positions in the company being open to all employees referred to a policy unusual in proprietary firms. Long before this anniversary, MacMillan and VanDusen laid down a policy that sons and sons-in-law of senior managers and executives could not be employed by the company. Apart from the fact that neither of them had any sons, their reasoning was that all employees should know any job in the company was open to them and was not being saved for a family member. As a result of this policy, there was no possibility of MacMillan's sons-in-law, John Lecky and Gordon Southam, being provided with a job. Having decided that he wanted his daughters living in Vancouver, MacMillan's solution to this problem was simple—and typical of his propensity to control and direct the lives of those around him. He bought two established companies and placed his sons-in-law in charge of them, Southam at Pioneer Envelopes and Lecky at Smith, Davidson and Wright, a local paper distributor.

The prohibition on nepotism did leave the problem of succession unresolved, and it was something upon which MacMillan realized he had to act. By the end of the war he had settled the question, at least in his own mind, and knew who he wanted to take over from him.

chapter 17

The good life

The problem confronting MacMillan in 1945, as he turned sixty, was a common one for entrepreneurs who built successful businesses from scratch. Their firms, even though they might be very large, as in MacMillan's case, were proprietorships in which the founder was the owner and quite often the chief executive as well. If the proprietor had an heir who assumed ownership as well as management of the company, the problem was not acute. If he did not, as in MacMillan's case—at that time daughters did not take over businesses—then management of the company evolved into the hands of professionals. They might be highly trained outsiders hired for their skills, or senior people in the company who had worked up through the ranks. In either case, they were not owners in any meaningful sense. For someone in MacMillan's

MacMillan and Major General Bert Hoffmeister at a Vancouver Club dinner, July 5, 1945. The dinner was organized to honour Hoffmeister as a Canadian war hero.

position, choosing such a manager was a difficult decision.

Under the circumstances, MacMillan's choice was a natural one. He picked Bert Hoffmeister, a thirty-eight-year-old company employee. In his eight years with the export company prior to the war, Hoffmeister had risen quickly through the ranks to become sales manager at Canadian White Pine. During the same period he attained the rank of captain in the Seaforth Highlanders. When war broke out, he went overseas in one of the first contingents from Vancouver. He soon acquired a devoted following among the troops, gaining a reputation as a hard-driving officer who worked and fought beside the men under him. He had an uncommon approach to fighting a war— win the battles without losing men. He quickly rose through the ranks, faster than any other non-professional soldier in the Canadian army, and eventually ended up a major general in charge of the Fifth Armoured Division, one of the five Canadian divisions in Europe. His legendary exploits as a senior officer who operated on the front lines, constantly exposed to enemy fire, earned him numerous decorations. All in all, he was one of Canada's most respected war heroes.

MacMillan remained in close touch with Hoffmeister throughout the war, dining with him on occasion in London. He admired him immensely and was proud of his achievements. As the war wound down in Europe, MacMillan was anxious for Hoffmeister's return to civilian life and his job in Vancouver, and made these feelings clear among his contacts in Ottawa. Prime Minister Mackenzie King, however, had different plans. When the war with Germany ended, Hoffmeister returned to Canada. MacMillan wired him, offering him a job as general manager of the company's operations. When Hoffmeister asked to be relieved of his duties, King prevailed upon him to take command of Canadian forces in the Pacific instead. Hoffmeister set up headquarters in Brockville, Ontario, and began mobilizing Canadian troops for action in the Pacific, reporting directly to King, with whom he met regularly in Ottawa.

The sudden end of the war with Japan, brought about by the dropping of atomic bombs on Hiroshima and Nagasaki in August 1945, eliminated the government's need for Hoffmeister's services. At his next meeting with King, he reminded the prime minister of MacMillan's offer, and asked to be relieved of his command. King agreed, and on October 1, 1945, Hoffmeister was back at work in Vancouver. For the people of the city, he returned as a conquering hero. In MacMillan's mind, he assumed almost the role of a son, and it was clear to all that young Hoffmeister would one day take over as head of the company.

In the meantime, MacMillan was firmly in charge of the company, making a series of high-profile decisions. Long before he had taken an interest in buying Victoria Lumber & Manufacturing, the BC government, in response to an active public campaign, had entered into negotiations with John Humbird over the fate of a small stand of magnificent Douglas fir, popularly known as Cathedral Grove, located near Cameron Lake on Vancouver Island. For almost ten years, pressure had mounted to preserve the stand as an accessible, surviving remnant of the fir forests which once covered much of eastern Vancouver Island. When MacMillan took over VL&MC, Premier John Hart was preparing to exchange the stand and some surrounding land for Crown-owned land elsewhere.

Shortly after the deal for VL&MC was completed, MacMillan wrote to Hart, informing him that on Taylor's behalf he was offering to give the 332-acre tract to the province for a park, without receiving any land in exchange. He did list some conditions he wanted met. Most of them had to do with such matters as taxes on the land, establishing its boundaries and so on. But his last demand related to the long-standing campaign he had been conducting against government grants to the BC Lumber and Shingle Manufacturers Association for overseas market promotion. He asked, in exchange for turning over title to Cathedral Grove,

> that there should be no more grants made by the Provincial Government to a particular section of the industry for market extension purposes, which in past years were used by the recipients to injure the business of their competitors, amongst whom we were numbered. . . .[1]

For some reason—either he had a change of heart, or the Premier rejected his conditions—MacMillan sent another letter three days later, offering to give the land unconditionally. Hart accepted graciously and announced the gift a few days later. The *Vancouver Sun* published an enthusiastic editorial, suggesting the area be named MacMillan Park. In a flurry of activity, the premier agreed, obtained MacMillan's consent, and passed an order in council naming the park—all within three weeks.

The announcement concerning the park was followed by another, this time from the University of British Columbia. It reported donations by MacMillan to the university of $45,000—half as a personal contribution to the forestry school, and the remainder from BC Packers to establish a course in fisheries. The combined effect of these gifts on the public was phenomenal. At this time Canadian businessmen were not noted for their generosity to the public, and editorial writers across the country pulled out all the stops, praising MacMillan to the skies. Letters of congratulations and thanks poured into his office.

His gifts to the university were not impulsive gestures. He had been discussing the university's needs with its president, Norman MacKenzie, for some time. Learning that the government was not about to provide the required funds, he took it upon himself to finance what he considered to be vitally important functions in two areas that were important to him. These contributions marked the beginning of a long involvement with BC's major university.

Since its establishment, MacMillan had been closely connected with the university's presidents, first with Frank Wesbrook, then with his old friend from Guelph, Leonard Klinck, and latterly, with MacKenzie, an ardent out-doorsman who had become a regular fishing and hunting companion. In the fall of 1945, MacKenzie accompanied MacMillan, Stanley McLean and Don Bates on a hunting trip to the Rocky Mountains in southern Alberta. With a large party of guides, horse wranglers and a cook, they packed into a camp at the base of Citadel Mountain. The weather was cold and snowy, and after a few days without success, MacKenzie and McLean left for home, leaving MacMillan and Bates behind with two guides. Two days later, Bates bagged a goat on a rock ridge at the top of a long, very steep, snow-covered slope. The rest of the herd moved along the ridge, where they remained until dark. The next morning, while Bates remained in camp, MacMillan went in search of the goats, taking his guide, a relatively inexperienced young Native who had just returned from military service in Europe. Tying their horses at the foot of the snow slide, they climbed over the rocks beside it up to the ridge.

At this point, the guide's lack of experience led them into trouble. He took MacMillan out across the top of the snow slope, which had frozen to ice in the falling afternoon temperatures. Within a few steps they both lost their footing and began sliding down the 60-degree slope. The guide managed to slow his descent by grabbing at trees growing through the snow. MacMillan tumbled a few times, then slid face first down the 2,000-foot incline and into the rocks at its base. The guide worked his way down the slope, and they man-aged to crawl to their horses, mount them and ride back to camp. When they arrived they were both in shock and incoherent. The guide had dislocated his shoulder and was in considerable pain. MacMillan's face was torn and bloody, the flesh hanging off his forehead over his eyes. Great chunks of his cheeks and chin were torn from the bones of his face. After giving them both a stiff shot of rum, which just about finished off the guide, Bates did what he could with their poorly stocked first aid kit. All he had was some disinfectant and part of a roll of adhesive tape. He carefully washed MacMillan's wounds and taped his lacerated facial flesh back into place. After letting them rest for the remainder of the day, Bates and his guide loaded the two onto their horses and led them on a long cross-country trip to the nearest highway. From there they caught a ride to Banff, and MacMillan was quickly removed to hospital in Vancouver. He spent the next few weeks in the Vancouver General Hospital where, for a sixty-year-old, he recuperated rapidly. Bates's first aid treatment worked beautifully; only under close scrutiny could any scars be seen. The following year, and for several years after that, MacMillan was back hunting in the mountains.[2]

His convalescence gave MacMillan time to think about broader issues, many of which found their way into long memos to VanDusen and other senior members of his companies. Sometime before the war ended, he had become preoccupied with the question of finding work for returning veterans. He wrote a booklet that was sent to the more than 1,000 export company employees serving in the armed forces, assuring them their jobs were waiting for them. Remembering the shoddy treatment World War One veterans had received, and the social upheaval which followed that war, he instructed his executives to do more than simply reinstate their workers.

> Employing additional returned personnel which is a load upon all companies—particularly to do what they can to absorb those who went into the war from schools or otherwise with no previous employment. If we can think this out to give a stronger lead to it, we should do so. A large body of disappointed and discouraged unemployed returned personnel will be not only a great injustice to those who assumed their obligations in the war, but will also be destructive in society. We might accomplish something by an apprentice system; also something by preference to returned personnel when new employees are being taken on, and also by deliberately undertaking during the next few months such expansion of the business as seems justified and such as we feel we can carry.[3]

Two weeks later he sent a letter of similar intent to the Vancouver Board of Trade, with copies to several leading businessmen. In it he argued that individuals and businesses should find employment for the hundreds of servicemen then scouring the city for work.

> If the individuals in this community fail to prove they are making their utmost effort to employ these men, the natural reaction will be that the problem will be left entirely to public bodies, which is another step in the direction of socialism. . . . The solution to this problem will require the greatest and unremitting efforts of each firm or person. It should be attempted according to the same principles as the Victory Loan Drives, which aim to get something from everyone according to the capacity of each.[4]

In December 1945, Chief Justice Sloan's much-awaited report was released. It made numerous recommendations designed to sustain a regulated annual harvest from the province's forests, along with proposals affecting education, research, taxation and related issues. One of the key recommendations dealt with the thorny question of forest finances. Since colonial days, public revenues from the forests had been placed in general revenues, with the costs of forest administration paid out of treasury board allocations. Typically, the government spent far less on the forests than it collected from their utilization. Sloan urged that forest revenues first be spent on forest protection and management, after which any surplus could be passed onto the provincial treasury for use elsewhere. On the question of forest tenure, however, Sloan was ambiguous. He suggested the formation of public and private working circles—sustained yield units, each established to provide a perpetual crop of a designated volume of timber—but he did not directly address either proposals to transfer more public land to private ownership, or Orchard's suggestion of Crown-owned Forest Management Licences. Consequently, those who studied the report were able to read whatever they wished into his sections on tenure.

When the legislature met in the new year, only a few minor, non-controversial changes in the Forest Act were introduced. It was clear that the government, like most governments considering Royal Commission recommendations, was going to be very selective in those it adopted. Contrary to Sloan's advice, the government did not intend to commit a portion of forest revenues to forest management. In fact, with the imposition of new charges and taxes on forest users, the Hart government intended to accelerate the established practice of milking the forests for the largest possible revenues without providing the funds for their proper management. The question of tenure was set aside until the following session.

After much consultation, a critique of Sloan's report and the government's initial responses was published by the Truck Loggers' Association, a new organization of small, independent coastal loggers who sold timber on the open market. Dedicated free enterprisers, they feared they were being squeezed out by the bigger companies, and they banded together in an attempt to influence government policy and fend off the large operators.

Curiously, the largest operator of them all, MacMillan, supported their struggle. He and Gilmour were regular speakers at TLA meetings and were sympathetic to its aims. Gilmour wrote the TLA-published document "A Discussion of the Sloan Report," and MacMillan wrote a foreword for it, which stated bluntly:

> The Sloan Report presented to the people of British Columbia a most thorough and statesmanlike recommendation for the management of forest land for permanent crops of increasing volume and value.
>
> Unless the people of British Columbia instruct their representatives in the Legislature to spend on forest protection and management a greater proportion of forest revenue, as is urged in the Sloan Report, the forest will become only a temporary, not a permanent crop, with greatly reduced future employment in British Columbia.

Gilmour elaborated on this theme, generally praising Sloan's report and urging the public to call on government to adopt different policies than those it appeared to favour. At this stage, both he and MacMillan felt Sloan's report supported their tenure proposals.

MacMillan knew he was treading on dangerous ground. In early 1946, the subject of forest policy was a volatile one, as evidenced by the strong emotions exhibited during the commission's hearings and the attacks on himself as a major industrial figure. At a managers' meeting the previous August, he had warned that politicians, the press and public bodies were sensitive to issues discussed at the commission and that "this company is to avoid collision with public opinion."[5] Having seen the government's intended course of action, he was now prepared to risk that collision.

Perhaps MacMillan's willingness to place his neck on the line stemmed from his discovery of the rejuvenative effects of the Mexican climate. Ever since his bout with tuberculosis thirty years earlier, he dreaded the wet, cold West Coast winters. Often he ended up in bed, and occasionally in hospital, with severe colds or influenza. In February 1946, after recovering from his tumble down Citadel Mountain, he and Bates travelled by train to Guaymas, Mexico. They rented a car and spent a week touring the small coastal villages, hiring local boats and guides to take them marlin fishing, a sport Bates had learned as a child.

It was an unusual and carefree holiday for MacMillan, who usually disliked vacations, considering them a waste of time. They spent their nights in seedy hotels and ate whatever they could find. MacMillan loved the hot, spicy Mexican food, and devoured it in great quantities, apparently immune to the intestinal ailments it often inflicted on tourists. The two aging gringos had the time of their lives as MacMillan, in the gregarious way he had visited logging camps, got to know the local people. On one occasion the wife of a guide they had hired died suddenly. The Mexican government was engaged at this time in one of its periodic feuds with the Catholic Church, and was discouraging church funerals. In order to avoid harassment, burial services for the woman

MacMillan with his first marlin, Mexico, March 1946.

were conducted at night. MacMillan and Bates found themselves serving as clandestine pall bearers, helping carry the body to a moonlit cemetery where it was laid to rest.[6]

His contact with Mexico's rural poor awoke him to conditions in BC that he previously had not noticed. A few months later, following his summer tour of the fish canneries, he wrote to Buchanan.

> In my memo of a few days ago I neglected to mention that I feel strongly that we should proceed with the building of housing units for the Indians and others, which will be as good as the ones you have built recently, and thus free us from the responsibility of forcing our workmen to live in conditions similar to those we saw in Mexico. I know we inherited a great load of bad houses and it will take some time to correct it, but I would like to see the low-grade shacks wiped out before too long.[7]

Near the end of their Mexico trip, MacMillan caught his first marlin, a 150-pound fish he insisted on hauling back to Guaymas in the trunk of the car

Don Bates and MacMillan with his catch at Guaymas, Mexico, 1946.

to have photographed. He, in turn, was hooked on Mexico—its people, its climate and, most of all, the fishing. He returned to Vancouver full of enthusiasm and energy. He had found a way to beat the debilitating Canadian winter, and an exciting new sport to enjoy.

It was a good thing MacMillan came back with lots of energy, because he had a busy year ahead of him. In March he gave a dinner in Vancouver for C.D. Howe, attended by 400 Vancouver businessmen, most of whom had in one way or another been involved with Munitions and Supply during the war. The following month he went to Toronto to meet Eddie Taylor; they were hatching another scheme.

He had another delicate task to perform in Toronto. He had to placate Mazo, who was furious with him, and had vented her anger in a scorching letter.

> We both agree you have treated us very badly—yet airily sign yourself "your affectionate cousin." For years you have expatiated on the benefits of our living in British Columbia. You have told us that we shall live five years longer if we move out there—even ten years longer if we hurry up! When we visited you last Spring you adjured us not to return to Ontario at all but to send word to have the contents of our large house forwarded to us without delay. The children and dogs could follow easily. [Mazo and Caroline had adopted two children.] Yet now, when we have quite made up our minds to go, have put our house into the hands of an agent, you tell us that there is no roof to give us shelter in Vancouver or Victoria. . . .
>
> I have looked earnestly at your picture, and I cannot believe that you would be easily balked. If you, with your stupendous achievements, are balked by this problem, I shall indeed think the matter hopeless. Would there be a house in New Westminster? Do you own a house you could throw the tenants out of? Would you rent me one of your aviaries? We could perch there while we looked about.[8]

Somehow he managed to convince Mazo and Caroline that wartime housing restrictions were still in effect in BC and that they should stay in Toronto. It is not clear what prompted his change of heart, although Edna and his daughters' lack of enthusiasm for his two offbeat cousins likely was a factor. In any case, they permanently abandoned their thoughts of moving west.

After leaving Munitions and Supply, E.P. Taylor had set up the Argus Corporation, ostensibly an investment company. Taylor was an early and accomplished practitioner of the leveraged buyout, using the assets of acquired companies to secure financing for further acquisitions. When the Argus board, which included British, American and Canadian directors, was appointed in late 1945, MacMillan was on it. Previously, during the purchase of VL&MC, Taylor had expressed an interest in direct involvement in the BC forest industry. MacMillan undertook to help him find some established operations with a good supply of timber. Initial overtures were made to San Francisco's Zellerbach family for the Pacific Mills pulp and paper operation at Ocean Falls, midway up the coast, but that deal fell through. Over the course of the winter, offers were made to several sawmill operators who also owned substantial timber reserves. It was to discuss these deals that MacMillan and Taylor met in Toronto in April 1946 to work out the details.

In exchange for an annual fee and a percentage of sales, MacMillan agreed to provide Taylor's new company, to be called BC Forest Products, with a general and assistant manager, to act as sales agent and to make available the advice of H.R. MacMillan Export's directors. MacMillan returned to Vancouver, where he discussed the proposal with VanDusen, Scott, Ballentine and Wallace, beginning with a memo headed "Thoughts on EPT situation." In it, he revealed one of his primary reasons for entering into such an arrangement.

> The new company if formed will operate—he will find management somewhere—who will probably raid our staff as being best known pool here—some of whom may feel opportunities with us look moving slowly now and unequal to their feeling of confidence and ambition.[9]

An advantage MacMillan neglected to mention was that most of the mills Taylor proposed to buy were members of Seaboard. Acquiring them and redirecting their sales through the MacMillan organization represented a real coup over Seaboard.

In May the deal was announced, possibly the largest single transaction in the history of the industry. Six of the larger lumber companies in the province—four on southern Vancouver Island, one in Vancouver and another on the Fraser River—were combined into one company with planned production of 200 million board feet of lumber a year. In view of growing public concern over the monopolization of timber, and to show this was not another VL&MC-type takeover, Taylor and MacMillan went to great lengths to separate their interests. No one from the MacMillan operation was appointed to the BCFP board, although MacMillan's longtime friend, Austin Taylor, was made a director. MacMillan resigned from the Argus board, while E.P. Taylor and his eastern partners each turned in their single share in MacMillan's Chemainus operation, and resigned from its board.

The relationship between MacMillan and Eddie Taylor was a curious one, and yet another example of MacMillan's contradictory character. They held a completely different set of business ethics. Taylor, for instance, had by 1947 acquired control of Massey Harris and voted himself and his associates unusually high dividends to pay off the bank loans used to buy Massey Harris stock. Bleeding operating companies dry to set up financial shells was not MacMillan's style. But while their methods were at odds, there was for many years a close personal relationship between them.

A major hurdle for BC Forest Products was to provide a co-ordinated management structure to merge the operations of six well-established, disparate companies which had evolved in fierce competition with each other. MacMillan selected as general manager one of his favourites from the export company, H.G. (Hec) Munro, an ambitious young man who had started out as a lumber salesman and spent the war in Ottawa in the timber controller's office. Taylor provided the company's financial controller, Trevor Daniels, a British accountant who came to Canada during the war for the British timber control office.

For MacMillan, the arrangement with Taylor offered more advantages than protecting his staff from raids. From the day it was established, BCFP had as voracious an appetite for timber as every other company at that time—including MacMillan's. But to the chagrin of BCFP's middle managers, the MacMillan organization was able to override potential BCFP timber purchases until the management agreement was terminated in 1953, giving the MacMillan Export Company a clear edge in the struggle to dominate the timber supply. Corporate relations between the two companies would never be cordial.

In the late 1940s, MacMillan began to extricate himself from the day-to-day operations of his business ventures. In the fall of 1946 he resigned as president of BC Packers, turning the job over to John Buchanan and assuming the position of chairman. During the fifteen years he had been involved with the fish company, MacMillan had become intimately involved with its operation, issuing a steady stream of suggestions, ideas and instructions to Buchanan. At every opportunity, he had acquired additional shares in the company, reselling a portion of them at cost to company employees and other BC business people. His strategy was to move control of the company to the West Coast, and

The *Marijean*, MacMillan's pleasure boat, in Mexico. He named the yacht for his daughters, Marion and Jean.

by 1947, every member of the BC Packers board was a BC resident.

In early 1947, MacMillan arranged for his son-in-law, John Lecky, to inspect some surplus US Navy vessels that were for sale in Seattle. Lecky found three boats, a 95-foot sub chaser he had picked out for MacMillan's personal use, and two larger minesweepers for BC Packers to convert to fish boats. When he saw the ships, MacMillan was taken with the minesweepers. They had more elegant lines and room for more spacious accommodations. He took one for himself, a 138-foot wooden-hulled vessel that he got for about a tenth of what it cost to build. Naming the boat the *Marijean* after his two daughters, he assigned them the task of furnishing and decorating the interior cabins and saloons after the boat was converted to a pleasure yacht at Burrard Shipyard. It had accommodations for eight, plus a full-time crew of six. When the conversion was completed, the *Marijean* took its place as one of the more elegant additions to the yachts tied up in Vancouver's harbour.

For the next twenty years, the voyages of the *Marijean* were a big part of MacMillan's life. From spring until fall, he cruised the BC coast, fishing for salmon regularly at Rivers Inlet and numerous other locations, holding business meetings, and taking weekend excursions or day trips with a load of the grandchildren his daughters were producing regularly. But the vessel's real purpose was the annual two-month excursion off the Mexican coast. After the first couple of years, MacMillan established a ritual that he stuck to for almost two decades. Immediately after Christmas, the crew headed south on the long voyage to Mazatlan. The expedition was divided into two-week segments. MacMillan flew south with his wife and stayed for the entire time, and Edna usually returned to Vancouver after the first month. MacMillan personally planned the *Marijean* trips down to the last detail of every family member and guest's arrival and departure. Dorothy Dee made all the arrangements, oversaw their execution while he was away and remained in constant communication with him. But she never set foot on the *Marijean*; that was Edna's turf.

Every two weeks the boat returned to one of several ports, where the guests would change. The Southam and Lecky families each came for a two-week stay, and often brought guests. Grandchildren were not usually included. After the first month, the family having spent its allotted time aboard, MacMillan spent another four or five weeks with a constantly changing mix of friends, cronies and business acquaintances. Bates and Buchanan were the most regular guests, scarcely missing a year. Norman MacKenzie and Wallace McCutcheon, Taylor's associate at Argus, Air-vice Marshall Leigh Stevenson and James Stewart, the Canadian Bank of Commerce chairman, were frequent guests.

Since it was MacMillan's yacht, life aboard the *Marijean* was not a lazy, relaxed vacation. His idea of a holiday was to pursue vigorously a different objective than he did the rest of the year. Everyone was up early, fed, and sent off on whatever activity had been planned the previous evening. The larger of two launches headed out to sea for six or seven hours of deep-sea fishing; the smaller one cruised along the shore for ground fish. After the first few years, a smaller boat was used to collect marine specimens. At 5:00 p.m., with everyone back on board, the cocktail hour commenced and everyone enjoyed their two-drink ration. MacMillan was not a drinker and had little time for those

MacMillan (at right) showing his catch, including a 410-pound marlin, to (left to right) C.S. Steen (captain of the *Marijean*), Peter Larkin, John Lecky, John Buchanan, Gordon Southam, Leigh Stevenson and Ian McTaggart-Cowan, at BC Packers' Steveston plant.

who were. Dinner was served at 7:00, and the evening was devoted to conversation. Guests who could not make a stimulating contribution to the discussion were not invited back.

MacMillan spent an hour or so every evening writing in the ship's journal, a series of large red log books in which he recorded the day's events—where they had travelled, fish they had caught, events he remembered from the past, and, if he felt it merited recording, the evening's conversation. Some of these entries were three or four pages long; as much as 2,000 words a night was set down in MacMillan's distinctive handwriting, in clear, concise prose. Among his remarks, he carefully taped photographs he had taken with a Polaroid camera.

Into the *Marijean*'s large walk-in freezer went a good supply of fish for eating back in Vancouver, and unusual or interesting marine specimens. Back in Vancouver, at the BC Packers freezing plant, MacMillan started a fish "museum." One of his favourite pastimes was to take friends, acquaintances and business associates to examine his collection. As they stood around shivering and shaking in the sub-zero temperatures, he went on at great length, describing the circumstances of the capture, the habits of the creature and other related information. He was fascinated by marine biology. After the first few trips, he asked Norman MacKenzie to find a biologist from UBC to accompany them on their annual excursion. Casimir Lindsay was the first scientist to take up this offer, then Peter Larkin, who became the ship's regular scientist during the last decade MacMillan owned the boat. Witty, convivial and a dedicated biologist, Larkin became something of a cross between court jester and scientific guru to the expeditions. He had at his beck and call a rotating crew of usually willing assistants, drawn from the ranks of Canada's senior business elite. Wallace McCutcheon was one of his regular helpers, setting nets in the middle of the night, poisoning fish and utilizing just about every other method known to mankind in the capture of marine specimens. Everything was meticulously sorted, counted, examined and prepared for freezing and shipping back to Vancouver. In time, a third biologist became part of the operation. Murray Newman's presence on board laid the groundwork for the establishment of the Vancouver Aquarium, which in turn provided an even better reason for the *Marijean*'s scientific expeditions. MacMillan began making trips to the Galapagos islands and through the Panama Canal into the Caribbean, gathering live specimens and shipping them back to the aquarium.

MacMillan could well afford to buy the *Marijean* when he did. From the last years of the war, well into the 1950s, the profits of the export company rose spectacularly, often doubling or tripling in one year. BC Packers had made money steadily since he rescued it from virtual bankruptcy, and it was paying good dividends. By the late 1940s, MacMillan was a very wealthy man.

It was not in his nature to live ostentatiously. He continued to reside in the house on Balfour Street in Shaughnessy that he had bought in the late 1920s. He drove a Bentley and had a summer home at Qualicum on Vancouver Island. His extravagances were lavished mostly on his daughters and their families, for whom he purchased houses, summer places and a steady stream of other gifts—basically out of a desire to make them happy, but also as means of keeping them near him. Beyond that, he derived most of his pleasure from reading and collecting books, and his annual hunting and fishing expeditions into the BC interior.

In mid-1947, as the *Marijean* was being refitted, he bought his second farm at Qualicum, a few miles inland from the summer house on the beach, at the edge of what was then a sleepy Vancouver Island village. A few years earlier he had purchased Arrowsmith Farm, a somewhat rundown operation, because it was such a bargain, but he had not become actively involved in its management. This time he bought the island's, if not the province's, most elegant agricultural establishment, a 3,000-acre model farm property adjoining the Arrowsmith farm, built at great expense by A.D. McRae, a former senator. Enormous barns housed a large population of purebred cattle, sheep and pigs, with elaborate ponds provided for ducks and geese. It was a highly mechanized operation, utilizing the latest scientific methods. And, of course, it lost money.

Although MacMillan was delighted to learn from Eddie Taylor that he could write off the losses on the farm against his taxes, he was determined to make it a successful operation that showed a profit. Part of his enthusiasm no doubt had to do with the fact that the farm provided him with something productive to do during the weekends he spent at Qualicum. His early morning rituals now included a tramp through the fields with the farm manager, and an examination of livestock, during which he picked out turkeys and beef to stock the larders on the *Marijean*, his house in Vancouver and the kitchens of his daughters and friends. He was very much the gentleman farmer, trying out the methods he had studied forty years earlier at Guelph.

MacMillan had begun a new phase in his life. During the war he had learned, albeit reluctantly, that his "outfit" could survive, and even prosper, without his constant attention. He had chosen and installed his successor. Now, he had the freedom to pursue some of his other interests.

The elder statesman

As his wealth increased in the late 1940s, MacMillan was faced with an important decision—what to do with his money. Once he had seen to his own needs and the needs of his daughters and their families, there was still a substantial income to dispose of. One option was to reinvest it in his companies, or in new ventures. This was the choice of many of his contemporaries, who used their wealth to extend the range of their power and influence. MacMillan took a different direction, a much more public-minded one.

Increasingly, as the dividends from companies flooded in, he began to give away money, systematically and in ever-larger amounts. Initially, the University of British Columbia was the main beneficiary of his generosity. MacMillan's renewed interest in forest policy had convinced him that the teaching of forestry in Canada badly needed improving. He had often expressed his opinion that Canadian forestry schools, including UBC, were devoting too much attention to training engineers to harvest timber and not enough on silviculturalists to manage forests, and he had a very low opinion of UBC forestry graduates. So he decided to use his new wealth to rectify the situation. Following his initial gifts to the UBC forestry school, in 1947 he provided more money—about $60,000—to pay the salaries of three forestry teachers, an entomologist, a mensurationist and the school's first silviculturalist. Out of the five foresters on the university's staff, he was financing three. Over the next year he also provided money for a student loan fund, bursaries, a series of lectures in forestry, $10,000 to beef up the forestry library, and another $5,000 toward development of a research forest.

Also in 1947, MacMillan developed some wider and, for him, unusual interests upon which he bestowed money. While he was hunting in the Bella Coola Valley, a local guide, Bert Robson, showed him his collection of Native artifacts and expressed a desire to give them to UBC. On his return to Vancouver, MacMillan contacted Ian McTaggart-Cowan, then curator in the zoology department, and arranged the gift. Performing this middle-man role sparked his interest in aboriginal culture, which soon grew to include a fascination with archaeology and anthropology. He began buying Native artifacts and giving them to UBC, forming the basis of a collection that eventually evolved into the Museum of Anthropology. In typical MacMillan fashion, once he became interested in a subject, he devoted his full mental and financial resources to it. Wishing to retain Native carvings and artifacts as an important part of BC's cultural history, he became one of BC's major buyers of these items. On one occasion, hearing that an American boat tied up at Nanaimo had a section of a totem pole aboard, he went into action, calling the police, customs agents and everyone else he could think of to prevent the removal of the pole across the border. He was not at all contrite when it turned out the man who possessed it was a highly respected collector who was taking it to a BC museum.

In 1947 Lawren Harris, the artist who was a leader in the Group of Seven, approached MacMillan with a request. MacMillan may have met Harris through his connections with the Argus Corporation, as the painter was the son of one of the founders of the Massey Harris equipment company, which Argus controlled. Harris convinced him of the need for an addition to the

Vancouver Art Gallery to house, among other works, the Emily Carr collection. MacMillan took on the task with his usual thoroughness, and made the initial donation of $10,000, which he persuded Stanley McLean to match. In the end he raised more than $50,000 for the building fund, despite the fact he had no appreciation of or interest in art, and rarely, if ever, set foot inside the gallery.

At the 1948 annual meeting of the export company, MacMillan announced a further diversification, this time into a pulp mill to be built on the east coast of Vancouver Island. He financed it out of the spectacular profits earned in 1947, more than triple those of the previous year. This mill, which opened in Nanaimo two years later, was one of his most astute and well-timed moves. Named Harmac, after the company's old cable address, it came into production at the beginning of a long rise in pulp prices. Unlike any of the other pulp mills built during the late 1940s and through the 1950s, Harmac's construction was not tied to a large allocation of Crown timber. It depended for its wood supply on waste wood from the MacMillan sawmills and low-grade logs from the company's Vancouver Island logging operations.

MacMillan and Bert Hoffmeister at the Alberni plywood plant, 1948.

Inevitably, all this activity attracted the attention of the media. In the late 1940s, MacMillan was the subject of several flattering magazine articles. He disliked this sort of attention, embarrassed as much by the fawning tone as by the crude, superficial techniques some publications employ in their attempts to reveal a subject's character. *Maclean's* magazine ran one of these articles in March 1948.

> The time is five minutes to two on a Friday afternoon and the place is downtown Vancouver.
>
> A big man in a loose grey suit steps from the solid, respectable doorway of the Vancouver Club. His powerful shoulders are hunched forward, he has an air of complete preoccupation. His objective is the doorway of an office building 100 steps away. He covers the distance in a quick, rolling gait and takes the elevator upward.
>
> H.R. MacMillan, one of Canada's top industrialists, is bound for his weekly directors' meeting.[1]

This sort of melodramatic description of him returning to his office after lunch, invariably illustrated with a Karsh photograph portraying him as a stereotypical captain of industry, made MacMillan cringe. He was appalled at the thought his friends and colleagues might think he had collaborated in these profiles, and after a few of them were published, he tended to avoid interviewers intent on discussing his personal life. Even Howard Mitchell, his publicist acquaintance from Wartime Merchant Shipping days, was rebuffed in his frequent attempts to publish a biographical profile of MacMillan.

MacMillan's major preoccupation during the last half of the decade was the ongoing forest policy debate and the political machinations in Victoria that were a part of it. Throughout 1946 there were numerous meetings of those concerned about forest legislation changes the government planned to introduce in the spring session of the legislature. Individually, and in groups, industry spokesmen, heads of companies and representatives of associations and interest groups met with Orchard, Forest Minister Kenney and other members of the Hart government. By fall, a rough outline of the various positions was clear.

Gradually it became apparent the government was taking seriously Orchard's proposal to create a new form of tenure, Forest Management Licences. Under pressure from powerful financial interests both in and outside the province, the government concluded there was need for a licensing mechanism to open up large tracts of timber for big forest companies. These developments precipitated a profound division within the Forest Service, culminating in the resignation of several senior members of the staff—including Orchard's second-in-command, Fred Mulholland—who were put off by the chief forester's arrogance. All these professional foresters ended up in industry.

This small group of company foresters, which included Gilmour, Mulholland and John Liersch, head of forestry at UBC, were basically opposed to the Forest Management Licence scheme because of the power over private forest companies it granted to the government and the chief forester. Most of them were familiar with the structure and organization of successful European forestry, which was based on privately held land, and believed this was the key to developing good forest management in BC. They were supported in their arguments by small logging and sawmill operators, who feared they would be squeezed out under Orchard's scheme. In September, Liersch circulated a confidential report of the group's first meeting, which addressed Sloan's recommendation for a Forest Commission to collect forest revenues and oversee financing of protection and management.

> It seems to all of us that the whole problem of advancement in private forestry rests upon the establishment of a [Forest] Commission. We realize that at present the Government is against such a policy. But we believe it is against a Commission largely for one reason. This reason being the Premier and the Minister have had the opportunity of hearing only one side of the problem, as presented by the Chief Forester.[2]

Mulholland, in a letter to Liersch, agreed that industry should present its views to the government, as a counterbalance to Orchard's influence.

> I do not think the Minister is hostile to private forestry; if he is a bit cynical about it and inclined towards State monopoly of forest ownership it is because his Deputy [Orchard] has not outgrown the Socialist theories which most foresters seem to pick up at College.[3]

With the blessing of their employers and the active support of MacMillan, the foresters then set out to convince the government—essentially, Premier John Hart and Forest Minister Kenney—that Orchard's plan would not provide the incentives required to persuade private industry to engage in managing forests. In November, MacMillan and other industry representatives met with Orchard and Kenney and tried to persuade the government to bend toward their views. In a seven-page letter to Kenney, reviewing the points made at the meeting, MacMillan did not bring up the question of land owner-ship directly, but instead addressed the matter of incentives for forest man-agement.

> One should not forget that no labourer, farmer, professional or small business man will embark on any important use of time or capital unless convinced that the end result will be in his interest. Similarly shareholders of forest-owning and using companies cannot be expected to enter upon or support policies that show no promise of benefit proportionate to the expenditures and risks involved.[4]

He concluded with a strong plea that the government reconsider the idea of placing the province's forests under the administration of a Forest Commission. It was a delicate request to make of government: it suggested that politicians and government bureaucrats could not be counted on to administer forests properly.

By early 1947, the people who owned and managed the larger compa-nies were getting nervous. The major concern of most of them was to secure more timber to meet their short-term needs. They were not opposed to forest management, but only if they could hold the line on spending. Most of them found it was still cheaper to buy timber from someone else, including the gov-ernment, than to grow it. To these people, with their short-term perspective, the great appeal of Forest Management Licences was that they granted exten-sive cutting rights on Crown timber. At best, the opportunity to use the land for growing timber was looked on as a necessary obligation. On the one hand, they were prepared to go along with their foresters and oppose Orchard, which they did through the BC Logging Association, but they were also keep-ing a close eye on each other to see who would apply for the new licences first. They became increasingly nervous as word spread that outside financial interests, which had never operated in the province, were sniffing around the corridors of the legislative buildings in Victoria.

Their nervousness was well justified. By early January 1947, American Cellanese, a New York pulp company with no prior interest in BC, had received a reserve of timber, for a huge Forest Management Licence in Kenney's riding near Prince Rupert. Approval had come from Kenney, appar-ently with Orchard's agreement, even though the amendment to create the licences had not yet been tabled in the legislature. Placing the area under reserve prevented anyone else from obtaining timber rights in the area until an FML was granted or refused. In late January, the Cellanese group was back in Victoria, this time accompanied by Bob Filberg from Canadian Western, demanding another reserve on prime timber lands on the islands in Johnstone Strait, between northern Vancouver Island and the mainland.[5] The timber in this area was of critical importance to market loggers and the independent mills to whom they sold. The delegation was turned down, but Filberg later returned with another US company, Crown Willamette, and was granted a reserve on the area.

At the end of February, the Canadian Society of Forest Engineers held its annual meeting in Vancouver, with Liersch as chairman. Its theme was "The Relation Between State and Private Forestry." After Mulholland's keynote address, an extensive defence of private forestry, MacMillan delivered a speech on the status of the forestry profession. It was a long, detailed talk, in which he documented the decline of enthusiasm for forestry since his days as a student. Pointing out how the federal forest reserves on the prairies had fallen into neglect, he went on to describe how the initial measures adopted in each province in the early years of the century had been abandoned or circumvented. He dealt at greatest length with British Columbia.

> The greatest handicap in the progress of forestry in British Columbia is political; the public would support the theory of forestry if it were put to a plebiscite; the Legislature would support it if put to a plebiscite for the Legislature; the Legislature has never failed to pass a Forest Act which stated implicitly and implied on every page that it was the intention of the people in this Province to place their forest resources under sustained yield management by a professional staff of trained men. The trouble has been, in spite of the intention of the public, the fight for the dollars that takes place every year in Caucus or elsewhere, the Forest Administration has been defeated by those members who can go back home to their constituencies under no necessity of saying that they have robbed the Forest Administration, have obtained enough money for public works and for various forms of needs that are demanded by the electorate of the particular district. If that fight could be brought out into the open, and the people could see the result of what has been so frequently in this Province an unfortunate division of money between all types of worthy expenditures and forest expenditures, I am sure that forest expenditure would, by the general demand of the people be allotted enough of the public funds to place the forests upon sustained annual yield management, because that is clearly, in my opinion, the desire of the people of British Columbia.[6]

MacMillan did not place all the blame on politicians seeking re-election, however. Much of the fault, he said, lay with foresters, who had failed to convince the public and demonstrate to the people of Canada what could be accomplished by managing forests properly.

As his speech showed, MacMillan had worked himself into a somewhat unusual, if not difficult, position. Once again he was thinking—and, at times, acting—as a forester. He saw clearly the failure of the profession over the past forty-odd years, and the mistakes in the direction the BC government was now headed. But, as the principal owner of the largest forest company in the country, he realized the virtue of discretion and the need to take whatever advantages offered themselves in the coming months—no matter what he thought as a forester.

In March, the government tabled its proposed legislation, giving the public, including industry, a scant two weeks to comment and respond. It was clear that the government, pressured by the deals it was already cutting with American Cellanese and other companies, was going to ram Orchard's plans through the legislature. An intense effort by industry extracted some minor reductions in stumpage for timber harvested from managed forests, but little else. Orchard had won. On April 3, the amendment passed in the legislature and Forest Management Licences became the only means by which private interests could get into the business of managing forests on the 93 percent of BC's forest land owned by the Crown.

Immediately the forest companies began jockeying for position. Even opponents of the legislation, including MacMillan, lined up with applications, knowing that any company which did not obtain a FML would find it difficult to obtain any Crown-owned timber. Canadian White Pine had requested a reserve in the same area of Johnstone Strait as Filberg had been granted one, and was refused. In August, Victoria Lumber Company requested a reserve on south Vancouver Island for a FML, and Alberni Pacific made a similar request in November.

MacMillan made a futile attempt to convince Orchard to establish some clear principles for allocating the licences, to replace the arbitrary procedures by which the minister of forests awarded them.[7] It was unclear, for instance, whether more than one application would be considered for a FML on an area placed under reserve. But MacMillan got nowhere. With the bit in his teeth, Orchard began issuing regulations for FMLs which made it clear the Forest Service was planning to closely control the forest management practised by private companies. It was exactly the type of bureaucratic triumph MacMillan and others had feared.

A further flaw in the FML allocation procedures became apparent when the government accepted Filberg's application for the Johnstone Strait FML. Filberg, in collaboration with his American backers, was proposing to use timber from this area to supply a new pulp mill in Campbell River, only a few miles across the Gulf of Georgia from the Powell River pulp mill, which had operated at its location for almost forty years. The Powell River Company, which had always supported a policy of leaving the Johnstone Strait forests for small market loggers, obtained its timber from farther north on the Queen Charlotte Islands and adjacent mainland coast. It found itself with a competitor setting up shop in its back yard. Both the Powell River Company and MacMillan protested the issuance of the licence. In response, Kenney dismissed MacMillan's objections and his arguments. "In this connection," he wrote, "it is interesting to note that interests represented by your company have already applied for more than nine percent of our estimated productive forest lands."[8] Kenney also dismissed MacMillan's request for a public hearing into allocation of the licence. Not only would failure to apply for a FML cut off access to all Crown timber, it appeared that making an application precluded the right to object to any other application.

During the first half of 1948, the trickle of requests for FMLs turned into a flood. When it was clear the licence was going to become an established form of tenure and that every company of any significance would have to acquire one if it wanted to maintain its timber supply, Gilmour resigned in disgust. "Recently I told you and Mr. VanDusen that I was leaving your employ in this province, because I could not see any professional opportunity for an industrial forester in British Columbia," he wrote to MacMillan in July. "It is a disappointment to leave the Province for these reasons, but I could not retain my professional self-respect any longer where I could not contribute anything worth while."[9] MacMillan regretted Gilmour's departure; he had lost his chief advisor and spokesman on the land tenure issue and, in the general climate of acquiescence to the government's plans, was increasingly reluctant to say anything critical himself. Along with Harold Foley and Doug Ambridge, a Toronto timber executive, he arranged to retain Gilmour on a long-term consultancy to continue researching, writing and speaking publicly about forest policy.

Relations between MacMillan and the forest minister had deteriorated to the point where virtually any suggestions he made were dismissed out of hand. Early in the summer of 1948, he wrote to Kenney in support of an Alpine Club proposal to protect the forested area through which the entrance to Garibaldi Park would one day be built. Kenney rejected the idea outright, saying there was too much timber being tied up in parks. It was a ludicrous situation: the province's principal timber user was asking for the enlargement of a public park, and the minister in charge was refusing in the name of the timber industry. MacMillan responded, backing away somewhat from his request, and suggesting it might be better if the federal government took over the park.[10]

By early 1949, MacMillan was saying very little about the province's forest policy. His comments were limited mostly to his correspondence with Gilmour, now living in Montreal.

> I am impressed by the fact that you are just about the only one doing any clear thinking respecting basic principles of forest management in this province. I have come to the conclusion that we are all too optimistic about forest management. People, particularly in British Columbia, are not ready for it and, this being a Democracy, nothing of fundamental importance can be done in advance of public opinion. Those who should give leadership have not thought out yet the home truths and governing factors, consequently they have not sized up the social reactions. Obviously, not having done their own thinking, they are neither equipped nor competent to lead the public in the right direction . . . the public is not getting any help from the foresters, who merely are semi-trained trade school graduates who have received their learning from men who preceded them by five or ten years through the same course of instruction.[11]

A few days later, he wrote to VanDusen, who was on vacation, about FMLs. At this point the MacMillan companies had only asked for reserves in various locations, but had not reached the point of actually applying for a licence.

> The argument still seems sound against this Company taking up a management licence or applying for one. There is some sign of criticism developing amongst some persons in the Forest Service. Therefore, to make up any deficiencies in our patch logging system and to get something else that can be used to the Company's advantage, we are working out a system of protecting the numerous suppressed and other trees that are left after logging so that they will not be burnt up in slash fires, thus resulting in natural reseeding for a decade or two at

slight additional expense. On examination it looks to be an attractive idea, particularly as the Forest Service shows signs of requiring that not all slash areas be burned. We have got to do something to show that we are not neglecting public welfare, and planting looks to be a very undesirable solution.[12]

The following month, Kenney delivered a blistering speech in the legislature, accusing opponents of Forest Management Licences of being against the basic principles of sound forest management. Opposition to government policy was tantamount, he and Orchard believed, to favouring a policy of forest liquidation.

By the end of the summer, MacMillan despaired of any change in direction, as he expressed in one of his regular letters to Gilmour.

Here in British Columbia I don't see any immediate correction for the Management Licence policy. It was put out without enough thought and consultation, it was endorsed by the top foresters and by the Cabinet Ministers, etc., and being ignorant they don't know where nor how to change it, and being stubborn to allow any change would be to admit a weakness, and any attack made on it would probably freeze them in their mental positions. Therefore, it possibly is best left alone for awhile.[13]

By the following spring, he had pretty well given up on the issue, writing brief notes to Gilmour, but otherwise keeping his thoughts on the subject to himself. "I am getting to the point where I think that forestry is not going to proceed any faster than Christianity, and that a person should not get in a sweat about it," he wrote. "This is a poor way to feel I know, but humanity seems to have to learn both the hard way."[14]

That year, 1950, Gilmour tried to prod MacMillan into action, urging him to come out publicly against the millions of dollars in capital gains that had been realized on Canadian Western's FML #2 since it had been awarded the previous year. MacMillan declined to engage in a public debate.

Governments make a good many errors, but, when one is conducting a business upon which thousands of shareholders and thousands of employees are more or less dependent, a continuing fight with the Government can turn out to be a managerial error.

There is not much point in such a fight if contracts have been signed by the Government, which has gone so far that it certainly will not and cannot withdraw. The situation becomes much more difficult when the only possible opposition to the Government is the Socialist Party.[15]

Gilmour, one of the few people willing to persist in an argument with

MacMillan, would not give up so easily, and wrote again a few days later. MacMillan again explained why he did not want to speak publicly against the government's actions.

> You will understand that this is a situation in which, speaking on behalf of our Company, I can say nothing; it is also impossible for me to speak excepting on behalf of the Company. It is obvious, of course, that any time I do speak it would be considered to be on behalf of the Company.
>
> We are in the midst of a very difficult period in this area at this time. It seems that everybody on this coast is too close to the situation to see the problems in perspective. It is now as it always has been in Canada: there is no body of persons who have enough information to enable them to see the problem in perspective and who have enough interest in the future welfare of the country to work unremittingly for the public interest.[16]

Soon after, correspondence between the two tapered off. Gilmour became intrigued with the single-tax theories of Henry George, and after MacMillan suggested he stick to writing about forestry, their association came to an end.

By this time, opposition to Forest Management Licences had all but disappeared. Seventeen of the licences had been awarded, including one to the Alberni Pacific Division of the H.R. MacMillan Export Company, providing a harvest of 75 million feet a year. Noting the MacMillan licence, the Victoria *Daily Colonist* commented: "This is evident support of the Provincial Government's policy. . . ."[17] For the time being the issue was laid to rest.

Throughout the course of the forest management debate, MacMillan had much else to keep himself occupied. Work on the *Marijean* was completed in time for its first trip to Mexico in early 1948. As profits in his companies continued to increase, he found himself devoting more time responding to attacks from the left. Then, in mid-1949, the driving force behind the export company's profits, the booming UK reconstruction market, took a severe blow. The British Board of Trade announced that, for the first time since the outbreak of war, it would resume major timber purchases from Russia, beginning with a huge order accounting for 10 percent of the country's annual imports. The market that year was already in bad shape, and this news caused the layoff of thousands of workers in BC.[18]

In anticipation of this decision, the province's two chief lumber

MacMillan with Montague Meyer at a lunch in London, commemorating the thirtieth anniversary of the founding of the H.R. MacMillan Export Company, 1949.

promotion associations—the Seaboard-dominated BC Lumber and Shingle Manufacturers Association and the MacMillan-led Western Lumber Manufacturers Association—set aside their differences and amalgamated. In the wake of the British announcement, MacMillan accompanied a large BC lumber industry delegation to London in an attempt to protect the province's export business there. His presence on the delegation attracted considerable attention in the British press, which characterized him as "the last of the lumber barons."[19] This was intended as a compliment, as MacMillan was widely known in the UK and was looked upon almost as a hero for his wartime work and public support of Great Britain. It was a successful expedition for the BC group, which returned to Vancouver with substantial orders.

Part of the reason for this success stemmed from MacMillan's introduction to the head of the British Board of Trade, Harold Wilson. The prominent Labour Party politician was interested in fostering increased trade with

Canada, particularly western Canada. Coincidentally, he was also a friend of Monty Meyer's. Subsequently, when Labour was voted out of office, Wilson worked in Meyer's office, and during this time he and MacMillan became close friends. The curious relationship which evolved between the arch-capitalist, MacMillan, and the socialist politician who would one day become Britain's prime minister, lasted until MacMillan died.

The Canadian lumber industry was saved that year by the adoption of a major housing bill in the US Congress. Overnight, a huge demand for Canadian lumber developed, and by the end of the year the US had replaced Britain as the country's primary lumber market, a situation that continued for the next forty-five years.

In the fall of 1949, MacMillan accompanied Don Bates on a highly successful hunting expedition into northern BC. After flying 400 miles north of Prince George into the Cassiar Mountains, they travelled into a remote area by pack horse where they bagged record-breaking sheep, caribou and goats. On his return to Vancouver, MacMillan announced his retirement as president of H.R. MacMillan Export Company. In making the announcement, he explained it was time for the older executives in the company to make way for a younger generation of managers: Bert Hoffmeister would become president and Ralph Shaw vice-president of marketing. MacMillan was not ending his participation in the company. He became the company's first chairman, with VanDusen— who had declined MacMillan's *pro forma* offer of the presidency—taking the position of vice-chairman. Control of the company's operations now lay in the hands of the finance and policy committee of the board of directors. This tight group ruled on policy changes, senior appointments, major expenditures and anything else MacMillan felt like introducing. In reality, he had not given up any authority to Hoffmeister, who found himself in a position more like that of a vice-president. Hoffmeister was clearly being groomed to take over, but as far as MacMillan was concerned, the transition was far from complete.

By now MacMillan was the elder statesman of the West Coast business community, and in recognition of this role, he instituted a kind of informal salon. Most weekday evenings, when he was in Vancouver, he hosted an intimate gathering of business associates for a cocktail hour at his Shaughnessy home. It began promptly at 5:00 p.m. and was held in

MacMillan's study, where regular guests regaled each other with a telling of the day's events. Often loud peals of laughter rang through the house, at least once provoking an inquiry by Edna from her seat at the piano: "I wonder who went broke today?"

At 6:45, the evening's guests were ushered out the door and dinner was served. Every night at 7:00, often during dinner, MacMillan listened to the evening news on the radio. He usually spent his evenings at home reading, rarely attending social functions.

MacMillan's Shaughnessy home at 3741 Hudson Street, Vancouver.

Business executives of MacMillan's stature receive awards and honours, and he was no exception. In 1950, noting his age, his retirement as president of his two major companies, and his honorary University of British Columbia degree, columnists and editors at Vancouver's two newspapers launched a campaign to have him appointed the province's next lieutenant governor. Privately MacMillan was appalled by the idea and thought it an absolutely ridiculous proposal. The suggestion was quietly dropped.

He received his honorary degree, his first, along with C.D. Howe. In MacMillan's convocation address, he lamented rising government expenditures, as well as the growing tax levies needed to pay for them. The politicians, he said, were only incurring these expenses at the request of the people. "Our voters throughout the Province are infatuated with the pleasurable feeling of constantly flowing public money irrigating all constituencies."[20] But the chief problem facing the province, he said, was not unrestrained public spending on education, health, public works and social services. It was the ability to maintain and improve the state of the provincial economy. This, he informed the graduates, would be their responsibility.

In June, the Harmac pulp mill was opened with great pomp and ceremony. It was a moment of pride for MacMillan, even if the mill's costs had spiralled upward from the projected $12 million to more than $19 million. Still, it was the most up-to-date mill in the world and—perhaps the real achievement—it had been built without the allocation of a Forest Management

Licence. No sooner had it opened, though, than the citizens of Nanaimo rose up in indignation. The smell produced by the mill, the first bleached kraft mill in the province, provoked howls of outrage. After initiating some studies to determine what could be done to solve the problem, MacMillan told a lunch for 200 of the city's leading citizens that the company would do whatever was possible to alleviate the stench. He then painted a vivid picture of the city's economic possibilities, which included at least an implied threat to take future investments to a more appreciative community. When he concluded, his listeners leaped to their feet in thunderous applause.

The following winter, when MacMillan was fishing off the coast of Mexico, he received an urgent summons from Howe, whom he had spoken to on several occasions since they received their degrees at UBC. It was almost identical to the one that had called him to Ottawa in 1940. A few months earlier, North Korea had invaded the south and the conflict was escalating. Howe wrote:

> You will agree with me that the war news grows worse each week. When I was last in Vancouver, I expected the situation to develop much more slowly, but that has not been our experience. I am busy setting up another Department of Munitions and Supply, and we are talking a defence budget for the next fiscal year of around two billion dollars, which is comparable with the scale of defence preparedness in all North Atlantic countries except the United States, which is on a considerably larger scale.
>
> I am sure that you will wish to take the same active part in this war as in the last, consistent with the fact that we are all ten years older than when the last war started. I am therefore calling your attention to an appointment that promises to be one of the most important that Canada will make, a job which you can fill to perfection if you will do so.[21]

Howe asked MacMillan to serve as the Canadian representative on the North Atlantic Treaty Organization's Defence Production Board. Once again, MacMillan was off to war.

Cold warrior

H owe's call to arms dramatically disrupted the 1951 *Marijean* cruise. By the time MacMillan received his letter, he had already spent a month on the yacht with his family and friends. Dorothy Dee had tried for a week before she managed to get a copy of Howe's message to MacMillan in Puntarenas, where he had come in to pick up John Buchanan, Don Bates and Blake Ballentine for four weeks' fishing.

MacMillan immediately cabled Hoffmeister, asking his opinion and giving him a reply to send on to Howe:

REALIZE SERIOUS EMERGENCY AND PRIVILEGE OF ASSOCIATION WITH YOU IN REPRE-SENTING CANADA STOP LEAVING HERE NOW NEXT ADDRESS MAJESTIC HOTEL IN ACA-PULCO IN ABOUT WEEK STOP MEANTIME CONSULTING MY ASSOCIATES RESPECTING SUCH AN ABSENCE DURING SERIOUS COMMITMENTS ONE OF WHICH KNOWN TO YOU AND ANOTHER OF MAGNITUDE IN NEGOTIATION STOP WHAT IS LATEST DATE I COULD REPORT OTTAWA STOP AM SOMEWHAT UNEASY RESPECTING MY HEALTH SUPPORTING

GOOD PERFORMANCE UNDER EUROPEAN WINTER CONDITIONS BUT AM INCLINED TRY IT IF CAN ARRANGE MY AFFAIRS VANCOUVER.[1]

Hoffmeister had already been in touch with Howe, and hurried to Ottawa to discuss the situation with his former military associates, all of whom expressed their belief that the political situation was deteriorating rapidly. Fears were growing that the Korean conflict might spread, possibly even leading to clashes between communist and western forces in Europe.

There was more behind Hoffmeister's trip than curiosity about MacMillan's potential role. The export company had recently committed itself to a major expansion of the new Harmac pulp mill, and Hoffmeister was concerned about the impact of these events on the company's plans. He dispatched a hurried letter to MacMillan in Acapulco.

> Harris, the Steel Controller, mentioned that he expected to have full control of steel production within three to six months, and thought that it might be accomplished in a little more than three months.
>
> Both Howe and Harris expressed keen interest in our proposal to complete the second unit and while neither was in a position to give a written guarantee each felt confident that they could see us through on our steel requirements. Harris mentioned that he felt quite sure that the USA would treat the project as a high-priority item and that some American steel would be available.[2]

MacMillan and his guests reluctantly carried on with their fishing trip, and missed a second telegram from Howe stating that the urgency of the situation was not so severe as he had thought a week previously, and that MacMillan was not needed for a month or two.[3] Disturbed by the news from Korea they were able to pick up on the ship's radio, and the tone of Howe's original summons, the fishing party was plagued by a feeling of uneasiness, as MacMillan related to L.R. Scott.

> This worried me a great deal and really upset the remainder of my trip to the degree that I ended it after two weeks more fishing and then flew to New York, Montreal and Ottawa. . . . In Ottawa I found the situation very much like where I came in 11 years ago. C.D. is older and gets tired, but possesses astonishing vigour and, of course, great experience, judgement and powers of discernment. He has a better looking organization now than he had 11 years ago. I found that I had to agree to help him until winter comes. He is sending me to represent Canada on the Defence Production Board for the twelve Atlantic Treaty Nations. I don't think that it will be onerous, but it is just being away and being so far away that concerns me. . . the opinion in New York and Ottawa seems to be that the

danger of a "shooting war" is receding, although I think they are all just as much in the dark as ever as to what Joe might do, especially in Yugoslavia, Turkey and Arabia where nothing looks too healthy for our side. If a "shooting war" comes at all, it will probably come about mid-summer this year.

The American preparations appear to be whole-hearted, in spite of political bickering. I think that Canada is finding it difficult to decide how to use her strength, which she can use only once but, having given care to selecting our field of operations, is determined to go ahead on a substantial scale. I believe that a universal draft will be adopted in this country, without much trouble, if necessary to maintain armed forces at whatever scale the government decides.

I have never seen this country on the eve of such vast development of natural resources as is now actively discussed and initiated from coast to coast.[4]

After meeting with Howe, MacMillan spent two more days in Ottawa, but was unable to learn much else about the political situation or his role in it. Circumstances were even more ambiguous than they had been when he had arrived in the capital as timber controller in 1940. He returned to Vancouver to ponder the matter and await further instructions from Howe.

In early 1951 the North Atlantic Treaty Organization was still suffering from birth pangs. NATO had been initiated by ten west European and two North American nations in response to the Soviet Union's aggressive military moves into eastern Europe at the end of World War Two. Canada, concerned the US would retreat into its pre-war isolationist position, was an enthusiastic advocate of the alliance. Canadian diplomats Escott Reid and Lester Pearson envisioned it as more than a military agreement, proposing far-reaching economic and industrial initiatives as well. NATO was not formalized until 1949 and, even then, little was accomplished until war broke out in Korea a year later, less than a month after Canadian Prime Minister Mackenzie King died and was replaced by Louis St. Laurent.

Abandoning isolationism with a vengeance, the United States took the leading role in mounting a United Nations force to confront the communist-led North Korean forces which were sweeping south, virtually unopposed. Under US General Douglas MacArthur the UN forces drove the North Koreans back across the 38th parallel, the line dividing the Asian country, and to the dismay of most of the world, pursued them north. By late 1950, the Chinese had intervened, and it appeared the world was once again being plunged into conflict. In this situation, the face-off with the Soviets in Europe took on a new complexion, and the rapid mobilization of NATO became a matter of extreme urgency. In this context, Howe had cabled MacMillan.

The treaty signed by the twelve Atlantic nations divided NATO into three parts. The political component was vested in a governing council, consisting of the foreign ministers of member nations; Pearson was Canada's representative on this council, which had no permanent chairman. The military section consisted of defence ministers—with Brooke Claxton sitting in for Canada—who oversaw an integrated military force under the supreme command of General Dwight Eisenhower. A financial and economic committee was presided over by the twelve finance ministers, including Canada's D.C. Abbott. The governing council, in turn, set up a series of permanent working committees and boards to perform the day-to-day work of the organization. One of these, the North Atlantic Defence Production Board (DPB), was assigned the task of reviewing the overall military supply situation, recommending ways and means of increasing available supplies, and promoting more efficient methods for producing and standardizing military equipment. An American, William Herod, the head of General Electric, was elected co-ordinator of the DPB, which operated out of offices in London. Initially there was no central headquarters for NATO; instead, its various components had offices in a number of UK and European cities.

MacMillan was Canada's representative on the DPB, with most of the ongoing work performed by Dana Wilgress, Canada's High Commissioner in the UK and the country's representative on the North Atlantic Council of Deputies, which performed the day-to-day work for the governing Council of Ministers. In Ottawa, Howe headed up a reconstituted version of Munitions and Supply, now called the Department of Defence Production. Essentially, it was a carbon copy of the World War Two body, with several senior people drawn from the private sector and given the task of overseeing each industrial sector. This group of controllers and a small army of bureaucrats prepared to mobilize and co-ordinate all domestic production to meet whatever military needs might arise. As MacMillan had noted, there was a linkage of postwar industrial expansion and the imperatives of global Cold War politics. He was describing the birth in Canada of what Eisenhower, after he was elected US president, deplored as the "military-industrial complex," a network that dominated western economic and political thinking for the next forty years.

Back in Vancouver, MacMillan spent the next few weeks attempting to make sense out of the flood of letters and cables he began receiving from

Ottawa and London. He deflected requests from newspaper reporters for interviews, and declined invitations to describe his new posting to various boards of trade and chambers of commerce. In response to an enthusiastic letter from a functionary in the London High Commissioner's office, he replied:

> I have had very little information yet. I do not know where the meetings are to be held and consequently do not know where I shall be going, to London or Paris. I believe there are some communications now on the way to me from Ottawa. I am looking forward to meeting Mr. Gill and you and to find out what the duties will be.[5]

MacMillan warned Howe that his participation might be limited.

> I appreciate the honour of the appointment you have offered me and, as I said to you in Ottawa, I will do the best I can until the weather in Europe becomes dangerous for my health, which I expect will be along about October or November.
>
> I mention this because this past winter has been the first in the past five in which I have not experienced a very bad cold, keeping me in bed two to four weeks and leaving me shaken. My doctor attributes these colds to my travelling in planes and trains and living in hotels during the months of November and December; therefore, I decided to terminate that particular kind of activity.[6]

Meanwhile, his attempts to elicit from Howe details of his role met with little success for several weeks. Finally, in mid-April, he received notification of meetings scheduled in London and Brussels. On his departure from Vancouver he told the press he had no idea what his duties would be and described NATO as a "long-term defensive contract to meet a threat, not a marching army."[7] He did not know when he would return, but the war news from Korea was not encouraging. Canadian troops had gone into action weeks earlier and were already suffering casualties in the grim winter conditions on the unfamiliar Korean terrain.

After a few days' briefing in Ottawa, MacMillan flew to London. Shortly after his arrival on April 21, he went to Brussels to meet with other members of the DPB, then returned to London, using Canada House as his base. He was joined by Dorothy Dee, who had come at his expense to assist him in London, as it appeared he would have several months' work there. For MacMillan, the city was a second home. Before the war he had often commented that he was better known there than in Canada, and he used this occasion to renew old acquaintances, revisit favourite restaurants and browse through antiquarian bookstores. For part of the time his oldest daughter, Marion, joined him, but most of his time, when he was not trying to comprehend the future of NATO, was spent enjoying the city with Miss Dee.

In early June, MacMillan returned to Vancouver to deal with some pressing business concerns, stopping off in Ottawa for talks with Howe and other government officials. Miss Dee stayed behind to keep an eye on DPB affairs at Canada House and enjoy herself exploring Britain.[8]

By now, the relationship between MacMillan and Dorothy Dee had endured for almost two decades. She was extremely competent and an indispensable factor in his business life. While she performed the functions of a secretary, she filled a much larger role and had far more authority than was usual for that position. She controlled access to MacMillan by outsiders as well as his associates and employees, some of whom learned to their regret that MacMillan would not tolerate attempts to bypass her. She scheduled and arranged all the details of his business life, as well as those involving his wide-flung personal interests. She had virtually no personal life of her own, apart from his, and devoted her full attention to his affairs. Her caustic comments, when the occasion demanded, were a match for his, and she had no reservations about telling people how they should address or relate to MacMillan, as well as when and where.

Dorothy Dee en route to London, 1951.

The nature of their personal relationship at this stage remains unknown. That Dorothy Dee was an intimate companion, and in certain ways the emotional centre of his life, seems quite clear. Whatever its nature, the relationship was discreet, unlike those of many of MacMillan's contemporaries. Naturally there was gossip about him and Miss Dee, but few apart from MacMillan's closest associates suspected that she was anything more than an exceptionally capable and devoted secretary.

Within his family, the closeness of their relationship was openly acknowledged. Miss Dee played an important role in the lives of MacMillan's daughters and grandchildren. She made their travel arrangements, was involved in the details of his relations with them as they grew up and, when circumstances warranted, interceded with him on their behalf. All of them regarded her highly and had no reservations about her relationship with MacMillan. Edna was not so sanguine, at times expressing her disapproval during arguments with her husband behind closed doors in the privacy of their home. She and Miss Dee maintained a reserved and polite distance.[9]

After returning to Vancouver in June 1951, MacMillan had time to assess his role and Canada's position in NATO. Quite clearly this was not, as Howe had suggested, a repeat performance of Munitions and Supply. Nor was it a rerun of World War Two. It was not a shooting war in Europe, and there was no immediate crisis, only a vague threat. For MacMillan, the task was not to mobilize an industry and redirect its efforts toward a clearly defined objective, cutting through bureaucracy to achieve the desired result. This time, at least in the early stages, the enterprise was cloaked in bureaucracy and protocol. The methods and sensitivities of a dozen countries had to be blended into a unified whole. Conflicting national views prevented the DPB from getting on with its assigned tasks until the diplomatic niceties had been observed. Canada, with its gung-ho proposal to integrate the members' economies, often found itself standing alone, waiting for the resolution of more subtle political issues.

MacMillan spent much of June and July in Vancouver, corresponding with diplomats and bureaucrats in Ottawa and London.

> I don't suppose there is much stirring respecting what comes in from London from the Defence Production Board. As I become a little more detached from the situation my feeling strengthens that the duties of the Board, although important, are rather nebulous and that some months will be consumed in obtaining a realistic and acceptable estimate of how much of each of these important items will be produced by each country up to the end of 1954 and annually thereafter, if the rate of production being sought is maintained.
>
> In a shooting war necessity freezes designs more quickly, items are produced more rapidly and disappear from sight. The compulsions and sanctions are possible which are next to impossible now. . . .
>
> I know nothing of military matters, but from what I read I judge that the stature

and authority and influence of the men supplied by the various countries to create the armed forces in Europe, is considerably greater than that of the men supplied to find out how much armament is needed, who is going to make it, what are the deficiencies, and how is the cost to be distributed and what is the time-table.

Perhaps the difficulty arises because the military people have authority and those other people, of the categories devoted to studying the production and financial problems, have no authority and also have not had a life-long training in that profession.[10]

It was not only bureaucratic procedures and delays in getting things underway that frustrated MacMillan. The nature and abilities of the people involved conflicted with his approach to management. "It seems very easy for these organizations to develop too much staff and for the members thereof to do as much work in a day as they would do in half a day if they were in competitive business or professional life," he complained to Evan Gill. "Also, I notice there is a tendency to keep too many helpers around. This appears to be inevitable, particularly with international organizations, if one is to judge by those who have sprouted off from UN."[11]

MacMillan with NATO colleagues Mitchell Sharp (left) and C.D. Howe (right) in Tokyo, c. 1955.

He was also cautious about the Canadian position in NATO, and mistrusted the enthusiastic pursuit of a leadership position pursued by the Department of External Affairs. He told Gill:

> Without ever having any convictions on the matter, I am instinctively against having Canada Vice-chairman and then Chairman when there is a great danger that Ottawa would be embarrassed by providing the Chairman of a body which is impelled chiefly by two forces over which Ottawa has no control. The one, the great American determination to set Europe up in their own pattern, and the other, the strong European inclination to conduct European affairs in their pattern.[12]

Senior members of the H.R. MacMillan Export Company just before the merger with Bloedel, Stewart & Welch in 1951. Seated, left to right: MacMillan, W.J. VanDusen, E.B. Ballentine, B.M. Hoffmeister. Standing, left to right: L.R. Scott, G.D. Eccott, R.M. Shaw.

During the two months he was in Vancouver, MacMillan had much more to concern him than the future of NATO. In April, following several months of preliminary discussions, he and Prentice Bloedel had announced their intention to merge Bloedel, Stewart & Welch with the H.R. MacMillan Export Company.

The Bloedel family had started a small logging operation on the coast, near Powell River, in 1911. In tandem with MacMillan's company, it had grown into one of the largest logging and sawmilling outfits in the province, pursuing a policy of aggressive timber acquisition, mostly on Vancouver Island. By 1950 it was MacMillan's primary competitor.

Taken together, the two companies complemented each other, like partners in a good marriage. Senior BS&W executives were at retirement age and willing to step aside as younger HRME employees took over management positions. BS&W had timber surplus to its mill capacity, while MacMillan was deficient in timber. In addition to sawmills, the Bloedel company ran shingle mills, while MacMillan operated two plywood plants. The Bloedel pulp mill was located in Alberni Inlet, while MacMillan's was on the east coast of Vancouver Island at Nanaimo. Merging the two operations allowed better utilization of their combined timber supplies. Because BS&W was a Seaboard member, the merger would also give the new company a great advantage over MacMillan's old competitor. The combined assets of the two exceeded $100 million, making it one of the largest forest companies in the world. It was the biggest corporate merger in the history of BC.[13]

More important was the personal compatibility of MacMillan and Bloedel. Bing Bloedel, as he was known to his friends, did not consider himself a particularly astute businessman. His interests lay more in the scientific and technical sides of the forest products industry. A quiet-spoken, gentlemanly person, he was in many ways the polar opposite of MacMillan. But they shared many ideas, one of them being that, practical and necessary though they might be in an era of corporate gigantism, big companies were not desirable in themselves. Bloedel later explained his views:

> As I looked at our place in the industry and the trends in industry it looked to me like it was necessary to have a property like ours—and it might have appealed to HR in the same way—in a bigger unit. I did not believe, from the standpoint of the ideal social good, in large units. In a practical way there was no alternative to that. If you were going to stay in the game at all you had to go big. These big units are undermining free enterprise. We are building ourselves a triple-lane highway right straight into socialism. In a practical sense, I was not going to be able to change that trend.[14]

With this in mind, Bloedel sought out MacMillan to propose a merger as an alternative to both their companies becoming part of other, less compatible units. Fifteen years younger than MacMillan, Bloedel was, as he said, "scared peeless" to raise the subject.

> I went to his house on Hudson Street at Christmas time, after supper. He was very cautious. For good reason he would think this might be a trick. He was receptive, that it was a good idea that might be worth investigating further. We met secretly a number of times after that, just the two of us. We went up to Qualicum. I told him I had no ambition to be the head of it, which I'm sure made a lot of difference to him.[15]

After a thorough investigation of the idea, the two agreed to go ahead. Each consulted his directors and, with their approval, MacMillan and Bloedel made their April announcement. The details were worked out while MacMillan was in London. He returned home to encounter opposition and a major obstacle to the merger—both from outside the companies.

Public concern with the concentration of ownership in the forest industry was reflected in a *Vancouver Sun* editorial titled "Road to Monopoly."

> By combining assets worth at least $120 million the MacMillan and Bloedel companies would create an industrial colossus. They would also put BC's free enterprise system in danger.
>
> The difficult practical arrangements of combining the two companies have kept the merger incomplete. It's still not too late to drop the whole thing. Nothing in law prohibits the merger. But there's a wholesome free enterprise principle that no corporation should become so vast that it must eventually swallow all its competitors. It's also a common experience that those who overgorge usually suffer acute indigestion.
>
> These dangers lurk clearly in the MacMillan–BS&W merger plan. In a decade, competition with this kind of domination is likely to concentrate the entire lumber industry in the hands of half a dozen firms. The top outfit would tend to swallow everything. As an able man, as a great Canadian and great British Columbian, H.R. MacMillan deserves a better fate than to become known as a great monopolist.[16]

The editorial touched one of MacMillan's sore spots. He was a staunch defender of the free enterprise system and of the smaller businesses that at that time constituted its largest portion. But, like Bloedel, he was not going to be dragged down by his principles. As well, he was a living example of the adage: Do as I say, not as I do. He encouraged Howard Mitchell, his publicist friend who occasionally was hired to write press releases, to draft a lengthy rebuttal, and he mentioned the editorial in several letters, including one to VanDusen.

> I see in the Sun editorial only a proof that the public, which can easily be aroused respecting destruction of the forest, has not been educated to the fact that small operators cannot maintain the forest as a crop because they will not live long enough, and, not living long enough, they cannot even be forced to put capital enough into the enterprise to do any more than give lip service to the growing of a new crop. It is the job of the large companies to teach the public that such companies are as necessary to maintain the forest, achieve complete utilization, and build up continuous management. . . .[17]

The merger took place via a simple share swap. After calculating the relative worth of each company, BS&W shareholders turned over their shares to the H.R. MacMillan Export Company, making BS&W a wholly-owned subsidiary. In exchange, they received 43 percent of the HRME shares. The company name was changed to MacMillan & Bloedel Ltd., with MacMillan as chairman, VanDusen and Bloedel vice-chairmen, and Hoffmeister carrying on as president.

At the last moment, a major tax problem almost scuttled the deal. BS&W held among its assets $16 million in undistributed profits. In a surprise move, the Canadian tax department ruled BS&W shareholders would have to pay taxes on this amount—to the tune of about $8 million—if the share exchange was consummated. Despite the opinion of almost every tax lawyer in the country to the contrary, the tax people refused to back off.

In the midst of this crisis, MacMillan was called back to London. He left Vancouver in mid-August, crossed the Atlantic and spent three weeks acting on behalf of a government whose revenue department was obstructing the biggest deal of his career. Accompanied by Dorothy Dee, he returned to Ottawa in early September for a full meeting of NATO. It was an inconclusive meeting inasmuch as the member countries could not resolve how to apportion costs and responsibilities for defence production, and they rejected a Canadian solution to the impasse. Rumours began circulating in the press that MacMillan was fed up with the situation and on the verge of resigning, which he emphatically denied in a letter to a Deputy Minister Max MacKenzie.

> I am quite clear in my mind that I have said or done nothing to give rise to such reports. They do not bother me, unless it appears to anyone in Ottawa that remarks of mine led to these reports, which is not the case. As I mentioned to you in Ottawa, I have no plans for resigning, but am willing to do whatever meets Mr. Howe's convenience to give him assistance or to adopt the course that will avoid embarrassment for him. I do not know what the next turn is going to be in the affairs of DPB.[18]

For both Howe and MacMillan, these reports must have brought back unpleasant memories of the controversy that raged around them during the early stages of World War Two.

At the conclusion of the conference, Bloedel went to Ottawa and joined MacMillan in a full assault on the tax department. After a week of lobbying and considering alternatives, they decided the best solution was to negotiate a settlement with the tax department. An amount of $2.3 million was agreed upon and eventually paid by the merged company, not the BS&W shareholders.

With this hurdle surmounted, the merger was completed in October and the new company was launched. This left MacMillan relatively free to devote his attention to the worsening mess in the Defence Production Board.

Many European members of NATO were still struggling to finance post-war reconstruction and were reluctant to undertake rearmament on the scale desired by Canada and the US. The latter was demanding conditions and a significant measure of control for bankrolling NATO, which irked the European members. The bureaucrats and diplomats, of course, revelled in this kind of situation, while dollar-a-day men like MacMillan and others on the board became increasingly frustrated at the lack of progress. In late October, Herod, the board's US co-ordinator, voiced a desire to resign.

MacMillan's misgivings mounted, and on October 30 he confided to Guy Smith, an official in the Canadian High Commissioner's office in London:

> I am deeply concerned about DPB. It seems to be one of those bodies which, after it gets going with a large enough staff, could continue to revolve in a maze of reports for possibly months with no accomplishment. . . . There is always the great danger when these organizations are put together that voluntary staff members supplied by the various countries, are persons not needed at home and are very likely to be those who can quite well be spared.[19]

Pleading business pressures, MacMillan decided not to attend the next general meeting of NATO, scheduled for Rome in mid-December. By this time Herod had resigned and the DPB was wallowing in confusion. At the Rome meeting a comprehensive report on NATO's operations was tabled suggesting, among other things, that the civilian agencies, including the DPB, be combined in one body under a director general, with a permanent staff and located in one city. The report was very critical of the accomplishments of defence production efforts of most member countries, including Canada. It suggested Canada had to drastically increase its financial contributions by about $200 million a year to meet its fair share of costs.

MacMillan, who largely agreed with the recommended changes, also declined to attend a January meeting in Lisbon. The reports and letters he received from London indicated the situation was not improving and most of the energies were focussed on reorganizing the operation. At the end of January, he met Howe in Vancouver and discussed the situation with him. The minister could see no reason for MacMillan to return to Europe, so he departed for a few weeks on the *Marijean* off Mexico. By the time he returned in March, the issue was settled. "Paris has been chosen as the centre for NATO civilian agencies," Guy Smith wrote to him. "This, of course, means that the Defence Production Board is definitely washed up."[20]

After obtaining a full account of the planned reorganization, MacMillan wrote to Howe.

> I have heard something of the re-organization of NATO. It is my opinion that bringing the whole organization together in Paris, getting rid of the Boards which spent much time arguing with one another, putting all the work under one executive head, and appointing an ambassador to live on the job and represent Canada, will bring about a striking improvement. There probably will still be some weaknesses, but this preliminary re-organization prepares the way to do the utmost with the resources provided by each country. Under these circumstances, I do not think you will need me any more.[21]

A month later, when the reorganization was complete, he added a footnote in another letter to Howe: "I did not succeed in accomplishing anything for you, but I was very glad to do what I could during the time the organization was sorting itself out."[22]

On his return to Vancouver from Mexico, MacMillan was confronted with a more immediate political crisis than that posed by NATO. In BC the coalition government, which for eleven years had kept the CCF out of office, had collapsed as the province headed for a June 12 general election. In March, writing to Gladstone Murray, a regular correspondent from Toronto who ran an anti-socialist organization called Responsible Enterprise, MacMillan correctly predicted no party would emerge with a clear majority. But he did not foresee the rise from nowhere of the Social Credit Party, which formed a minority government with one seat more than the CCF. He was not particularly impressed with the new government, as he indicated in another letter to Murray following the election.

> I think that the election of Social Credit in British Columbia is an expression of the degree of distrust which the electors have built up over long years toward political manipulations. Apparently they have reached the point where they would vote for anything rather than vote for the two old Parties.[23]

As historian David Mitchell has noted, the Socreds were a "motley crew: railway workers, garage mechanics, restaurant owners, bus drivers, teachers, farmers, loggers and clergymen."[24] None of them were from the big business or professional communities. But within a couple of months, like many skeptical British Columbians, MacMillan found his opinion of the new government led by W.A.C. Bennett had begun to improve. He wrote to Murray:

> I think that our crowd here may be settling down somewhat. It is too soon to tell anything. . . . They seem to be sincere people thrown into power by voters who had become disillusioned by the old parties.
>
> I do not think that the old parties know the degree to which they have lost the confidence of the people—a large number of whom voted for the old parties because they did not like to vote either socialist or Social Credit. However, next time many people, who last time voted Liberal or Conservative, will vote Social Credit.
>
> It looks at present as if the government here is for free enterprise, and will do work for the public without any animosity against business. Of course, as yet we really do not know what they will do.[25]

For the next year, BC was in a state of political ferment with another general election slated. The entire province was polarized around the diametrically opposed options of the CCF and the Social Credit parties. MacMillan, like most business leaders, feared the CCF might triumph. Along with Dal Grauer, head of BC Electric, and Howard Mitchell, he took a leading role in organizing and raising funds for an ad hoc group called the Canadian Foundation for Economic Education. This group reflected the business community's deep suspicion of Social Credit and devoted its time and money not to electing the new party, but to defeating the CCF. In the run-up to the 1953 election, MacMillan expended considerable effort soliciting funds from business leaders and corporations across Canada to fend off the socialists. He wrote to Ross McMaster, the chairman of Stelco:

> As you are probably aware, we are undertaking a bitter fight here against Socialism which Party, of the four Parties running in the provincial election here in June, could easily win the largest number of seats and might even constitute the Legislature. If this happened they could do tremendous damage in this province; therefore, we must defeat them.[26]

Although the Foundation did not throw its support directly behind Social Credit, Bennett's party triumphed on June 9 and formed a government that lasted for the next nineteen years. MacMillan and his group did not slack

BC Premier W.A.C. Bennett and MacMillan, c. 1953.

off after the first clear victory by the Socreds; they continued their anti-social-ist campaign until the CCF suffered a decisive defeat in the 1956 general elec-tion. In 1954, MacMillan and Harold Foley, president of Powell River Company, conducted a fundraising campaign for their anti-socialist group, as outlined in a letter to Bill Mainwaring at BC Electric Railway.

> Harold Foley and I got together and divided up the names in order to get the money in, upon which we agreed last Thursday evening. You are to take the B.C. Electric at $6,000. M&B Ltd. and Powell River have each paid in their money, $6,000 each. I have spoken to the B.C. Forest Products for $2,500, and also the Canadian Forest Products for $3,000.
>
> Harold is taking the Columbia Cellulose Company at $3,500, and you might get Dal to take the Canadian Chemical & Cellulose at $2,500. They are a going concern now, and should at least come in for as much as the B.C. Forest Products. Harold will also take Crown Zellerbach for $6,000 and the Alaska Pine & Cellulose for $3,500.

I will take the Tahsis Company and will speak to Frederiksen when he gets back from Europe. I will also speak to Gordon Farrell for $5,000 from the B.C. Telephone Company and $6,000 from the Consolidated Mining & Smelting Company.

Harold will take the Aluminum Company at $6,000. Harold will take the three big banks at $2,500 each, and the remaining smaller banks at $500 each. He will work with the Raikes. You might mention to Dal that the Royal should come in for $2,500, the same as the Bank of Montreal and the Commerce.

I will speak to Ross MacKenzie re all the Breweries at $3,000. Harold will arrange with Alan Williamson for the Trust Companies at $200 each, and the insurance companies for $200 each.

You might take on the Dominion Bridge at $1,000—if you think that is right— and the United Distillers for whatever you think they ought to do. You might also take on the three big oil companies and the Trans-Mountain Oil Pipeline at, say, $2,000 each, or for whatever you think they ought to give.

I hope that you approve of what we have done; we just tried to push things along. If any of these figures are out of line, or if anyone has been omitted, let's get together on it.[27]

Freed from his obligations to Howe at NATO, MacMillan accepted an appointment to the International Pacific Salmon Fisheries Commission in June 1952. In the early 1950s, conservationist awareness began to focus on the West Coast fishing resource. The previous month, MacMillan had delivered a stirring, hour-long speech to the largest-ever meeting of the Canadian Fisheries Council, held in Vancouver. He described BC's salmon-spawning streams as one of the most valuable resources in the country and said there were large numbers of jobs available in caring for and enhancing their productivity. The greatest threat to the salmon, he warned, was from power-dam construction.

His warning anticipated by many years the covetous desires to dam some of BC's major salmon-bearing rivers by the state-owned BC Hydro Corporation set up by Bennett in 1961. But even in the early 1950s, it was apparent to perceptive observers what the Hydro people had in mind. One of the most vociferous advocates of unlimited hydroelectric expansion was Gordon Shrum, a UBC physics professor who chaired the Fraser River Hydro and Fisheries Research Project. Shrum achieved a certain notoriety when he claimed it was inevitable the Fraser River would be dammed. In a letter to one

of the original founders of the BC fishing industry, Henry Doyle, who forwarded a copy to MacMillan, Shrum stated:

> I think even the most ardent commercial fisherman in British Columbia realizes
> that eventually the Fraser River will have to be used for power. What they are
> insisting upon is that this be the last resort and that other sources of hydro elec-
> tric power within economic range of Vancouver be developed first; furthermore,
> when it does become necessary to utilize the Fraser, that every possible step be
> taken to minimize interference with fisheries.
>
> With our continued industrial expansion and the consequent increase in our
> standard of living, the need for an ever increasing supply of energy becomes fair-
> ly obvious. In the final analysis the problem will be one of economics; which is
> more important to the people of British Columbia, adequate sources of power or
> commercial fisheries? The ideal solution, of course, would be to have both, but
> personally I am very sceptical about this possibility.[28]

From the outset, MacMillan set himself against this clearly defined intention,
and became—from behind the scenes—one of many opponents to the
damming of the Fraser River. In 1956 MacMillan noted on a letter from Peter
Larkin about the Fraser dam controversy:

> Peter: Don't let them put the fish people to sleep. Their tactic could be to act like
> they don't need Fraser, then find Columbia too slow and costly. Put on a few
> brown outs, say the only cure would be a couple of quick low dams on Fraser. And
> the girl would be only a weeny teeny bit pregnant.[29]

The destruction of salmon stocks caused by the damming of the
Columbia River was evidence enough for those concerned with the future of
the salmon fishery that dams on any of the province's major salmon-rearing
rivers were unacceptable. But the hydroelectric enthusiasts, led by people
like Shrum, were highly regarded by the development-at-any-cost Bennett government. Eventually BC Hydro proposed a specific project on the Fraser, the
Moran dam. What finally defeated it was a broad coalition which included the
federal government and a wide cross-section of British Columbians. Much of
the opposition in BC came from the fishing industry, through which
MacMillan, along with his colleague and friend John Buchanan, exerted their
influence.

Bennett's Social Credit government, in spite of its idiosyncratic ideas
and unorthodox members of the legislature, had an obsession with economic
development that was consistent with the unrestrained industrial expansion

occurring throughout the western world during the 1950s. If the pace in British Columbia appeared to be more frenzied than elsewhere, it may be because there were more unrealized opportunities in that province, requiring only financing and additional workers to realize their development.

The province had become a magnet for eastern Canadian and European immigrants, who began arriving in a flood. MacMillan, with his extensive business contacts in the UK, received a steady stream of letters on behalf of young men who wished to secure jobs in Canada before they left Britain. His usual response was to advise them to take the plunge and find a job after they arrived, but occasionally his patience wore thin. "Dear Bertie," he wrote to Lieutenant Colonel Harewood Williams, who had written asking for a job for the son of a friend.

> As a matter of fact, I am a little weary of the number of young men from the UK who seek the promise of a job before they come out. If they are good enough to come out, they ought to be good enough to find a job for themselves when they get here. We do not promise employment to the young Canadians who come here from eastern Canada, nor to the Americans coming into this province, and it seems to me that young Englishmen coming to this large growing country, should be prepared to take the same chance. If they are not, they ought to stay under the umbrella of social security.[30]

MacMillan's views on industrial growth and development, as he began to express them in the early 1950s, were curiously inconsistent with prevailing attitudes. It was a period of unrestrained economic expansion often entailing government participation, during which the idea that there might be, or should be, limits to growth was heresy. His reservations about such developments would soon earn him considerable criticism.

The second Sloan Commission

In the mid-1950s, with a new government in power which showed no inclination to replace the forest tenure system, MacMillan suspended his criticism of Forest Management Licences and turned his attention to company-owned lands on Vancouver Island. Prentice Bloedel's presence on the board of the newly merged company provided an additional spur to investments in forestry. Following the merger, M&B launched an aggressive program of land acquisition on Vancouver Island and, with an expanded forestry department, began implementing the most intensive program of forest management in the province. Angus MacBean, the company's new chief forester, directed policy and handled relations with the Forest Service. Keith Shaw was manager of lands and Ian Mahood manager of forestry operations. A forestry and timber committee of senior directors, including MacMillan, Bloedel, Hoffmeister and VanDusen, oversaw the program, insuring it had the status and financial resources it required.[1]

With some trepidation, the company included its extensive private land holdings in the Alberni area in its applications for two Forest Management Licences on the west coast of the island. A third application on the northeast coast of the island also had been submitted. As MacMillan later pointed out, the proportion of private land committed to the FML scheme by M&B was much higher than that of any other company.[2]

When MacMillan explained the FML applications to shareholders at the 1955 annual meeting, he was misquoted in the press as criticizing the FML system and its administration by the Bennett government. He quickly obtained a verbatim transcript of his remarks and sent them under a covering letter to the premier. "Again the newspapers have represented me as criticiz-

MacMillan and Prentice Bloedel (right) at the time of the merger of their companies in the early 1950s.

ing your Government," he wrote Bennett. "In this instance they print statements which I did not make and which do not sound like any statement I made."[3] Two days later he wrote to Socred Forest Minister Robert Sommers, enclosing a copy of the transcript and suggesting a meeting the following day.[4] He was at great pains not to alienate the government, which he realized was the only alternative to the CCF, and which—whatever his reservations about it—was going to retain the forest tenure system then in place.

By this time a veritable flood of FML applications had been submitted to the government, including a contentious one from BC Forest Products. The ten-year management agreement between Taylor and MacMillan's companies had been terminated in 1953, three years before it expired. The two companies had been in direct competition for timber, and Hector Munro had requested, on BCFP's behalf, an end to the arrangement. He left MacMillan's employment and became the president of the Toronto-based company. BCFP immediately submitted an application that included an already-established Public Working Circle near Tofino, on the west coast of Vancouver Island, adjacent to areas under application by M&B. By now Chief Forester Orchard's vision of a few hundred modest-sized FMLs had degenerated into a handful of very large licences, including in some cases lands reserved in Public Working Circles, which were areas managed by the Forest Service for small operators. The Tofino application set off a storm of controversy. Newspapers once again began publishing editorials deploring the allocation of large tracts of forest land to the big companies and a hue and cry arose, demanding another royal commission on forestry. On January 7, 1955, the government bowed to this pressure and appointed Gordon Sloan to conduct a second enquiry into BC forest policies.

MacMillan threw himself fully into overseeing preparation of the M&B brief. In a four-page memo to Hoffmeister, he outlined his ideas.

> Although I spent a lot of time on this in 1944–45 and have thought a lot about it, solely from the interest of this Company, I am not satisfied with my thinking. . . . Very cautious and thorough thinking must be done before we present views, or become connected with any group. . . . It should not be overlooked that several of the large companies . . . are not too Canadianized. Our Company has been Canadian for just about 36 years, and the background of our whole active staff is Canadian. We might naturally therefore have different viewpoints from the others. This may be why Foley [an American] suggested someone from this Company as a front for the Pulp Committee. . . . We might differ from some other companies

in the desire to maintain public working circles, and thus keep open the biggest possible field for independent loggers.[5]

Within a week, MacMillan had drafted a long memo for the Forestry and Timber Committee, outlining M&B's strategy for its presentation to Sloan. The company would support the allocation of more FMLs in undeveloped areas such as the north coast and the west coast of Vancouver Island, known as the Western division. The already-developed area of the Eastern division—inside the island and along the protected mainland coast—should be left in Public Working Circles to maintain the supply of logs which could be purchased competitively on the open market. There would be no criticism of the Forest Service and no arguments that would alienate the forest minister or the premier.

Having laid out the guidelines and entrusted preparation of the brief to Mahood and Keith Shaw, MacMillan departed on his annual trip on the *Marijean*, which had been dispatched through the Panama Canal to the Caribbean. He was appalled at the social conditions he encountered on some of the islands. In Jamaica, after lunch at a luxurious country club, he hired a car and toured some of the rural areas, recording his impressions in the ship's log that evening.

> It is conspicuous that the native population now crowded into the roadsides, gullies, steep hills and poorest land, and increasing rapidly, cannot be held off the large areas of best land in non-intensive use. Something will give and my guess: concessions will be forced on the large land owners, chiefly impersonal and non-resident.[6]

The parallels to BC were obvious to him: the province's forest land could be made available to individuals and small companies for the purpose of growing timber, as he advocated; or it could be turned over to big corporations, leaving a large population of landless and, someday, poor workers, as was the case in Jamaica. MacMillan came home more convinced than ever that the FML system had to be resisted, and he began to dispose of other obligations on his time. In July, he submitted his resignation as chairman of BC Packers, retaining a position as director, and was succeeded by John Buchanan.

By the time MacMillan got back to Vancouver, Sloan's public hearings had commenced and the entire FML system was under intense attack, along with the Forest Service and the big forest companies. Gordon Gibson, a Liberal MLA whose family had pioneered logging on the west coast of

Vancouver Island, had been ejected from the commission for arguing with lawyers representing the big companies. He followed this up by charging in the provincial legislature that the government was covering up wrongdoing in the issuing of FMLs. When he refused to retract his statement, he was ejected from the legislature. The next day, the government appointed a one-man commission under Judge Arthur Lord to look into Gibson's accusations. Gibson declined to appear at the hearing, which opened March 7, for fear that without his parliamentary immunity he might be vulnerable to slander charges. The only witness who testified was Chief Forester Orchard. A few years later, in his unpublished memoirs, Orchard claimed there were irregularities in the approval of BC Forest Products' FML #22 and that he had refused to sign the documents, as required by the Forest Act. But when Orchard appeared before Lord, he provided no evidence of wrongdoing, and the hearing concluded that same day, ruling that nothing illegal had taken place.

In MacMillan's view, it was clearly a time for prudence. By early summer it was obvious to him and others with an intimate knowledge of the industry that the system of allocating FMLs had been corrupted. He had no desire to become involved in any way in the scandal he could see was about to erupt.

On November 3, MacMillan presented his company's brief to Sloan. Once he had dispensed with the niceties of praising the Forest Service and the chief forester, he launched into a lengthy articulation of his vision for the development of coastal British Columbia. He painted a picture that managed to encompass the dreams and aspirations of a large proportion of coastal residents in a manner that has kept it relevant even forty years after it was delivered. MacMillan began by defending the big companies as necessary entities under current global economic circumstances. They were required in order to build and maintain the large efficient pulp mills needed to fully utilize the timber harvested. But, he argued, they should not be allowed to proliferate and assume dominance over the timber supply at the expense of small, independent loggers and sawmill operators. He defended the idea of Forest Management Licences, but said no more should be issued in the Eastern division, where numerous applications had been made by several large companies, including M&B. Then he went on to describe how large and small companies could survive, and thrive together.

> If no further Forest Management Licences are granted, we, in common with the other owners of converting mills in the Eastern division, will draw a portion of

our log and pulpwood needs from the open market which we will help to create and maintain. Similarly, the other applicants, if disappointed in their efforts to establish a preferred position, will have the same opportunity to compete for an undiminished supply of wood. . . .

It will be a sorry day for the Eastern division or elsewhere in British Columbia when forest industry here consists chiefly of a very few big companies, holding most of the good timber and good growing sites to the disadvantage and early extermination of the most hard working, virile, versatile and ingenious element of our population—the independent market logger and the small-mill man. . . .

There is every reason to avoid legislating or regulating the market logger out of existence. The Eastern division now has all the big mills and big companies that can be supported from its forests. There is no apparent good and sufficient reason why any of the large companies should be aided by Government policies to grow bigger at the expense of the smaller. Our forest industry is healthier if it consists of as many independent units as can be supported.

This is particularly true if these units, which run in number to thousands, are not dependent upon any of the larger companies. In a healthy forest society they would cover the whole range from fuel cutters, shakemakers, pole and piling producers, market loggers, small shingle and sawmills to the largest integrated multiproduct units. The division of this vast and important industry into two groups, on the one hand a few large units protected partially or fully for raw material by Forest Management Licences, and on the other hand a thousand or more smaller units who already see the end of the timber available to them, would furnish the motive for drastic political overturn. . . .

If there is to be a free enterprise system, there must be room left in which it can operate. Otherwise, we get a system of "share croppers". . . .

There is no visible applicant whose merits are such that he should get a Forest Management Licence at the expense of destroying very soon the existing investment and employment of other estimable and responsible residents of this district, many of whom were born and brought up here in the industry.[7]

In a series of specific recommendations, MacMillan elaborated on the social objectives of such a policy.

The object of this suggestion is to encourage the maintenance of permanent communities, roads, schools, hospitals, electric power and other services. The allowable annual cut will not decrease. As management improves, it will in the distant future increase somewhat. This crop will give employment every year to a continuing population, which can become as dependable an element in upcoast settlement as a farming community in the Fraser Valley. Many coastwise settlements which were started by fishing, mining, logging, have shrunk or disappeared. These coastwise population centres are most valuable to our society. Public Working Circles offer an opportunity to maintain permanent villages. The residents will know the business and know the characteristics of their home Public Working Circle. They will work hard to prevent fire and will be on hand to suppress and fight the common enemy.[8]

His alarm at the voracious appetite of the big companies for timber rights caused him to back away from the position he had taken at the first Sloan Commission on private ownership of forest land. "The consequences of a sale policy," he declared, "would be that the large companies would get most of the young Forest now left in Public ownership."[9]

MacMillan's proposals constituted a direct challenge to a gathering coalition of politicians, bureaucrats, technocrats, corporate fixers, union leaders and other vested interests trying to wrest control of the forests from the people who lived and worked in them. His presentation provided a dramatic moment in the forest policy debate that was rivetting public attention and commanding front-page newspaper coverage. For a month, it shifted the focus from the brewing political scandal to the real substance of the issue—who would get the right to manage and harvest the province's forests.

The response was instant and enthusiastic. Overnight, MacMillan became the hero and champion of thousands of people living along the coast, from Victoria to Prince Rupert. He, of all people, the biggest operator of them all, had provided the most cogent delineation of their aspirations ever presented on a public platform. Forty years later, his name and his arguments would still be evoked in defence of the "little man's" never-ending struggle against big monopolistic interests.

Those interests, as represented by the large companies with applications submitted for FMLs in the Eastern division, were apoplectic. The counterattack was led by Gordon Wismer, an elegant Vancouver lawyer and former attorney general in the Liberal-led coalition government, who represented an association of sixteen forest companies, thirteen with FML applications under consideration. He questioned MacMillan's motives.

> I put it to you, Mr. MacMillan, that in your position of advantage, sitting in your acres, that you are in the position of a benevolent monarch of the Vancouver forest district with a few crown princes, and you are throwing the crumbs to these other fellows.[10]

During several days of cross-examination before the Commission, MacMillan shrugged off these accusations and used the opportunity to elaborate on his vision of a sustained yield forest economy that was inextricably bound up with a large, diverse population of independent loggers, forest managers and processors, working in tandem with a few large companies. In the days that followed, the briefs of several large companies were quickly amended to oppose his views directly. Poldi Bentley, a vice-president of Canadian Forest Products, which had submitted a FML application for land on the northeast end of Vancouver Island, levelled the strongest attack. He said MacMillan's defence of the small logger was only a pretence. "His real purpose is to weaken the competitive position of everyone else but Mr. MacMillan," Bentley charged. "The number of truly independent loggers is so insignificant that I don't know what Mr. MacMillan is talking about when he says he wants to protect them."[11]

After MacMillan made his submission to Sloan, the commission hearings took on the legalistic complexion of a criminal trial. Argument and counter-argument continued for more than a month, until they were driven off the front pages by the sensational appearance on December 16, 1955, of a Vancouver lawyer, David Sturdy, who asked permission to submit evidence that Forest Minister Robert Sommers had accepted bribes to issue Forest Management Licences. Sloan refused to allow Sturdy to testify, and outside the hearing room the lawyer told reporters he had first taken the evidence to Attorney General Robert Bonner, who had refused to act on it.

The province was in an uproar. A few days later, Sommers sued Sturdy for libel, effectively stifling discussion of the charges. A relieved Sloan recessed the hearings for a two-week Christmas holiday. When they resumed in early January 1956, MacMillan was back before the commission for further cross-examination, and the major companies pressed their attack on his views. He was saddened in the midst of this debate to learn that John Gilmour, his chief supporter in the forest policy debate over the previous decade, had died in Montreal.

Meanwhile, in the legislature, both opposition parties launched a sustained attack on the government, delicately skirting the boundaries of the legal dispute between Sommers and Sturdy, and focussing on the role of Attorney General Bonner in the affair. The scandal was beginning to widen and threatened to bring the entire Socred government under suspicion.

In the midst of the growing uproar, MacMillan & Bloedel held its annual general meeting, at which MacMillan delivered a long talk on the state of the

Left to right: BC Premier W.A.C. Bennett, MacMillan and Forest Minister R.W. Sommers at the Alberni pulp mill, July 1955.

industry and the company, without reference to the Sloan Commission. At conclusion of his address, he startled shareholders by announcing his retirement as chairman, informing them that Hoffmeister would replace him, with Harry Berryman, a longtime HRME employee who had begun as a fireman on an Alberni Pacific locomotive, moved up to president. Close observers of the company noted, however, that MacMillan was not actually leaving; he continued to serve as a director and, more to the point, retained his position on the key committees, including a seat at the newly created Finance and Policy Committee. It was from this strategic post that he continued to exercise control over the company, at times relegating Hoffmeister and Berryman to what amounted to vice-presidential roles.

On January 30, MacMillan's protégé, Hector Munro, delivered the BC Forest Products brief to Sloan. It too took direct aim at MacMillan's proposals, leaving him alone among officials of big companies in opposing the issuance of more Forest Management Licences in the Vancouver district. The following day, under cross-examination, Munro was asked if his company had bribed Sommers for a FML. He denied it before Sloan cut off questioning.

On February 1, MacMillan departed on his annual trip south, hoping to leave behind the increasingly acrimonious debate in which the pros and cons of the new licensing system were becoming enmeshed in the growing political scandal. After cancelling the family component of the trip, he flew to Mexico City with Norman MacKenzie, Don Bates and Peter Larkin. A few days' sightseeing included a bullfight, which, MacMillan noted in the ship's log, he did not enjoy and left halfway through. They flew on to meet Wallie McCutcheon, John Buchanan and the *Marijean* at Lima, for another voyage to the Galapagos Islands. A few weeks later, when they returned to Lima—bringing with them eight live penguins for Vancouver's Stanley Park zoo—a telegram from Hoffmeister was waiting.

> SOMMERS RESIGNATION ACCEPTED BY BENNETT TODAY STOP WILLISTON NEW MINISTER LANDS FORESTS STOP BITTER DENUNCIATION OF LIBERAL PARTY AND STU KEATE [*Victoria Times* Publisher] BY SOMMERS WHO STILL HOLDS SEAT IN LEGISLATURE STOP TENSE ATMOSPHERE PREVAILS WITH RUMOURS EARLY ELECTION[12]

As the finger of suspicion began to point at other members of his government, Bennett had been forced to replace Sommers. But this move merely added fuel to the conflagration consuming the political life of the province and threatening to overwhelm the Sloan hearings.

In July, MacMillan appeared again in front of Sloan to deliver part of M&B's response to questions put to the company by the commission. In his presentation he addressed, one by one, the criticisms levelled at him by the heads of other companies seeking FMLs.

> A lot of effort has been spent and time taken up during the sittings of this Commission in an effort to persuade you, Mr. Commissioner, to a conclusion that public timber should be parcelled out to the selected few, selected on one basis or another, but all with the sole objective that the few should survive and the many struggle for their existence until they can struggle no more and cease to trouble. To this we are unalterably opposed. . . .
>
> Some of our large companies [he was referring to Canadian Forest Products] . . . do not even have to pause to think that this province gave them the opportunity which has enabled them in a few years and not entirely on their own merits to rise from small beginnings such as they would now deny to others. . . . They look upon this Commission as a heaven-sent opportunity to enable them to take a tow instead of continuing to work their passage. What they want is most favoured company treatment, favoured by the government, with the objective that they will be awarded all the timber they could ever need even though the result will be that the timber available for their present competitors will be

correspondingly reduced and other wood consumers less powerful, less crafty, but just as good Canadians, will be put out of business.[13]

In conclusion he issued a dramatic and, as it turned out, accurate prediction of the consequences of issuing more FMLs in the district.

> A few companies would acquire control of resources and form a monopoly. It will be managed by professional bureaucrats, fixers with a penthouse viewpoint who, never having had rain in their lunch buckets, would abuse the forest. Public interest would be victimized because the citizen business needed to provide the efficiency of competition would be denied logs and thereby prevented from penetration of the market.[14]

This, his last appearance before Sloan, effectively ended MacMillan's participation in the decade-long forest policy debate. For several months, the participants and the public waited for Sloan's report. By the time it came out, MacMillan, then in his seventieth year, was caught up in other, more pressing events.

Given the uniqueness of MacMillan's position during the two Sloan hearings, and through the intervening years, several questions arise concerning arguments he and like-minded participants advanced. They come down to two issues: was MacMillan right in his predictions of what would occur if the Forest Management Licence system was adopted; and was he sincere in his support of the small, independent business sector?

Considering the state of the BC forest industry today, after operating for almost fifty years under the policies adopted in 1947, the answer to the first question is an unequivocal yes. Following a period of growth and expansion, the industry has fallen into a steady decline. The timber supply, which was drastically overcut for many years, is now insufficient to sustain the capacity of existing mills. The number of jobs the industry provides is decreasing. And public revenues, which have been used for decades by political parties of all stripes to bribe the voters into electing or re-electing them, have declined. In fact, between 1982 and 1987, it cost $1.1 billion more to fund the Forest Service bureaucracy than was collected in forest revenues.

MacMillan, along with Mulholland, Gilmour, Liersch and a few others, understood the economic dynamics of the forest sector. Orchard did not. What MacMillan and his associates knew when they began criticizing

HR MacMillan

Orchard's plan was that without the security of ownership, no one would invest their own money in managing forest land. Events have shown they would have been stupid to do so. Governments change every few years, politicians change their minds even more frequently, and bureaucrats have an infinite capacity to create more rules and regulations. The opportunity to grow timber on Crown-owned land in Forest Management Licences did not appeal to anybody who acquired one. Apart from the fact that licensees had no idea what would become of their licences, it was cheaper to buy timber or cutting rights from someone else than to grow it.

MacMillan argued that the future of forest management in BC depended on the willingness of the private sector to reinvest a portion of its revenues back into the forests. After many years' experience working in the public sector, he knew governments could not be relied upon to collect the full value of the stumpage and return an adequate portion of it to the forests. In spite of widespread public perceptions, forest management by Canada's provincial governments has been woefully inadequate.

MacMillan had a pretty good idea what both sectors would do with the vast wealth the forests were capable of generating. Governments would spend it on popular non-forest programs. Industry, particularly during good market periods, would use its profits and its ability to raise capital to build bigger mills and to buy each other out. He and his associates predicted, quite accurately, that forest company profits would be used to expand by acquisition. They knew that in countries where forest land was privately owned, profits were routinely reinvested in the owners' forests, because it was a safe place to store surplus funds and would provide a good rate of return. Lacking that opportunity in BC, the obvious choice for a forest company wanting to stay in business was to buy out a competitor.

It is reasonable to assume that if the policy alternative proposed in great detail by MacMillan, Mulholland and Gilmour had been adopted, the present condition of the BC forest sector would be quite different than it is. Two types of financial activity would have taken place to a much greater extent than they have. Large amounts of capital would have been put back into the forests, and BC would now have several thousand people employed as stewards of the new forests that would have grown up after logging. The future timber supply would be much bigger than it is, and there would be less need to log areas the public would like to reserve for other uses. As well, there would

318

have been a tendency to extract more value from the available timber supply by manufacturing the most valuable products possible, instead of trying to compete in world markets by selling the largest possible amount of timber at the lowest possible price.

Concentration of control over cutting rights has occurred to a degree which even MacMillan could not imagine. In 1954, the ten largest companies in the province harvested 37 percent of the timber cut in BC; in 1990 the ten largest companies logged 69 percent. The performance of this narrowly controlled industry is relatively poor. In 1984, the value added through processing to a cubic metre of timber in BC averaged about $56; the average in the rest of Canada was more than $110, and in the US it was $174. In BC 1,000 cubic metres of timber created 1.05 jobs, compared to 2.2 jobs in the rest of Canada, and 3.5 jobs in the US.[15] If anything, MacMillan's predictions about the consequences of the FML system were understated.

The other question—was MacMillan sincere in his support of the small, independent business sector?—relates to the charges laid by MacMillan's critics, such as Poldi Bentley, who accused him of hypocritically supporting small business in order to protect his own interests. There is no question that MacMillan was trying to protect M&B in his presentations; he would have been derelict in his duty to shareholders and employees to have done otherwise. But it should be remembered he offered to withdraw the M&B application for a FML on the inside of Vancouver Island if his competitors agreed to do the same and buy timber from independent loggers.

MacMillan's position was fundamentally different than his critics. His concept of the best industrial structure for the BC forest sector was one that included a healthy range of firms, from small to large. He had a genuine belief in a system of free enterprise that enabled an individual to enter the business at the lowest level, in terms of size, and proceed from there according to his abilities and desires. He believed attempts by government and state bureaucracy to regulate and control this enterprise were at best unproductive, and at worst an evil of the sort that lay behind German and Russian totalitarianism. He also believed the rise of big corporations was a dangerous development. His opponents, including Orchard and the heads of many of the big forest companies, disagreed. Most of them, including Orchard, were not socialists but nascent corporatists. They believed in a province dominated by large corporations, big government and—eventually—labour unions. Almost from the outset, they co-opted W.A.C. Bennett's Social Credit government, which had been put into power by the province's small-business and working sectors. Two generations of Bennetts stayed in power by promising to protect free enterprise interests, while acting at the behest of the big multinational corporations which emerged after World War Two.

If MacMillan erred, it was in thinking that the real threat to free enterprise in BC came from the "socialists." But it does appear that he also understood the true nature of the Social Credit Party. Despite its determined and effective opposition to the CCF, which he supported, he never was enthusiastic about Bennett or any of his Socred colleagues.

MacMillan's articulate support of what he liked to call "citizen business" is consistent within the context of his life. He had the vision, the public platform and the financial security to say what he thought. The fact that what he said contradicted his image as a buccaneer capitalist should not be surprising. MacMillan was full of contradictions.

While Sloan prepared his report, the case of Robert Sommers dragged on, unresolved. By sheer stubborn stonewalling, Attorney General Bonner managed to delay the inevitable step of hauling Sommers into court, even after a police investigation into Sturdy's accusations concluded that the former lands minister should be charged. Bonner held off until well after the September 1956 provincial election, in which the Socreds, including Sommers, were returned to power. Apparently the electorate had paid scant attention to the FML scandal and were prepared to give Social Credit the benefit of any doubts.[16]

From MacMillan's point of view, the scandalous aspects of this farcical affair were headed toward tragedy, and he spent most of the winter travelling outside the country. After a trip to Japan and China, he returned to Vancouver for a brief stay during which he advocated recognition of Red China—well before the idea gained common currency. After a few weeks in Europe with Edna in early 1957, he made his annual pilgrimage to Mexico, spending the first month there with his family. Much of the year's trip was spent at Acapulco, where the Southams learned to water ski, much to MacMillan's amusement.

His family had grown considerably. Marion had borne two more

MacMillan and his wife Edna on the *Marijean*.

children and Jean seven, giving him ten grandchildren. It was rare for any of the children to be included on the *Marijean's* trips off Mexico, although they often were included in day trips on the yacht around Vancouver. Johnny Lecky was the only grandchild MacMillan paid any particular attention to, and during the 1950s he occasionally included the youngster on fishing and hunting trips. Seven of the grandchildren were girls, which he found of little interest, and Jean's two sons, Gordon and Harvey, were too young to be included in their grandfather's activities. His daughters were both devoted to him; even after they had acquired husbands and children of their own, he continued to be the central figure in their lives. They both travelled abroad with him regularly and placed his interests first among their priorities.

The MacMillan smile, often seen after he had bested a competitor.

Although MacMillan refrained from expressing disappointment at not having a son, it was clear in his relationships with younger colleagues that he gave a great deal of thought to choosing successors and agonized over not having a male heir. The subject of succession was much on his mind, and occasionally a matter of discussion during the 1957 Mexico trip, as the log of the *Marijean* revealed.

Conversation took us back to the Cromies, General Stewart [the railway-building co-founder of Bloedel, Stewart & Welch] and A.D. McRae [head of Canadian Western Lumber]. The Stewarts and McRaes led Vancouver society when Shinnie and Prentice [Bloedel] came to Vancouver about 30 years ago, and where are they now—no survivors. And many others of that era who were on the eve of the greatest opportunities but reaped before they sowed and neither equipped themselves nor learned to work. Prentice remembers Bob Cromie the elder from 1911 when he looked after General Stewart's business affairs and, at the beginning of the business, the office affairs of BS&W when they opened up at Myrtle Point. Then the PGE trouble and Stewart, to fight a political battle to get out of his contractual liability to build and pay for the PGE, bought a newspaper and turned it over to Cromie temporarily to independently support Stewart. After the battle was won and John Oliver, Premier, released Stewart from his liability, Stewart expected the newspaper back, but he had no proof of title and Cromie, a tough guy, kept it. Now the very large and prosperous Vancouver Sun. Stewart as the political fixer and front man for Foley, Welch & Stewart got as his share over $25,000,000 and lost it all in the next years due to poor judgement, the wrong kind of friends and lack of business ability. He died insolvent. Aristocratic Vancouver of those days, at least the most flamboyant members, left nothing but a bad example.[17]

Now in his early seventies, MacMillan was becoming concerned about the example he would leave. The lack of a son in MacMillan's life offers insight into the relationships he established with some of the bright, hard-working young men in his companies. And while the rules agreed upon with VanDusen may have precluded Crown Princes following the route to top positions, there were certainly court favourites, including Hoffmeister and Shaw.

One of the favourites had been Hector Munro. Starting with the export company, then accompanying MacMillan to Ottawa for the war years, Munro had returned as MacMillan's hand-picked manager of BC Forest Products. Electing to stay with that company when the management agreement was terminated, he was implicated in the growing FML scandal. When and how much did MacMillan know about Munro's activities? MacMillan's personal and corporate papers are of no help in illuminating the matter. Given his deeply ingrained habit of communicating his thoughts and ideas in letters or memos, it is hard to believe he never wrote about it, even in passing. Written references to the affair may have been purged from his files by Dorothy Dee after he died. The fact that Wallie McCutcheon, one of his closest friends, was intimately associated with BCFP and must have known how it obtained FMLs, suggests MacMillan must have had some knowledge of events as well. That he would approve of such tactics is another matter.

When MacMillan returned from Mexico, the legislative session had resumed and the opposition was back flaying the government over Sommers's activities and Bonner's refusal to charge him. The issue dominated the political agenda through the spring and summer.

In July, Sloan released his report, a lengthy two-volume compendium in which Forest Management Licences were endorsed unequivocally. The report was notable primarily for what was not mentioned. There was no reference even to the possibility of irregularities in the allocation of FMLs. Nor was there the slightest acknowledgement of any of the major points made by

MacMillan, who wryly noted later that it was if he had never appeared before the commission. His arguments that no more licences should be issued in the Vancouver forest district were not referred to once, even though his ideas represented a large body of opinion and were a perfectly credible option. Instead, Sloan recommended no more FMLs be issued for five years to allow for the completion of administrative details on the licences already awarded. The only exception to this rule, he suggested, was to allocate licences to some of the large coastal companies in the Vancouver district—the very thing MacMillan had resisted.

Sloan ignored almost entirely the plight of small independent loggers. Not only did he dismiss their claims for timber, but he said Forest Management Licensees should be allowed to obtain timber in Public Working Circles—the government-managed areas reserved for independent market loggers—if they agreed to hire independent loggers as contractors. It was as if all of the independent loggers' appeals, and MacMillan's eloquent defence of their position, had never been uttered.[18]

By now the royal commission was inextricably bound up with the Sommers affair. Sloan's attempts to separate forest policy issues and political

MacMillan during a CBC television interview at his home in Vancouver, September 1956.

corruption were further undermined on November 1, 1957, when he accepted a request from Bennett to serve as a one-man royal commission to investigate the Sturdy allegations which had precipitated the controversy almost two years earlier. This commission ended in a fiasco after thirty minutes of hearings, when C.D. Schultz's lawyer successfully argued the government had no right to set up a royal commission to investigate criminal charges. Three months later, Sloan left the bench and become the government's special advisor on forestry—at more than three times his judicial salary.

Finally, ten days after the collapse of Sloan's ill-fated investigation, the police arrested Sommers on charges of conspiracy and bribery. They also charged Wilson Gray, Schultz, and, after further investigations, the companies using Schultz and Gray's services, including BC Forest Products. After a lengthy, sensational trial, Sommers and Gray were convicted and sentenced to five years in prison. During the trial, prosecutors questioned Hector Munro and Trevor Daniels, the BCFP controller. Trained in the austere traditions of the British accounting profession, Daniels was unable or unwilling to give the answers the company's senior officers had instructed him to deliver, and admitted BCFP had put up $30,000 to bribe Sommers. A few days later, on December 2, Munro took his own life.

The death of one of his longtime favourites was undoubtedly a severe blow to MacMillan, although no evidence of his reactions can be found. It came near the end of a particularly traumatic year, the disappointing Sloan report being one of the low points. M&B's profits dropped by almost half during the year, and share prices fell from a high of $48 to $22.50. Yet for MacMillan, the year was far from over.

While Munro had been a favourite, MacMillan's all-time fair-haired boy was Bert Hoffmeister, who by all accounts became the son MacMillan never had. MacMillan had gone so far as to tell him so, earlier that fall during one of their regular overnight hunting trips to Kirkland Island. When Hoffmeister revealed he had recently been offered the position of Canada's RCMP Commissioner, MacMillan advised him not to accept, saying his future was assured and he was needed at M&B.

Eleven days after Munro's death, MacMillan called Hoffmeister into his office and, with tears running down his face, fired him. Everyone in the company, most of the local business community and the business press, not to mention Hoffmeister himself, were stunned. MacMillan never offered an explanation.

Because the consequences of this action for the company were enormous, numerous theories have been offered to account for it. Almost forty years later, Hoffmeister himself could not fully explain the event, and handwritten notes MacMillan used to steer himself through the ordeal only refer to the chairman's "health problems."[19] But Hoffmeister had no health problems, and was quick to remind MacMillan that the company's medical advisor had just given him a complete physical examination which showed he was in excellent health. One theory for the firing is that MacMillan learned from his daughter, Jean, about rumours Hoffmeister was having a clandestine affair with a prominent Vancouver woman during his regular Wednesday afternoon ski sessions at Grouse Mountain. In their final discussion, MacMillan made reference to Hoffmeister's skiing as evidence of ill health. Hoffmeister pointed out that skiing was what kept him in good health. When asked many years later, he dismissed rumours of an affair.[20]

A second theory is that his dismissal had something to do with the Foley family's desire for a "neutral" chairman if and when the off-again-on-again merger discussions between M&B and the Powell River Company came to fruition.[21] Others think cost overruns for the Alberni pulp mill expansion, which had been kept from MacMillan, contributed.[22] Hoffmeister's own belief is that a significant factor was an explosive encounter between himself and MacMillan at a Finance and Policy Committee meeting a few weeks previously. He had been in the midst of requesting that committee members check in with divisional managers before visiting company operations, when MacMillan interrupted. His face ashen and his fists clenched, MacMillan shouted at the chairman, saying he had no right to tell him when he could visit "my outfit." From then on, MacMillan never dropped into Hoffmeister's office for a casual talk, as had been his habit. But this sort of outburst was not out of character for MacMillan, whose temper was well known within the company, so it does not seem credible that this incident alone was the cause.

Ralph Shaw has a simpler explanation. He feels the firing was the result of several factors, including MacMillan's reluctance to give up authority. "When H.R. and Van told Bert and me we were now in charge, I didn't take him seriously," Shaw says. "Bert did."[23]

Whatever MacMillan's reasons, Hoffmeister was finished at M&B. He declined the offer of a seat on the board and, after receiving a generous severance payment, took a long vacation and went on to other prestigious jobs. For MacMillan, and for MacMillan & Bloedel Ltd., the immediate problem was not so easily resolved. The precipitous firing of Hoffmeister left the company without a successor for the position of chairman. Berryman had resigned as president for health reasons a few months previously, and been replaced by Shaw. The senior directors—essentially MacMillan, VanDusen and Bloedel—were faced with a major difficulty. They were the last of the proprietary managers in the company, and none of them had heirs who had been groomed for the job. Although an entire generation of managers, of which Shaw and Hoffmeister represented the cream, were brought up through the ranks and meticulously trained to run the company, none of them were professional managers in the modern sense of the term. And while several of these home-grown managers were capable of filling the president's position, that of chairman required qualities other than administrative ability and a lifetime of experience in the company. The real role of the chairman was to replace MacMillan as the company spokesman, its representative and ambassador in the wider world of politics and public attention. Hoffmeister, the war hero, had the potential to fill that role, although with MacMillan around he never had the chance. The search for his successor was extended outside the company and, eventually, the industry.

Bloedel and MacMillan's final choice astounded the business community. On January 9, 1958, it was announced that J.V. (Jack) Clyne, a BC Supreme Court judge with no business experience, was the new MacMillan & Bloedel chairman. His appointment was as inexplicable as Hoffmeister's dismissal. MacMillan would later admit it was the biggest mistake of his life.

chapter 21

Succession

Why MacMillan picked Jack Clyne to head MacMillan & Bloedel in 1958 is one of the great mysteries of Canadian business history. That Clyne also became chief executive officer is even more puzzling, considering his complete lack of business experience.

Unquestionably, Clyne was a colourful character with a powerful personality. Like MacMillan, he had lost his father at age two. Raised in Vancouver, he was educated at the University of British Columbia and in England. In university he was an accomplished athlete, actor and student leader. After returning to Vancouver from studies in London, Clyne quickly established himself as the city's top marine lawyer. It was in this capacity that he came in contact with MacMillan while performing legal work for Canadian Transport Company. At the end of World

War Two, he headed a commission overseeing the Canadian shipbuilding industry. He turned down the presidency of Canadian National Railways and accepted a position on the British Columbia Supreme Court. Clyne was witty, erudite and an avid reader. He had never had a great deal of money and, having come in contact with the wealthy and powerful, was very ambitious. His formidable reputation was exceeded only by his own high opinion of himself.

A factor in Clyne's appointment undoubtedly was his friendship with Prentice Bloedel, with whom he regularly played poker.[1] Still, it was not Bloedel for whom a successor was needed, it was MacMillan; and it was MacMillan who made the choice that was later approved by Bloedel and VanDusen. The company needed a distinguished speaker and prominent citizen to fulfill the position of chairman, and in the mind of Bloedel, and perhaps to a lesser extent MacMillan, Clyne fit the requirement. There is no indication they expected he would ever serve as chief executive officer.

Clyne's own version of events is not particularly reliable. His autobiography, *Jack of All Trades*, is more self-serving than most and is full of errors,

Left to right: J.V. Clyne, W.J. VanDusen, Ralph Shaw, Prentice Bloedel and MacMillan, at Comox wharf, 1958.

distortions and outright untruths. For instance, in his account of the initial discussions with MacMillan that led to his appointment, Clyne says he first turned down the offer and over a period of three months was reluctantly persuaded to accept it. In fact, slightly more than three weeks passed between the time Hoffmeister was fired and Clyne's appointment was announced. Clearly, he jumped at the offer.

Two weeks after Clyne's appointment, MacMillan left on his annual Mexico excursion on the *Marijean*. In his absence, Clyne began to flex his new muscles. At the time the West Coast industry was bogged down in a pulp and paper strike, the first major labour dispute in the history of this sector, which had begun the previous November. It had been a bad year for strikes in BC generally, and this one was perhaps the most bitter dispute of them all. Although negotiations for the pulp and paper companies were being conducted by the newly formed Pulp and Paper Industrial Relations Bureau, and were at a delicate stage, Clyne chose labour relations as the subject of his first speech as M&B chairman. He told a Vancouver Board of Trade lunch meeting that the real problems in the pulp and paper dispute were the fault of the union leaders. His antagonistic approach made it quite clear to labour leaders M&B was adopting a drastic new style of labour relations.

Only days later, Clyne made one of his first major corporate decisions, incurring the long-distance wrath of MacMillan in the process. It was their first, but certainly not their last, significant disagreement. The company was in the process of borrowing $50 million for capital expenses. Until his departure for Mexico, MacMillan had been handling negotiations. He had given his proxy to Clyne to vote on the matter at the directors' meeting which would decide the deal. One of MacMillan's longstanding principles was to borrow money only in Canada. In his absence, Clyne decided to raise $10 million of the required funds from US lenders, who were offering a one-half percent lower rate than Canadian financial houses. MacMillan's reaction, when he learned of this decision, was to cable Clyne, cancelling his proxy and insisting the money be borrowed in Canada. He cut short his vacation to attend the directors' meeting. VanDusen met him at the Vancouver airport and told him Clyne was determined to have his way and would force the issue at the meeting the next day.

It must have been a long, difficult night for MacMillan. As a major shareholder he could easily have mustered the power to overrule Clyne. But as he well knew, Clyne would have no choice but to resign and, on the heels of Hoffmeister's abrupt departure, his resignation would damage the company's reputation immeasurably. MacMillan backed down. Clyne now had the upper hand, which he proceeded to use.

It was Clyne's contention that from this point on his relations with MacMillan were friendly and agreeable. Except for a maudlin description of his last meeting with MacMillan, the night before he died, Clyne's autobiography essentially dismisses MacMillan's presence and his role in the company from this time on. In fact, the situation was not as straightforward as Clyne wanted the world to believe. MacMillan lived for another eighteen years, and he spent much of it keeping a sharp watch on M&B's new chairman. He retained his seat on the board and served on various committees, including the critically important Finance and Policy Committee, along with VanDusen, Bloedel, Shaw and Clyne. Accepting the fact he could not confront Clyne, he set about to educate and influence him, in part through a steady stream of memos on a wide variety of subjects.

One of these, an eleven-page memorandum written in December 1958, titled "Looking Back on This Business," is perhaps the most comprehensive analysis of the company he ever produced. It began with an historical overview of the principles and objectives of the company, followed by some lessons learned, and then a long section on employees. By this time Clyne had locked horns with many of the company's senior staff. MacMillan's ideas on handling personnel were an elaboration of his longstanding admonition: "analyze, organize, deputize, supervise"—to which he occasionally added—energize and excise.

> Select beginners and other new employees carefully by selecting from the ranks, or bringing in new and appropriately qualified persons. Rugged constitutions important as well as other qualities. Occasionally, when opportunity offers, bringing in experienced persons from other organizations cross fertilizes the home grown.
>
> Maintaining healthy competition for promotions. Seniority to be disregarded except with candidates equal. Otherwise frustration set up against brightest of the younger and staff average age rises too fast. A clear road ahead for open competition for top jobs in HRM Export Co and M&B was made by owners agreeing their sons and sons-in-law would not be taken into the company. This was to avoid nepotism, encourage competition amongst rising professional management, encourage selection of, and rewards for, those selected as the best qualified.

Avoiding appointments in apparent permanence to important positions before appointee has proved himself. If appointee unsuitable, correct quickly.

Promoting or moving sideways or otherwise without fear, affection or favour. The Company or function should not be weakened and only rarely bent to fit the man.

Moving the most promising looking occasionally from function to function so as to broaden their experience and to cross check on qualities.

Deputizing a duty does not bury it or put it out of sight. It is necessary to learn how the deputy is performing and, if needed, direct and help him. Otherwise, too much policy will by default come from the bottom instead of from the top. In other words, the company sometimes will be run from the bottom up. The higher executive still carries his responsibility and authority to be exercised to maintain standards. Study Monty of Alamein. He was a successful leader and manager. After he analyzed, organized and deputized, he supervised to the end that his standards prevailed.

Beware of nepotism, apple polishers, office politicians and pets. I have run into many and know when I did wrong and when I did right. I did right when I got rid of them.

Some companies are handicapped by having field jobs to fill, bush jobs where wives complain. If a wife complains, never yield. If the family can't stick it out where the job is, let the man quit. If he succeeds in politicking to get moved to head office, or to a good town before he has normally earned the promotion by good performance, as his contemporaries must do, the fat is in the fire and the disease will spread.

Only the very best men should be selected and retained in managerial positions anywhere, and emphatically so at head office. The prominence of strikingly competent persons in these positions radiates immense leadership and influence. There is no equal way to build up morale.

Supervision and contact must be continuous top to bottom to produce reliable, productive, brainy performance, good habits and quick communication of news, threats, market trends and action accordingly. Otherwise, insufficient authority will be exercised at some levels and too much at others. The tendency to sit back in head office, or other offices, and await reports from subordinates, should be supervised out of existence. . . .

The top men, including the Chairman, the President, as well as responsible VPs, should show up around the operations without laying on a do, and without any of the drill as in a super-organized institution like an army. The local top man is sufficient to meet and go around with. Otherwise, our "tops" will have only hearsay, will be slow in sizing up "comers", and practices, will come to rely too much on the typed word. So the organization will slow down and stiffen. A few years can easily see walls built around jobs, defensive thinking, which I have observed more than once in other companies—a detriment to growth. Emergencies sometimes demand that at any level in an organization, from the top down, an executive, without violating protocol, has to have the road open for quickly getting ideas and facts and bringing about action, thus strengthening rather than weakening the fabric. This is the art of supervision—perhaps a difficult but also an essential art.[2]

The memo concluded with a long section on the board's Finance and Policy Committee, which MacMillan saw as the key committee because it represented the directors. They, in turn, represented the shareholders whose interests, MacMillan believed, must be of prime importance to management. The F&P Committee, as it was known, directed management and, now that MacMillan was not part of the executive hierarchy, it was his major avenue of influence. Judging by subsequent events, much of his advice appears to have gone over Clyne's head.

By the time of Clyne's arrival, M&B had become the largest integrated forest company in Canada, and one of the biggest in the world. It produced 550 million feet of lumber a year, 320 million square feet of plywood, 400,000 doors, 390 squares of shingles, 200,000 tons of newsprint, 300,000 tons of pulp, 60,000 tons of kraft paper, and 4,000 tons of paper bags. Including its own output, the company marketed a third of BC's lumber exports. It controlled close to a million acres of forest land, operated close to 200 miles of logging railways and 400 miles of logging roads. Most of these operations were concentrated on Vancouver Island and the Lower Mainland, and the company maintained sales offices throughout the UK, Europe, the US and Asia. Its subsidiary, Canadian Transport Company, shipped M&B products, and those of others, all over the world. In 1956, the company had earned $20 million in profits, although poor markets had reduced earnings to $8 million in 1958.

Throughout Clyne's settling-in period, merger talks with principals of the Powell River Company continued at a subdued pace. The idea of combining these two companies had been in the air for years. In 1953 MacMillan had instructed his manager of forestry operations, Ian Mahood, to make a discreet investigation of Powell River's timber supply—and was surprised and upset when Mahood reported it was inadequate.[3] In 1956, Powell River's president, Harold Foley, approached Hoffmeister to propose merger discussions. Extensive talks were held by the Foley brothers, VanDusen and MacMillan, leading to a decision in 1957 not to merge. However, the two companies did agree to joint ownership of Martin Paper Products, a Winnipeg-based box manufacturer.[4] Clyne later claimed he initiated the merger talks, but this is not true.[5] The discussions he entered into with Harold Foley in 1958 were merely a continuation of a long courtship.

The Powell River Company was the oldest pulp and paper producer in the province, and its newsprint mill on the mainland coast northwest of Vancouver was the largest in the world. It was a closely held family company that became a public corporation after World War Two. Harold Foley was a nephew of Joseph Scanlon, a company co-founder, and had brought an atmosphere of Southern grace and elegance to the rough-hewn Vancouver business scene when he moved up from Florida in the late 1930s. His patrician looks concealing a tough, astute business capability, Foley was known as "the Silver Fox." He and his younger brother, Joe, who joined the company later, were highly regarded on the local business and social scene. The Powell River Company was run as a family concern. The owners spent a great deal of time at the mill and logging operations, talking freely with employees who were treated in a familiar, paternalistic manner. The Powell River management style was the opposite of that at M&B. A manager was given a job to perform and left alone to do it, whereas MacMillan's approach was to hire good people, then ride them hard.

Three-quarters of Powell River's business was in pulp and newsprint, with the rest in some recently acquired Vancouver-area sawmills. The company's timber supply was extensive but scattered, much of it far up the coast on the Queen Charlotte Islands. Given the nature of pulp and paper markets at the time, it was clear the two companies needed either to merge or to square off for serious competition, with M&B likely to be the winner. Both were holding discussions with other potential merger partners in the fall of 1958 when Clyne and Harold Foley resumed talks. MacMillan played a distant role in the discussions. He was good friends with Foley, as were his daughter Jean and her husband Gordon. He made it clear to Foley that in a merger, M&B would be the senior partner, with Clyne as chairman.[6] In most respects, however, he worked through Clyne, advising him and supporting his decisions.

By the spring of 1959, discussions had become serious and on June 30, Clyne and Foley signed a memorandum of agreement outlining the mechanics of the deal, the redistribution of power, the naming of senior officers and a new corporate name. To effect the merger, Powell River took over

M&B, its shareholders exchanging seven shares for three M&B shares. Each group elected an equal number of directors, with Clyne appointed an additional director and chairman. It was agreed he would act as a neutral chairman, and would not favour one group over the other. Harold Foley was vice-chairman, Joe Foley president, and Ralph Shaw dropped back to executive vice-president. MacMillan was to be honorary chairman for a one-year term. The cumbersome character of the merger was reflected in the new name—MacMillan, Bloedel and Powell River Limited.

The choice of a new name was memorialized the following winter off the coast of Mexico in an entry in the *Marijean* log by biologist Peter Larkin.

Said M to B one sunny day
The time has come to merge
Together we can weather out
The economic surge.
Now wooden you and wooden me
A splendid splicing make
With each providing half the work
And each gets half the take.
Said B to M, I am agreed
A merger would be swell
Permit me to suggest the name
MacMillan and Bloedel.
The years went by, the friendship grew
And times were roly-poly
If two are grand, three are superb
And so they took in Foley.
It all sounds grand but hear me out
There's one thing gripes my liver
Couldn't they get a shorter name
Than MacMillan, Bloedel and Powell River?

By early October, both companies' shareholders had approved the deal. In November, a new executive committee replaced the M&B Finance and Policy Committee. Members included Clyne, the chairman, with MacMillan, Bloedel, VanDusen and Shaw from the M&B group, while the two Foleys, Conley Brooks and George O'Brien represented the Powell River group.[7] Conflicts appeared almost immediately, even before the first meeting of the new board in January 1960. Clyne and Harold Foley disagreed openly over the respective roles of senior people within the two groups. Clyne, instead of being the neutral chairman he was intended to be, took a strong, confrontational stance against many Powell River people, particularly Foley.

Disagreements festered and grew throughout the year. Foley and Clyne were barely speaking to each other, with Clyne accusing the vice-chairman of trying to run the company. On one occasion in August, he told Foley there was not room for both of them in the company and suggested he, Clyne, might as well return to the bench where he would be happier. Foley convinced him this would have serious consequences for the company.[8] By the end of the year, several senior Powell River employees had resigned. Early in the new year, John Liersch, the Powell River group's most senior employee, quit when Clyne decided he would, henceforth, report to a lower-ranking member of the M&B group. Sometime after he quit, Liersch sent Foley a long, personal, handwritten letter, full of sadness for what he saw as the company's lost opportunity to fully utilize the talents of both groups. His assessment of Clyne was remarkably lacking in bitterness.

> I have known Jack Clyne for a long time as you have. I respect his ability and knowledge—but the ability and knowledge learned in one line of endeavour cannot be always applicable under different conditions. Nor can the secrets and methods of industry be acquired overnight. Jack wants—and is rushing it too fast—to be known as a "Giant of Industry, Formerly one of Canada's Outstanding Jurists." His ambition is laudable, but a smile and a wave once in a while might contribute more than he realizes to its attainment.[9]

Joe Foley, who had expressed his concerns to Clyne about the handling of Powell River staff, quit as president in mid-April. He was convinced a plot had been hatched the previous winter on the *Marijean,* aimed at undercutting the power of the Powell River group, and that "HRM is calling the plays."[10]

A further point of contention developed between Clyne and Harold Foley over ownership of the Powell River Sales Company, which was owned by the main company's major shareholders. At the time of the merger, Clyne claimed, Foley told him the sales company was held in trust by the shareholders and should be treated as an asset of the company. When Clyne asked him to expedite transfer of ownership of Powell River Sales to the merged company, Foley said he would urge the new company's directors to pay the sales company shareholders $4 million for having held the shares. Clyne raised the issue at the next directors' meeting, recommending Foley be fired

MacMillan with Harold Foley at the Powell River Company newsprint mill.

as vice-chairman and removed from the executive committee, leaving him with his seat on the board of directors. With several members of the Powell River group absent, the measure was approved. During May, several more Powell River directors resigned, along with more senior staff.

At the end of May, Harold Foley sent Clyne a letter in which he laid the problems within the company on Clyne's failure to act as an impartial chairman. After listing several of Clyne's decisions, he concluded:

> Your prejudiced attitude and actions since the amalgamation, contrary to all our understandings, leave me no alternative but to conclude that I can be of no further use to MacMillan, Bloedel and Powell River while you remain as chairman of its board of directors. . . . I do not now consider myself a director of MacMillan, Bloedel and Powell River Limited."[11]

He sent copies of the letter to the Vancouver newspapers, which gleefully printed it on their front pages. The city's business and social elite were shocked by such naked animosity. Clyne named himself chief executive officer, Ralph Shaw was appointed president, and another M&B man, Ernie Shorter, became the sole executive vice president, leaving the company's management totally in the hands of M&B people, and its board dominated by M&B directors. The Foleys had only one card left to play, which they were to sit on for almost a year.

It is difficult to determine MacMillan's role in these tumultuous events. There is little to indicate he was actively engaged in purging the Powell River people; nor is there any evidence he took steps to restrain Clyne. The same can be said of Bloedel and VanDusen. The three of them appear to have sat back and let events take their course, at least until this point. Subsequent actions by the Foley brothers would require their active participation.

Throughout the merger, MacMillan had numerous other matters claiming his attention. The implementation of the Forest Management Licence system by the Forest Service bore watching. Apparently the government viewed these licences primarily as a means of attracting investors and raising revenues, and was not particularly concerned with the long-term management of forests. In early 1958, MacMillan wrote to VanDusen, who was on vacation:

> Bonner [the Attorney General] tells Clyne that he intends to reduce the tenure of existing FMLs to 21 years. This is still a secret. We regard it with horror, and must oppose it. I think that it would have very serious consequences on the good faith of provincial governments here.[12]

This decision, when implemented, did more than undermine faith in government; it erased any doubts about the real purpose of Forest Management Licences. From that point on, all licensees treated them as a means of acquiring mature timber and not as an opportunity to grow the next crop. MacMillan was one of the few who objected to this and other ongoing forest policy developments of the Bennett government, which had little interest in forest management but a lot in rapid economic development. He wrote to VanDusen:

> There is a strange unreadiness on the part of the big companies to talk about any of these things, which bothers Foley and bothers us. We are inclined to put it down to Crown [Zellerbach's] and Rayonier's confidence in their own special "pipe lines," and lack of leadership at the moment at BCFP. During the next two months, Clyne and Ralph will do what they can. Bennett needs more money to expand his high expenditure programme, and he can only get it from those who have got it—unless he raises the consumption tax still further, which I do not think he will do yet.[13]

It was increasingly apparent that MacMillan's stance before the Sloan commission had cost the company a measure of its influence in Victoria, and that its competitors were gaining ground. He urged Clyne and Shaw to protect the company's interests aggressively, including the right to compete for timber within Public Working Circles—in apparent contradiction to his earlier contention these areas should be left for independent market loggers.

> We must be thorough and strong in our efforts before Sloan [now working as a government advisor] to maintain our right to bid on Timber Sales in Public Working Circles. We must move thoroughly and energetically to prevent any part of any PWC going to Powell or other FML application. I would oppose Powell on this point with all the arguments at M&B's command. If PWC raided now a precedent is created to open to any—friend or thug. M&B's future depends on competitive wood supply. If chief competitors get 100% of their requirements underwritten by Government [in FMLs], if PWCs are raided, if M&B can't compete in PWC for wood, M&B, with only 50% or so of current annual consumption in Western District within M&B control, will be cramped for all time vis-a-vis Crown, Rayonier, Powell, who work cleverly all the time and seem to be winning.[14]

MacMillan's seat on the Finance & Policy Committee, while strategic, kept him at a distance from many company activities he regarded as important, and it took longer for information to reach him through the informal network he had built up within the company over the years. Within a year of Clyne's arrival, he was alarmed to discover one of his treasured programs, forestry and timber, was faring poorly. For most of the 1950s he had devoted a great deal of his own time and effort to the acquisition of forest land on Vancouver Island and the development of a management plan on the land. He spent many weekends at his Qualicum house, driving his Bentley over backwoods logging roads, accompanied by Forestry Manager Ian Mahood, inspecting the young forests and passing on his ideas of how to manage them for the future. On one of these excursions, while Mahood cleaned a deer one of them had shot, MacMillan sat on a stump gazing out over a stand of young fir trees. Mahood noticed tears running down his cheeks, and asked if he was feeling unwell. "No," MacMillan answered abruptly. Then, after a few moments, he added: "Ian. Remember my words. When the pulp and paper people get a hold of my company it will be like a blight on the prairie wheat. They do not understand the forest and they will waste and abuse the very foundation of the company."

MacMillan delivering a speech, West Vancouver, c. 1955.

It soon became clear to everyone at M&B that Clyne was favouring the pulp side of the business at the expense of the timber and solid wood divisions. The forestry division was drastically downgraded and within a short period of time, following an explosive encounter with the new chairman, Mahood quit.[15] MacMillan raised this question in an April 1959 memo headed "Thoughts—for reflection, stimulation, criticism and to avoid oversights." He pointed out that a timber and lands budget had not been presented for half a year. "To a degree I can't measure, staff has taken over, seniority inbred, promotion of nearest and best known, a kind of 100% inbreeding has weakened us."[16]

At this stage he was not usually blaming Clyne for the problems he saw emerging, but pointing them out to him so they could be corrected. As M&B headed into the merger with Powell River, he wanted problems dealt with before they became incorporated into the new company.

It is doubtful that MacMillan realized the profound differences in corporate philosophy between himself and Clyne. One of the fundamental principles MacMillan had instilled in the company at its inception, and which it had adhered to for forty years, was to keep its operations within British Columbia. Clyne, by his own admission, chose to abrogate that principle almost as soon as he took over. He told the author of the company's official history, Donald MacKay:

> MacMillan had said that the company achieved its strength because of the fact it was the lowest-cost producer in the world and what we should do would be to continue to manufacture in British Columbia; all our manufacturing facilities were on salt water so we could sell throughout the world and we could maintain a high profitable position by remaining in B.C. because all our forests were accessible, our mills were accessible, and we could ship.
>
> But once I had been in the company for about three months I came to the conclusion that we had all our eggs in one basket and we should expand. So that was the philosophy I worked under.[17]

Clyne's grand scheme was to use the assets MacMillan had accumulated to create a multinational corporation engaged in whatever looked like it might turn a profit. Within the first year or more of Clyne's tenure, it is doubtful MacMillan could have comprehended the colossal arrogance it took for someone with so little business experience to embark on such a grandiose course of action. Rather, he put his energies to educating and assisting Clyne, first with overseeing the Powell River merger, and then with consolidating the expansion. By the time he figured out what was actually going on, it was too late to do anything about it.

But all that lay in the future. In 1959, MacMillan saw only adjustments to be made, and other, personal matters to attend to. The highlight of the year was a visit from the Queen. M&B hosted a lunch for her on July 16 in Chemainus, attended by about fifty of the company's senior people, including the Foleys, with whom negotiations were progressing well at the time. MacMillan, along with Edna, and the Clynes, Buchanans and Howard Mitchells, had gone over to Vancouver Island the previous day on the *Marijean*.

After docking, MacMillan walked into town to see what changes had been wrought since he had lived there forty years previously, and to have a haircut at the local barber shop.

> While waiting I studied intently an old gentleman whom I first decided must be the ghost of old Bill Horton, boss stevedore, who I thought had died a few years

ago. I listened to conversation which gave me glimmers, emboldening me to ask: Was his name Horton? The answer, yes, led a companion to say "this is Bill Horton."

This old man, a little crippled, huge frame, gnarled hands, considerable hair, black streaked with grey, heavy black eyebrows, all his teeth sound but two upper fronts knocked out, was born in Moodyville in 1865, began work at Hastings Mill in 1877, was on the roof of Hastings Mill wetting it down during the fire of 1886, moved to Chemainus as boss stevedore in 1903, worked there until 1952, established a coastwise reputation in loading lumber ships for quality of work and speed.

He talked about the old timers he knew, Sue Moody, John Hendry, Arthur Hendry Senior and Junior, R.H. Alexander, E.J. Palmer, Captain Gibson and many others.[18]

This encounter, judging by the space he devoted to recording it, ranked as high with MacMillan as meeting Queen Elizabeth. The following day he sat with her during lunch, under the trees at the Chemainus golf club. He had a pleasant conversation with her and Prince Philip, and on their departure MacMillan returned to Vancouver on the *Marijean*. It was a significant day for him, impressed as he was by titles, royalty, pomp and ceremony; the Queen and Bill Horton, all in one go. It was, he wrote in the ship's journal, "a red letter trip."

The next winter's trip on the *Marijean* was memorable for another reason. Just before his departure he met with his old friend, Harold Wilson, who was visiting Vancouver. Wilson was accompanied by Sir Arthur "Bomber" Harris, the Royal Air Force commander famed for the controversial bombing raids he launched on German cities at the end of the war.

MacMillan invited Harris to join the expedition, which that year was making its third trip to the Galapagos Islands. Harris jumped at the opportunity and agreed to meet the ship in Lima. After he came on board, they headed out to sea. Early the next morning, Harris arose and, for some reason, presumed to order the captain to make a change of course. When MacMillan arose and learned of this, he ordered the captain to take the ship back to Lima. Harris was put ashore, and they departed for the Galapagos without him. No one but MacMillan gave orders on his ship.[19]

The year 1961 was a melancholy one for MacMillan. After her return from the annual *Marijean* trip, Edna went into a long, debilitating decline.

Doctors eventually discovered she had a hereditary disease, idiopathic steat-
orrhea, resulting from the body's inability to absorb certain foods. Through
the year she weakened slowly, with occasional,
brief revivals of energy and spirits from blood
transfusions.

In June, MacMillan received a letter from a
Toronto friend telling him Mazo was also in poor
condition, bedridden, in great pain, her ability to
speak gone. She died the following month.
Montague Meyer died the same year.

The end came quickly for Edna on January
21, 1962, with all the family present. Following a
heavily attended funeral, MacMillan left with one
of his daughters and her family to meet the
Marijean in Mexico.

The relationship between MacMillan and
his wife of fifty-one years can best be summed up
in one word—loyalty. Beginning in the days of

Edna MacMillan, c. 1960.

their childhood, when he left rosebuds on her school desk in the mornings,
Edna patiently waited for him. She stood by him through six years of absence
when he was at university, and another two and a half years while he was in
sanatoriums, much of the time on the edge of death. As MacMillan pursued
the opportunities presented to him, and grew into a powerful, dynamic citizen
of the world, Edna remained quietly at home, in spirit if not always physically. She did not fit the model of the powerful wife behind the successful man.
She was a quiet, reserved woman, whose husband went off on his own.

Throughout their long marriage and the vast changes of circumstance
he experienced, MacMillan also observed a dedicated loyalty to his wife. No
matter what the nature of his relationships with others, he was careful never
to cause her public embarrassment. During the last decade and a half of her
life, some of their most enjoyable times together were spent on the *Marijean*,
and the numerous photographs he took of her on the yacht reflect an affec-
tionate concern for her well-being. Although her death had been expected for
a long time, it saddened him immensely.

The philanthropist

B y the time Edna died in 1962, the annual *Marijean* expeditions had become more than a winter diversion for MacMillan. Probably because of his inability to engage in purposeless activity, the *raison d'être* of the southern excursions became the advancement of marine biology in British Columbia. During the early 1960s, the task of gathering display specimens for the Vancouver Aquarium had been added to the *Marijean*'s annual itinerary. MacMillan was a major donor to the institution, and after it opened in 1956 its director, Murray Newman, was a frequent guest on the winter trips.

On the 1962 trip Newman came along to collect specimens, and early one evening he set out a light gillnet, hoping to catch a few small fish. Late that night, before retiring, he checked the net and discovered an extraordinarily large manta ray in it. He and two crew members managed to pull the net alongside the ship and secure it. Their efforts awoke MacMillan and his guests, who lined the rail, watching. Newman explained to MacMillan the unique nature of his catch and the opportunity it offered, so the reluctant captain was requested to haul it aboard. After a great struggle it was hoisted onto the deck and, the next morning, stuffed into the food freezer. The creature weighed more than a thousand pounds and had a span of twelve feet. MacMillan quite happily agreed to cart it back to Vancouver and stick it in a freezer at BC Packers until some technicians from the Smithsonian Institution in Washington could be hired to make a replica, which was hung in one of the aquarium's galleries.

Midway through his vacation, MacMillan flew back to Vancouver to speak at a dinner in Victoria marking the fiftieth anniversary of the BC Forest Service. Soon after his arrival, he received a call from Conley Brooks, one of the remaining members of the Powell River group on the MB&PR board,

**MacMillan in Mexico,
April 1959.**

urgently asking to meet him. Brooks arrived with an offer from Harold Foley. A few weeks later, MacMillan described the encounter in a letter to his grandson, Johnny Lecky.

The first we learned of HFs actions was on February 28th last when Conley Brooks, carrying out HFs plan offered me $22 US per share for two and one-half million . . . shares of MB&PR on a day when the market price of these shares was under $20 Can. At that time, Conley told me that his family group, including the Foleys, would sell to St. Regis [Paper Company of New York] two to two and one-half million MB&PR shares at the same price. St. Regis wanted to get a total of five million shares.

Conley made this offer to Messrs. Bloedel, VanDusen and myself, and it was left to me to arrange in concert with these partners how many shares would be supplied by each. The offer was conditional upon my finding two and one-half million shares. I asked Conley how many shares his group would provide, and he said "from two to two and one-half million, if needed."

We have since learned that from December to the end of March, they had made firm offers to an unknown number of large shareholders and other directors. This offer was not made, nor was it intended that it would be made to the ordinary shareholders of MB&PR, who number about 16,000. The offer was not made to all the directors of MB&PR, although there had been great activity between Harold Foley—supported by some of the directors who were friendly toward him—negotiating with the St. Regis Company and whoever else was going to finance the transaction, which would run into $110 million US.

Neither the Chairman of MB&PR [Clyne] nor any of us here knew anything of the negotiations until Conley Brooks sent word that he wanted to see me, and he did not disclose why. He told me about it when we met at breakfast at my home on February 28th.

Immediately after my conversation with Conley, I told the Chairman. He had heard nothing about it from any other source. I assume that the conspirators did not intend to tell him until they had the five million shares in hand, and were thus enabled to make their next move, whatever that might have been.

If St. Regis had got five million shares, and had got one-half of them by reducing the share holdings of Messrs. Bloedel, VanDusen and myself, they would have been the largest shareholders in the company. They would have been on firm ground to proceed with their plans, which, because of the great activity of the Foley and Brooks families and relatives in the conspiracy, undoubtedly would have been to oust the current Chairman. Actually, he would have resigned of his own accord, because he would not be associated with them for more than a day or two, or perhaps even a few hours. If our Chairman had gone, some of our best men would have gone. Many of them have been doing their utmost for more than two years to correct the errors made by Foley management of Powell River

Company, and also have been working hard to build up the merged Company.

In the past two weeks we have learned that the Consolidated Paper Corporation of Montreal, in which St. Regis Paper had 13% ownership, was working with St. Regis to take over MB&PR. We have also learned that these two companies were prepared to buy up to ten million shares of MB&PR, if their initial scheme had worked. Undoubtedly HF in his efforts was activated by the thought of getting "back into the saddle" here and elsewhere.[1]

Clyne's response, once it was clear none of the M&B group were interested in the scheme, was typical of his style. At the 1962 annual meeting of the company in April, he asked for the removal of all the "conspirators" from the board, and was supported by 85 percent of the shareholders. The only survivor of the Powell River group was Anson Brooks, Conley's cousin. The merger had, in the end, become a takeover.

As he indicated in the letter to his grandson, MacMillan fully supported Clyne in his actions, as he apparently had done throughout the stormy consummation of the corporate union. At what point in the proceedings the merger began to resemble the mating ritual of the praying mantis, in which one partner eats the other, is difficult to determine. Equally unclear is whether this strategy was deliberately engineered by Clyne with the agreement of MacMillan, Bloedel and VanDusen after the fact, or something they planned together.

The personal animosities aroused by the affair were unusually intense for people accustomed to drawing sharp distinctions between their business and personal lives. Most of the actors lived within a few blocks of each other in Vancouver's Shaughnessy district. They routinely encountered each other at dinners and parties, and discussion of the ongoing squabble enlivened many a dull social event. Some time after the events described, Clyne and Harold Foley met at a Shaughnessy cocktail party. After engaging in a conversation that became more heated as it continued, the bald-headed Clyne poured his drink over the elegant grey head of Harold Foley and walked out.

The final act of the tempestuous saga, ending with the general expulsion from the company of all but one of the remaining Powell River directors, shocked many in the staid Shaughnessy crowd. When, at another party, someone raised the issue with MacMillan, he merely smiled, adding enigmatically: "The dogs bark, but the caravan moves on."[2]

But there were limits to his support for Jack Clyne. The following winter, the Clynes were guests on the *Marijean* at the same time Jean Southam

J.V. Clyne and MacMillan aboard the *Marijean,* c. 1963.

was on board. One evening at dinner, Clyne, in imperious Supreme Court tones, informed Jean she would have to stop seeing Harold Foley socially, as he was an "enemy of the company." With MacMillan observing, somewhat amused, she informed Clyne there was only one person whose orders she obeyed: her father. MacMillan made no comment, and shortly thereafter retired to bed. Clyne spent most of the night pacing the decks of the *Marijean*. At one point, through the thin cabin walls of the ship, Jean overheard him telling his wife he wanted to leave. It was said in such a manner to indicate he would also resign his position at MB&PR. The next morning Jean attempted to mollify him by saying she had not intended to upset him. Clyne, choosing to interpret this as an apology, decided to stay. Nothing more was said about the incident until much later, when MacMillan asked his daughter what had transpired during the night and why she had spoken so diplomatically to Clyne in the morning. When she told him of Clyne's threat to leave, MacMillan shook his head in disbelief. "Oh, no!" he said. "Why didn't you let him leave? You'd have saved me so much grief if you had let him go."[3]

By mid-1962, the Powell River situation in hand, MacMillan was free to devote his attentions to distributing the wealth he had acquired. His first consideration was to care for his family. He had no dynastic pretensions and did not intend to burden his daughters and their families with a corporate empire. Early in his career, he had observed the difficulties encountered by second- and third-generation family businesses, and long since decided to avoid this option. By the time Edna died, he had established substantial trust funds for his daughters and all of their children.

The largest portion of his wealth was destined to be given away, mostly within Vancouver and British Columbia. This process had begun years earlier, but now that his energies were no longer focussed on building an industrial enterprise, he could concentrate on distributing the accumulated assets of forty-five years in business, most of which were in MacMillan Bloedel stock.

One vehicle MacMillan utilized was a creation of his partner. Twenty years previously, VanDusen served a term as the endowment committee chairman of the Vancouver Welfare Federation, the precursor of the United Way of the Lower Mainland. In 1944, a shy and reserved woman with no close relatives, Alice MacKay, gave the federation $1,000, with the stipulation that only the interest it earned be used each year. At about the same time, VanDusen had organized a body called the Vancouver Foundation to support the Welfare Federation's operating expenses. He conceived the idea of using Alice MacKay's $1,000 to start an endowment fund to support various charitable activities. Realizing the interest on her donation would accomplish little, he went out and persuaded nine others, including MacMillan, to join him in making a $10,000 donation each to the fund. This was the beginning of the Vancouver Foundation, which within fifty years became Canada's largest community foundation, with endowments of more than $350 million and annual grants of more than $22 million. Although MacMillan was an early and major supporter of the Vancouver Foundation, it was really VanDusen's creation. Over the course of his life, VanDusen was to donate to it and the related VanDusen Foundation more than $100 million.

MacMillan had become a regular supporter of local charities during the depression through contributions to the Vancouver Welfare Federation, starting with $150 in 1931. Through the depression, his gifts grew to a total of $2,000 to $3,000 a year, including a regular contribution to Chalmers United Church, which his mother had attended. By the end of the war, his annual tax-deductible contributions to community activities had increased to $40,000. In addition, he made numerous non-deductible donations—he never did claim a

HR MacMillan

MacMillan lands a fish in the BC interior, 1961.

Left to right: MacMillan, Frank Rush and Mike Crammond in the BC interior, 1961.

MacMillan at Kenny Lake.

deduction on his initial $10,000 jump-start for the Vancouver Foundation. By the early 1950s, he was giving to a wide array of causes, with the bulk of his funds going to various departments at the University of British Columbia. In 1955, the annual total had increased to about $60,000, in addition to $500,000 he gave to establish an education fund at the Vancouver Foundation on the occasion of his seventieth birthday. It was used primarily for donations to UBC. By about 1960, he had established a policy of giving away at least 10 percent of his net annual income, with additional donations from personal or corporate funds.[4]

MacMillan favoured a hands-on approach to giving away his money. Occasionally he would make a donation at the request of a close friend, although many such solicitations were turned down. He preferred to have detailed knowledge of the activity he was supporting, and spent considerable time and effort informing himself about the organizations, projects and activities he helped finance. He favoured direct, one-shot grants or declining funds, over long-term or permanent endowments, which he believed had less of an impact.

After the Powell River merger, MacMillan began to sell shares in the new company to support charitable causes, thereby diluting his holdings. In 1963 he gave 50,000 shares, worth about $20 each, to the Vancouver Foundation to establish a family fund, in which the donor advises the foundation on allocation of the money. MacMillan also favoured BC in his gifts. Occasionally, he gave money outside the province, but most often he turned down requests from eastern and American universities with the explanation that West Coast institutions like the University of BC did not have as long a period of development and their current needs were greater than, say, Toronto or Yale. His only significant contribution to the latter, in spite of repeated requests, was in 1964 when he gave $10,000 in support of a request from his old friend and classmate, Nelson Brown.

His major funding interest was UBC. Within that institution, the largest amounts were given for fisheries, scholarship funds, libraries and the Anglican and Union theological colleges. His contributions to forestry were substantial, but far less than other areas of interest, considering his connection to the profession and the industry. Most of his contributions to the forestry school were to cover salaries for a silviculture professor and the Dean of Forestry, although he also set up a number of scholarship and fellowship programs. One, the H.R. MacMillan Prize, went to the top student in the forestry graduating class, with the first award given to Peter Pearse, who later became one of Canada's prominent forest economists. On other occasions he turned down forestry school requests for assistance.

> I find that I am so far out of touch with Forestry and the Forest School, that I can neither give good advice nor do I feel like undertaking any more financial responsibility. I have no knowledge of the UBC Forest School plans, nor do I expect that these plans will be discussed with me.
>
> You may wonder why I undertook the responsibility for more money. I did so because it seemed to me that the School was becoming weak, and that perhaps it was because money was not available to maintain it at strength. Therefore, I decided that I would provide the money to give them freedom. Whether I was doing the right thing or not I do not know.[5]

In 1962, when he discovered his annual donations to the forestry school—about $23,000—accounted for more than one-fifth of its budget, and that the only other corporate contribution was a $1,500 gift from Rayonier Canada, he drew the line and turned down further requests.

> Financial support over the years to the Forest School of the UBC from MB&PR Limited and its predecessors, and by me personally, has been considerable. But it has not been accompanied by much help from any other forest industry. . . . This brings up the question in my mind as to why the prosperous element of the very large forest industry in this Province is not doing more financially for the Forest School at the UBC. It could be possible that MB&PR Limited and some of the names related to it have done so much that others have been deterred.[6]

His interest in fisheries was far more sustained, and more enthusiastically reciprocated. In part this was because of the role Larkin and other UBC scientists played in the work of the *Marijean*, and MacMillan's direct involvement in fisheries through the collection of about 80,000 specimens for the university's fish museum. He was a major supporter of UBC's Institute of Fisheries from its inception, and in 1965 he gave the institute $750,000 in the expectation

> that this money would be used primarily for attracting and keeping the best teaching staff in the Institute of Fisheries . . . [and] that any portion of the money used for buying teaching equipment would be kept at a minimum—my thought being that our governments, which spend millions on equipment in biological stations and research vessels, should provide for the University such equipment as is necessary to train the scientists.

He went on to explain his interest in the subject to UBC President J. B. Macdonald.

> Under present conditions of exploding world populations in the midst of food shortage, particularly proteins, we British Columbians apparently have an unusual opportunity in the unknown and unexploited fisheries within our reach, concerning which there is so much to learn.[7]

That same year he gave the UBC library system $3 million, the largest contribution anyone had made to a university library in Canadian history. This was directly related to his intense interest in books and his firm belief in their importance. He counted several librarians as his friends, and employed one to care for his private collection for a brief period.

MacMillan's generosity was not always unconditional, and he did not hesitate, on occasion, to make use of the influence it gave him. In the fall of 1963, he wrote to UBC recommending an honorary degree be given to Loyd Royal, the highly regarded director of the International Pacific Salmon Fisheries Commission, who is credited with bringing about the restoration of the Fraser River sockeye salmon runs. Royal, a Washington state fisheries biologist, was a rough-hewn, down-to-earth character of the sort MacMillan admired. He was a little bit too earthy for the university's honorary degrees committee, which politely refused to include his name on the list forwarded to the university Senate, saying it had been "unable to place Mr. Royal sufficiently high in the order of priority." Arming himself with a list of honorary degree recipients since UBC began granting them, and enlisting the support of John Buchanan, MacMillan renewed his campaign. It took him more than two years, but at the 1966 spring convocation, Royal received an honorary Doctor of Laws degree.[8]

MacMillan was also quick to respond when he felt the university was taking him for granted or was pushing him too hard. He was highly indignant when he received a form letter from the university's Wills and Bequests Committee, suggesting he might want to include the university in his will. His response to this thoughtless request was brief and to the point. "I do not know what I shall do, therefore I suggest to your Committee that they contain themselves in patience until, if ever, they find out."[9]

MacMillan with his daughters, Jean and Marion, c. 1960. He is wearing Lord Amherst's Collar, which he purchased in London.

After Edna's death, MacMillan made numerous trips abroad. He spent the following Christmas in Hong Kong with Marion and her family. A few months later he was back in Tokyo with Jean and the Clynes to visit the Keswicks and plan a partnership with Jardine Matheson. He planned to spend Christmas 1963 in London with Johnny Lecky, revisiting some of his favourite restaurants and bookstores. But early in the month he came down with a severe cold he could not shake. He cabled Lecky at Jesus College, Cambridge, suggesting they spend Christmas in Trinidad instead. Lecky, who was twenty-two at the time, agreed, and the two of them spent ten days touring the Caribbean, including a trip to Venezuela. MacMillan returned to Vancouver full of energy and with a renewed interest in travelling.

After catching up with various items of business, he headed south for what looked like an extra busy two months on the *Marijean*. The first few weeks went well, but his plans for the last leg of the trip disintegrated as, one by one, all of his guests were suddenly unable to come. Cutting his trip short, he came home and made plans for a month-long visit to Europe.

On May 1, 1964, MacMillan and Jean Southam flew to Athens where he attended a meeting of the World Association of Newsprint Manufacturers. He gave an animated dinner speech—the last of his life—which drew a standing ovation, and retired early to a hotel room adjoining his daughter's in anticipation of a morning flight to London. In the middle of the night, awakened by a noise, Jean went to check on her father. He was not awake, but one of his legs was flailing wildly. Eventually, he lay still and appeared to be sleeping quietly, so she returned to her room. The following morning, MacMillan complained he was not feeling well and wanted to stay in bed. Jean cancelled their flight, and then their London engagements as the days went by and he was not improved. After two weeks' rest in Athens, he was flown to Rome in Aristotle Onassis's private plane. There he was met by the local MB agent and his four sturdy sons, who carried him in his wheelchair across the tarmac to an Alitalia jet to Montreal. Dorothy Dee was waiting for him there, and after a few days' rest in Montreal, they returned to Vancouver where he was confined to his house for several weeks.

To his friends and acquaintances, MacMillan explained variously that he had contracted influenza, pneumonia or a bad cold. In fact, he had suffered a mild stroke. This incident brought home to him the realization that he was getting old. That September he turned seventy-nine.

Early that fall, MacMillan and Leigh Stevenson, accompanied by two US naval officers who had helped a consortium of BC forest companies obtain a fleet of water bombers, flew into Whitesail Lake in Tweedsmuir Park. MacMillan was fishing from the bank of the fast-flowing, icy-cold Lindquist Creek when he hooked a trout. The two officers came to help him; both managed to fall in, and they were swept downstream. While the pilot went to their rescue with the float plane, Stevenson returned to the stream bank just in time to find MacMillan, still playing his fish, slowly sliding into the water. Hooking his jacket with a stick, Stevenson pulled him in to shore. MacMillan looked at him sheepishly. "You might not realize it, but you just saved my life," he said. "I wouldn't have lasted five minutes in that water." Holding up his rod, he added, "I've still got the trout, too."

This was one of his last fishing trips, apart from excursions on the *Marijean*. Although his hunting expeditions into the mountains had long since ended, he still shot regularly at Kirkland Island, bagging geese, ducks and pheasants every fall. Each year he stocked the island with 250 pheasants, and he and his guests shot about fifty, the rest falling to poachers or escaping to adjacent islands.

On one of his trips to Kirkland Island, MacMillan was accompanied by Charles Camsell, Canada's deputy minister of mines, an old friend who had come out for lunch and an afternoon of shooting. They were walking along a trail, with Camsell in the lead, when a duck flew up. Camsell shot and it came down in a field. He went over and picked it up, commenting to MacMillan, "I've never seen a duck like that before." MacMillan laughed. "You'll probably never see another one again. I just imported a pair from Russia. They cost me $150 each."[10]

His life had settled into an orderly pattern that he fully enjoyed, as he explained to Johnny Lecky.

> My routine is made more bearable by great freedom in the use of my time, which I confess has developed into a habit: breakfast at 8:00 to 8:30 a.m., driven to the office by Hilding [his gardener and driver of more than thirty years] where I arrive at 9:30 to 10 a.m. I lunch with someone at the Vancouver Club, or elsewhere in the City, between twelve and one o'clock, afterwards I return to the office, almost daily, for a short time, then leave for home about 4:00 p.m., where I make myself comfortable and read until dinner at 7:00 p.m., which I usually have alone, and read again after dinner until I go to bed at 8:00 to 9:00 o'clock. Three days out of four one or two of a small list of friends come in for a chat between five and six o'clock. This list includes your mother and father, your Aunt Jean and Uncle Gordon. Occasionally one or two of the family of [your] cousins comes in, however, they are all very busy with their own affairs so are not frequent visitors. One or two friends from another list also come in such as Fred Auger, John Buchanan, "Stevie" [Leigh Stevenson], Ritchie Nelson and Larry MacKenzie. My list of acquaintances is quite wide, but those who call on me are very few in number. I see others around the Club more or less daily.
>
> I find it an easy life which shows what it is to maintain one's interest and also have formed the habit of hours of reading daily. Other people of my age say that they get tired and find time hangs very heavily. As yet, I do not.[11]

In the spring of 1965, this peaceful existence was rudely shattered one morning when MacMillan answered the doorbell at his home. He later described the incident to Johnny Lecky.

> These two young fellows, one about 20 and the other some years older, well dressed in the latest spiv costumes, arrived at 9:13 a.m., and nervously stated that they had had a motor accident, had hurt a boy and wanted to 'phone the police. Two minutes later, by their not using the 'phone and by their producing a meat cleaver, I found their errand was of a different sort. They promised me immunity if I co-operated. The cook was doing some work in the upper part of the house and the house keeper was busy with something in the basement, so I was alone on the main floor.
>
> They demanded $150,000. I told them that I did not have that amount of

Visiting a logging show in Squamish, BC, 1965. MacMillan chats with Cyril Fitch (right).

money in the house. They suggested that I could go to the bank and get it. I then told them that if I did go to the bank for the money someone would have to drive me because I did not drive a car, and afterwards would have to drive me to a spot where they could collect the money. This being the case, it would not be a secret. The only other way would be for them to come to the bank with me. I told them that I would be willing to do this if they were willing to go to the bank with me. They had no answer to this, so, at the point of the cleaver, shoved me in the main-floor closet and ran away. I have not heard from them since and the police have not found them. I was fortunate that I was not hurt. . . .

This incident is further evidence that our society is developing more and more people who do not want to work for a living and who are without conscience. However, when I read of the time of the Stuarts, the Plague, and the Fire, I am not sure that it is worse now than it was then.[12]

It was the second attack on MacMillan in his home. At one point during the Powell River affair, a few hours after news of Harold Foley's resignation had appeared in the local papers, MacMillan was sitting reading when a fake bomb, complete with a sputtering fuse, came crashing through the window and landed at his feet. An attached note read:

Dear Harvey: This should be real. Why do you have to get your hatchetmen to do your dirty work? Another thing, maybe if you smiled (which would be a new experience) life would seem better, but I guess it would hurt. Your many money-grubbing friends.[13]

Police later apprehended two teenage boys, one the son of a former Powell River Company employee. As well, about this same time, MacMillan had received some vague threats concerning his grandchildren.

After the attempted holdup, he hired two watchmen to patrol his and his daughters' properties each evening. This lasted for more than a year and eventually was discontinued when no further incidents occurred. The primary usefulness of the watchmen was to report on the comings and goings of his several teenage grandchildren to their parents.

As he had written in the letter to his grandson, reading had become MacMillan's passion. Books had always been important to him; he had amassed one of the largest collections of rare books in the city and now was able to indulge himself fully in reading them, particulary histories and biographies. During the 1960s he was also instrumental in bringing several books to publication. In a practical sense, British Columbia at that time had no book publisher. Most books written about the province or by people living there, were published in Toronto or London. Consequently, very little about BC's history found its way into print.

One of MacMillan's first publishing endeavours was to pay Bruce McKelvie, a *Vancouver Province* reporter, to write a biography of Eustace Smith, a well-known coastal timber cruiser. MacMillan admired Smith and enjoyed his accounts of coastal history. He persuaded his friend, Fred Auger, publisher of the *Province*, to give McKelvie time off from his duties in the legislative press gallery to complete the Smith biography. Unfortunately, just as it was finished, Smith died, and his family asked that the manuscript, titled "Sign of the S," not be published. It has lain in the BC Provincial Archives, virtually unread, ever since.

When MacMillan discovered that the diaries of Dr. William Fraser Tolmie, who had come to the West Coast in 1832, were resting unread in the archives, he financed their publication by Howard Mitchell. The production of this book, combined with a large loan MacMillan gave Mitchell, led to the expansion of Mitchell Press, a printing house, into one of the province's first book publishing firms.

For several years, Mitchell tried to persuade MacMillan to agree to publication of his own biography, and sent him various outlines and mockups of the proposed book. MacMillan always declined. Then, in the mid-1960s, he commissioned G.R. Stevens, a former Canadian Trade Commissioner to

Australia, who had published several Canadian history books, to write his biography. After reading Stevens's first draft, MacMillan paid him off in exchange for all copies of the manuscript. It was a sycophantic portrait, subtitled "A Study in Impenitence." Henceforth, he declined all suggestions for a biography, including a proposal from Margaret Murray, one of BC's best-known and most colourful newspaper publishers.

MacMillan also helped finance the publication of a history of the Chemainus district, *Water Over the Wheel* by H.W. Olsen, and provided a detailed reminiscence of his experiences at Victoria Lumber & Manufacturing for inclusion in the book. He encouraged a cousin, Georgina Sinclair, to research and write a genealogy of his mother's family, the Willsons, from the arrival of the first members in Whitchurch Township. He also paid for publication of the book. During the mid-1960s, MacMillan enjoyed the friendship of a young journalist, Pat Carney, who shared his interest in Martin Grainger's writing, as well as the early history of the forest industry. He tried without success to persuade her to write a history of BC coastal logging, one of his favourite subjects. Along with John Buchanan and Norman Hyland, he financed the publication of Cicely Lyons's *Salmon: Our Heritage*, an exhaustive history of the BC fishing industry. Near the end of the decade, he helped finance the writing of a biography of one of his first friends in Vancouver, UBC President Frank Wesbrook. The book was written by another friend, William Gibson.

In the fall of 1965, MacMillan had another idea, and wrote for advice to R.S. McLaughlin, founder of General Motors of Canada.

> I notice that you have been thinking about a Planetarium for the City of Toronto. Oddly, I have been thinking about something similar for the City of Vancouver. I think my lawyer, Mr. Charles W. Brazier, may have been in touch with you about it. I have had in mind to pay for the cost of a planetarium for the City of Vancouver if some institution could be found to pay the upkeep. This is as far as I have got. If you have any information about this subject, I would greatly appreciate your sending it to me.[14]

After months of exchanging information with McLaughlin, researching the subject and conducting arm's-length negotiations with the city, Brazier wrote to Vancouver Mayor W.G. Rathie: "I am instructed on behalf of an undisclosed client to advise you that he is prepared to make a gift to the City of Vancouver in a sum not exceeding $1,500,000 for the purpose of constructing and equipping a planetarium."[15] The offer was accepted and the planetarium, which MacMillan eventually allowed to be named after him, opened on October 26, 1968. A few months later, MacMillan noticed a letter in a Vancouver paper from a teacher in Clearwater, a small town 350 miles northeast of the city in the BC interior. It described how a class of Grade Seven students had come to Vancouver by train to visit the planetarium and been refused admittance because they were one minute late. MacMillan lost no time letting the planetarium's management know of his displeasure. He wrote to the director:

> I suggest that you inform everyone on your staff to keep in mind that people who come from out of town to Vancouver usually have to make some effort to get here, and also that most of them are not as well informed about conditions operating around such institutions as the Planetarium as you and your staff are. . . . I would be glad to pay such expense [of bringing the children back to Vancouver] if the situation would be corrected by my doing so. . . . Referring to your second paragraph wherein you state: "Nobody is more sorry than we are. . . ." I think you must admit that the disappointed children are more sorry than you are. Again, I suggest that everyone connected with the Planetarium try not to disappoint children in this manner.[16]

There were definite limits to what MacMillan was prepared to support, and they were exceeded when he was asked by several friends and acquaintances to help fund an endeavour launched by the soon-to-retire governor general of Canada, Georges Vanier.

> I told John Buchanan a few days ago that I don't approve of each Governor General trying to find a cause to which he can pin his name, and then call upon people to support. Taking up the family—through the Vanier Institute of the Family—with a view to improving it, is like taking up the cause of God and motherhood, and I am not enthusiastic about it. Massey set an example by the works to which his name is attached. I hope this example is not going to be followed by each Governor General as a sort of "command performance."[17]

MacMillan's charitable activities stand in apparent contradiction to his views on public assistance. Beginning in the 1920s, he was a consistent opponent of the growth of the welfare state, and there is nothing to indicate he changed his views in later years. Yet, as he found himself easing into old age, a wealthy man, he systematically gave away a great deal of his money. It

is in keeping with his earlier opinions that he contributed little to social causes. During the depression, he was an important contributor to these sorts of
programs, but for the most part he preferred to support educational and cultural institutions, most of them in Vancouver. Generally he did not like to
respond to fundraising appeals, and by no means could he be counted on to
support the numerous causes his friends and acquaintances assisted. He
much preferred to give on his own initiative, especially when his gift determined the fate of a venture. He liked to see the effect of his endowments, as in
the case of the planetarium. This was especially true in his contributions to
UBC, where most of his gifts went and to which he was, until recently, by far
the most generous donor. Occasionally he would respond favourably to a
request to fund a university program, although he was just as likely to refuse.
His preferred method was, through his many contacts in the university, to
identify a need (as in the case of marine biology) and, through his own investigations, to determine precisely how and where the money would be spent.
He rarely dealt with the university's fundraising bureaucracy, relying instead
on his relationships with people in the institution.

One of his greatest fears was that if he financed some endeavour,
other potential donors, particularly government, would fail to fulfill their
financial obligations. Although he never did spell out in detail what he
believed those obligations to be, they were clear in his mind, as were the
obligations of people like himself. The motivation of MacMillan's charity was
his acceptance of his responsibility, as a person of wealth, to finance public
institutions and undertakings. At the same time, in keeping with his character,
he reserved the right to determine what he would finance. His choices and
decisions, which have had such an enormous influence on the cultural and
social development of BC in general and Vancouver in particular, are a testament to his judgement—as well as his determination.

Losing control

Throughout his life, MacMillan's relations with and feelings about politicians were ambivalent. He could not entirely avoid them, but it was as if they were a separate species. Although he knew and corresponded with most BC premiers and federal prime ministers, for the most part he did so in his capacity as the chief executive of his companies, to protect or further their interests. Apart from working to ensure that socialists—by which he meant the political left in general—never gained power, he did not hold any consistent partisan political views. When he did support individual politicians, either financially or politically, it tended to be because of a personal liking for them. This was certainly the case in his ongoing friendship with Harold Wilson, to whom he regularly cabled Christmas wishes.

MacMillan was impressed with displays of personal honesty in politicians, even if they were socialists. When Frank Howard, a CCF member of parliament from northern BC admitted to a criminal conviction earlier in his life, MacMillan wrote to commend him:

> I have not the privilege of knowing you. I write to compliment you on the character which led you to make a public statement concerning an episode in your early life. Most persons would have issued a denial rather than speak the truth.[1]

He was an enthusiastic backer of Wallie McCutcheon, who served briefly in Diefenbaker's Conservative cabinet and later the Senate. This was clearly based on his friendship with McCutcheon, as MacMillan could never hold Tory politicians in high esteem after observing R.B. Bennett in action. He did correspond regularly with Diefenbaker during his term as prime minister, but was never close to him.

He had more of an affinity for Liberal party politicians, not so much because they were Liberals, but for personal reasons. He greatly admired Gordon Gibson's courage in taking on the government and the big forest companies during the Sommers affair, and when Gordon Gibson, Jr. entered politics, MacMillan backed him with one of his rare financial contributions to an election campaign.

MacMillan and Prime Minister John Diefenbaker with Lord Amherst's collar, which MacMillan purchased at a public auction in London and gave to the Public Archives of Canada, c. 1960.

The leader who, for a time, received more of his support than any other was Lester Pearson. MacMillan had come to know Pearson during his stint on the NATO productivity board, while Pearson represented Canada on the NATO governing council. In spite of Pearson's lack of support for Britain during the 1956 Suez crisis, MacMillan held him in high regard. Shortly after Pearson was elected Liberal leader in 1958 he visited Vancouver and had lunch with Clyne and MacMillan, as later recalled by Clyne:

> I do want to thank you again for sending me Volume II of Mike's memoirs. I enjoyed the first volume very much indeed and am looking forward to reading this one.
>
> It reminded me of the time when we had Mike at lunch in the Vancouver Club while he was leader of the opposition. You will remember that he told us that he did not have any money at that time to employ a research staff and we furnished him with funds for that purpose for several years.[2]

MacMillan never approved of Pearson's choice of a new Canadian flag, and when sailing the *Marijean* outside Canadian waters, continued to fly the Red Ensign. In the end he was sadly disappointed in Pearson's government, in part because of its increased expenditures on social programs, Pearson's obvious reservations about US foreign policy, and a variety of other perceived misjudgements.

> I would like to see the PM pay more attention to Canada and to building up our friendship with the U.S., rather than using this important position to badger Johnson about the war in Vietnam. I do not think we are responsible for how the Ten Commandments are practised by other people. I have not sufficient knowledge of what the Government does, however it appears to me that the PM should develop his attention, as I said above, to building up our friendship with the U.S. so that we may escape the imposition of regulations which might put a barrier in the way of our selling our goods to the US.[3]

Although MacMillan corresponded regularly with Henry Cabot Lodge throughout his tenure as US ambassador to Vietnam, there is no indication he either supported or opposed the American position in that disastrous conflict. Primarily, he was concerned about the effect on business relations of Pearson's views about the war. His opinion of the government's social policies, as well as his views on another, future Liberal prime minister, were no higher.

Perhaps the new trend toward spreading the benefits of this country amongst all the people would be better for everybody, but also perhaps it will not. It will have to be tried out. We see the new trend in the young, inexperienced Minister, John Turner, whose aim it is cynically to buy most of the votes by promising that a government containing him will pay everyone a salary.[4]

He was characteristically blunt in his assessment of the Pearson government in a letter to his old colleague from the wartime years in Ottawa and Montreal, diplomat Hugh Scully.

I am one of those who does not feel very happy about the government the Liberal Party has been giving us. It might be the worst we have ever had in Canada. Perhaps the only worse one was the one that disappeared in 1896.[5]

Although Pearson stopped in to visit MacMillan at his home a few days after this was written, their conversation did not improve his opinion of the prime minister, and he did not mourn Pearson's departure when the latter retired in 1968.

Although I never really knew him, I feel I can afford to be pleased that Mike has gone. I know nothing about Trudeau, but he looks better to me. Being French may be an advantage. The serious, important Frenchmen whom I have met had good ideas, particularly about what to do with money.[6]

MacMillan had an opportunity to meet Trudeau a few weeks before the 1968 leadership convention.

I had a chance last Sunday to be one of a small number of people who spent a couple of hours with Trudeau in the sittingroom of his hotel. I had never seen him before. He did not make any speeches but in a gathering of about 20 people he answered questions and conversed about his political views. I was very favourably impressed by him. I do not expect that he will win the leadership of the Liberal Party in April, but I would expect that later, when he has had more experience, he would make a good leader.[7]

Closer to home, he displayed a curious indifference to W.A.C. Bennett, who served as BC's premier from 1952 until 1972, spanning a period of great importance to MacMillan. For many years prior to the election of Bennett's Social Credit government, it had been MacMillan's habit to make regular trips to Victoria. He would check into a suite at the Empress Hotel for two or three days, meeting with politicians and cabinet ministers. This practice ended with the advent of Social Credit. Initially, he seems to have been wary of their populist rhetoric. After the Sommers scandal, he viewed them with distrust. And when they began turning the province's resources over to large, multi-national corporations, he apparently wrote them off as failing to represent the best long-term interests of BC. But he was very careful not to be observed criticizing them.

During the mid-1960s, following the final purge of the former Powell River Company people from MB&PR, Clyne pursued his plan for corporate expansion outside BC. One of his first moves was to bring in a team of US management consultants, McKinsey and Company, to plan a corporate reorganization. Many within the company scoffed at the idea of Harvard Business School graduates, with no knowledge of the forest industry, telling experienced managers how to run their business. MacMillan was among those critical of the McKinsey role, as reflected in his letter to the management company.

> Your letter brings to my mind a quotation from Goldsmith: "and still the wonder grew that one small head could contain all he knew." Of course, any "wonder" is quickly dispelled by the knowledge that so many of the best intelligences in the world are continuously spending money in large sums to educate McKinseys.[8]

The scoffing, along with the wonder, disappeared quickly when Clyne began to implement the McKinsey recommendations and appoint to senior positions some of the people the consultants recommended. Among the first of these was Charles Specht, an executive from an American mining company. Clyne appointed Specht president, shunting Ralph Shaw—whose first loyalty he believed lay with MacMillan—into the position of vice-chairman.[9] Clyne ordered Shaw to report to Specht; instead Shaw chose to resign. Although Shaw had more than thirty years' dedicated service to the company, having worked his way from the lowest rung on the corporate ladder right to the top, Clyne took advantage of a technicality in company rules to cancel all of Shaw's pensions.[10]

MacMillan's acquiescence in this brutal treatment of one of the company's most capable and loyal executives reveals one of the less pleasant aspects of his character. It was not the first time he had made or co-operated in a decision to dispense suddenly with a colleague and a friend. He believed it was necessary to distinguish clearly between business and personal relations—when the occasion required it. He had behaved similarly in his break with Percy Sills more than twenty years previously, and again with Bert

Hoffmeister. It was an attitude which occasionally surfaced in his conduct toward some employees with whom he had worked closely. When Keith Shaw died of a heart attack after overworking himself for months preparing the brief for the second Sloan commission, MacMillan curtly rejected a suggestion that he offer financial assistance to the forester's widow. These actions did not grow out of a dislike for people, or a lack of generosity, but from an obscure and flint-hard corner of his mind. After Ralph Shaw's dismissal, he and MacMillan, who never discussed the matter, continued to be good friends, and MacMillan left Shaw a bequest in his will.

In 1964, MB&PR acquired a one-third interest in a Dutch paper company. Shaw had initiated this move, but Specht took over in the midst of it and oversaw an aggressive expansion program. The following year the company, in partnership with United Fruit Company, built a large, integrated forestry complex in Alabama. Around the same time, MB&PR acquired a variety of paper, container and lumber companies in eastern Canada and the US. Overseas, company sales agencies were set up to replace the local agents MacMillan had used since he began operating in 1919. The only exceptions were agency partnerships with Jardines in the Orient and John Meyer, Montague Meyer's son, in the UK.

Prime Minister Harold Wilson, MacMillan's guest at the opening of a new MacMillan Bloedel wharf at Newport, England, September 8, 1967.

In BC, the company began a $110-million modernization and expansion plan at the Powell River mill—a boost to morale among survivors of the merger. There were a few comments but no serious objections when, in 1966, "Powell River" was dropped from the company's name, which became "MacMillan Bloedel Limited."

By this time—the spring of 1966—MacMillan, although still a member of MB's executive committee, had lost most of his influence over Clyne. He was assiduously courted by the chairman who, at the same time, was carefully ushering him out the back door. The previous fall, Clyne had hosted an eightieth birthday party for MacMillan at the Vancouver Club. Instead of a speech, Clyne read a poem he had composed for the occasion, which, in addition to being a terrible piece of doggerel, managed to miscalculate the guest of honour's age by ten years.[11]

MacMillan had taken to penning complaints in letters to Bloedel—who had vacated his house in Vancouver and moved back to Bainbridge Island near Seattle—and sending querulous memos to Clyne. Firmly in command of the company, Clyne began buying up some of MB's customers, including the largest US lumber wholesale firm, Blanchard Lumber, prompting MacMillan to question the policy. "This sits uneasily in my mind," he wrote.

> I should have asked these questions at the meeting. Why should we buy out our wholesale customers—to increase our profit? To increase our volume? To strengthen our hold on the business? Any other reasons? Respecting these important points, should we not have the written opinion of our most experienced men in this trade in which we have been important factors for over 40 years? My memory is that although some of our wholesale customers have been as good as we could hope them to be, none of them has made enough retained profit to tempt anyone into buying them out.[12]

One of the new chairman's policies that particularly irritated MacMillan was the growing amount of revenue devoted to corporate expansion outside BC, while shareholder dividends were declining. This issue was raised at the 1966 annual meeting by Gordon Gibson, who challenged Clyne directly from the floor, warning that the company was headed for financial difficulties. Clyne brushed off his arguments, and Gibson unloaded his shares.[13] This and similar incidents were not lost on MacMillan.

In May, sixty years after he graduated from the Ontario Agricultural College, its successor institution, the University of Guelph, granted MacMillan an honorary degree. He travelled to Ontario to accept it; but pleading tiredness, he avoided most of the convocation social events and returned to Vancouver a day or two later. On June 30, he joined an inspection tour of BC Packers' Imperial Cannery in Steveston. It was an arduous morning, ending with a lunch in the company cafeteria. During lunch he suffered another, more serious stroke. An ambulance rushed him to the Vancouver General Hospital, where he spent the next four months recuperating. For the first few weeks, he was unable to speak and his left arm and leg were paralyzed. But with the aid of private nurses, he slowly improved, and by September he was going out for drives with his daughters. In November he came home, accompanied by three nurses. He had lost some of his energy and vitality, and needed help walking because of an impaired sense of balance, but otherwise he had made a good recovery.

Typically, one of his first actions after returning home was to provide $150,000 financing for a ten-year research study on strokes at the UBC Faculty of Medicine. It was not until a few months later that he asked Dean J.F. McCreary for some personal advice. "I would like to know what eventually happens to most people who have strokes," he wrote. "For me this is an interesting subject. Have you any records?"[14]

He was advised to lead the most productive and active life possible, including travel. It was all the encouragement he needed to proceed with that winter's trip on the *Marijean*. The vacation was relatively subdued, with little fishing and a lot of relaxing, and conversation with his daughters, scientists, friends and a steady stream of doctors who relieved each other at two-week intervals. He returned home in mid-March much improved and in great spirits. "I have no plans at the present time for anything but staying here and trying to improve my walking and my ability to look after myself," he wrote to Johnny Lecky.

> When I went away a little short of two months ago, I had three full-time nurses during the twenty-four-hour period. I am now going to try to get along with only one—not exceeding eight hours, which I have discussed with my doctor and which he thinks should work out all right.[15]

Within a few weeks of his return, he settled into a domestic routine that he maintained for the rest of his life.

> I am down for breakfast by eight-thirty each morning. I then walk for half an hour with the nurse, because I can not walk alone on the street yet. I must have someone beside me in case I stumble and fall and perhaps break something. I can get around the house quite easily with a cane, and even without one, because I can usually reach a piece of furniture if I become unsteady on my feet. Miss Dee comes up about 11:00 a.m., and we attend to the letters, appointments, etc. Then

lunch and after lunch a sleep, and later in the afternoon a drive or visitors come in. Dinner is at seven and bed about nine or shortly after. After I get into bed I read for awhile but try to have the light out by ten thirty. I sleep very well.[16]

As time passed, and MacMillan's friends and acquaintances withdrew into their own old age, or died, Miss Dee maintained a steady stream of visitors to his home. If there were no requests to visit him from those he wanted to see, she would call and suggest it was time they come by for a talk. She still devoted all her energies to MacMillan's welfare, and their close personal relationship continued into their old age.

He maintained an office in the MacMillan Bloedel building, which he visited less and less often. Miss Dee still worked there as his personal secretary, answering his mail, arranging appointments and sorting his papers for posterity. She was his primary connection to the outside world and the guardian of his legacy within the company. Eventually, without consulting MacMillan, Clyne assigned his office to someone else. Next he moved Miss Dee into a smaller office and, later, stopped paying her salary, which MacMillan then paid himself. It was as though Clyne was attempting to stamp out MacMillan's last connections with the company, to eliminate all vestiges of his authority and influence. In this subtle battle of wills, Miss Dee gave no quarter.

In June 1967, with John Buchanan presiding as chancellor, MacMillan opened a new forestry and agricultural complex named after him at the University of BC. As he explained in his brief address, he had turned down a request to pay for the new building, believing it was up to government to finance construction. But he continued to provide a large portion of the forest faculty salaries, grants to the forestry library and contributions to specific research projects.

That fall, accompanied by Miss Dee, he made a brief trip to London. With Harold Wilson, he took part in opening a new MB wharf on the Severn River near Bristol.

> I had an opportunity of spending a few hours with the Prime Minister, who is an old friend of mine. I have always liked him and I find him very interesting. He is carrying a terrible load, which I could see is wearing him out. Although he is the political leader of the Socialists, he is too intelligent to support their extreme policies, which one can see from the newspapers are troubling him greatly these days.[17]

MacMillan at the opening of the Newport, England wharf, September 8, 1967.

Apart from the official excursion, and a quick trip to his tailor, MacMillan spent the rest of his time in London in his hotel suite, reading, resting and entertaining visitors.

In spite of his medical advice that travel was beneficial, he no longer had the energy for it. He took only one more trip—to the Caribbean, immediately after Christmas 1968, to get away from Vancouver's cold, damp winter. After spending three days in Barbados, where he encountered former prime minister John Diefenbaker at the Coral Reef Club, he realized he was homesick and returned to Vancouver—arriving just in time for one of the city's rare heavy snowfalls.[18] After this trip, he was content to stay at home. In early 1969, for the first time in seventeen years, MacMillan did not go to Mexico. The

Marijean stayed in Vancouver, and the following summer it was sold, along with the farms at Qualicum and Kirkland Island. His world was now reduced to his home in Shaughnessy.

One of the effects of MacMillan's less active life was a renewed interest in his far-flung relatives and the family which, for the most part, he had ignored since his youth. While recovering in the hospital, he had learned that Caroline Clement, still occupying the large, well-furnished home where she and Mazo lived in Toronto, was short of money. He wrote immediately.

> The precipitous rise in all costs, which inevitably will get worse, must be bringing great grief to many people on fixed incomes. I feel that this must not be allowed to happen to you in the closing years of your life, and that you must be protected in the dear home in which you have lived so long. Therefore, in order to safeguard you against any unwelcome changes, I am enclosing my cheque for $250 and shall send you a cheque for the same amount monthly hereafter. I hope you will permit me to do this. After all we are all one pioneer family from York County, and I am only too happy to be able to help my little cousin—as I am doing for other relatives.[19]

Two years later, for reasons he never explained, he instructed Miss Dee to stop sending Caroline cheques. Suggesting it was cruel to cut her off, Miss Dee argued against terminating the payments. In the face of this staunch opposition, MacMillan maintained Caroline's cheques—and wrote her long, chatty letters—until she died three years later at age ninety-four.

Throughout the late 1960s and early 1970s, he corresponded with several distant cousins at various locations in North America. His mother had ten siblings, whose children, MacMillan's cousins, numbered almost fifty. He established contact with many of them and in several cases, when their financial circumstances merited support, he sent regular or occasional cheques. He developed a great curiosity about his family, including the few remaining members on the MacMillan side.

He enthusiastically supported Georgina Sinclair's work on a Willson family genealogy, and he anxiously awaited her letters informing him about her latest findings. The more he corresponded with distant members of the family, the more news he had to pass on to them all. In the summer of 1967, his first great-grandchild was born—Jason, the son of Jean's eldest daughter, Carol—which news he happily conveyed to numerous cousins.

Giving up the *Marijean*, the farms and the travel was one thing for MacMillan; letting go his hold on MacMillan Bloedel was something quite different. By the end of 1967, he was seriously concerned about various developments, particularly the rapidly increasing role of management consultants in the company. Early in the New Year, he drafted a letter of resignation to Clyne.

His ambivalent feelings about Clyne were evident in the contrast between this letter and a telegram he had sent the chairman on New Year's Day. "I CONGRATULATE MYSELF FOR HAVING GOT YOU TO JOIN US TEN YEARS AGO AND I CONGRATULATE YOU ON ALL THE TREMENDOUS THINGS YOU HAVE DONE DURING THAT DECADE WARM REGARDS."[20] Ten days later he wrote:

> I hereby submit my resignation as Director and Member of the Executive Committee, effective today. I feel that I am not making the contribution I would like to make. Also, I am not in sympathy with the extent to which the McKinseys are taking part in the formal organization. I take this step with both regret and unhappiness. . . . I say again that you have performed a magnificent service for MacMillan Bloedel Limited. McKinseys would not have found you.[21]

Left to right: J.O. Hemmingsen, J.V. Clyne and MacMillan, May 2, 1968.

It appears, however, that MacMillan had second thoughts and did not send the letter to Clyne, as he continued to serve on the board and executive committee. In July 1968, he cabled Clyne, who was in Toronto. "WOULD IT BE AGREEABLE TO YOU IF THE COMPANY PLANE TOOK ME TO VANCOUVER ISLAND AND FLEW OVER OUR TIMBER HOLDINGS I WOULD LIKE TO HAVE A GENERAL LOOK AT THEM."[22] It was his final look at the forests upon which he had built his industrial operation. For MacMillan, corporate growth and expansion were based first on careful use and management of the forests, and second on the best and wisest use of the timber extracted from them. He was not unaware that this understanding was being forgotten in the upper reaches of MB management.

In March the following year, he wrote to Clyne: "On May 6th this year fifty years will have elapsed since the H.R. MacMillan Export Company was incorporated. I intend to resign from all connections with the present successor company on that anniversary."[23] Clyne wrote back a long, chatty letter expressing regrets at the decision, then quickly moving on to the mechanics of MacMillan's final departure from the company, which he arranged with great dispatch.[24] On April 22, MacMillan attended his last executive committee meeting. Three days later he wrote to Clyne:

> I have your letter of March 24th and thank you for your kind feelings. The Company has meant so much to me during so large a part of my life that I do not feel right about severing my connection with it. In fact, I feel that I could not be happy outside of it. Therefore, I am withdrawing my intention to resign which we can discuss further upon your return from your projected trip.[25]

MacMillan did not wait for Clyne's return, but cabled him in London two weeks later.

> I FEEL THAT I COULD BE USEFUL TO THE COMPANY ALSO HAVING A LARGE INVESTMENT IN THE COMPANY AND BEING FRIENDLY WITH ALL OTHER OWNERS I THINK IT WOULD BE HELPFUL TO THE COMPANY IF I WERE RETAINED ON THE EXECUTIVE COMMITTEE STOP I WISH BETTY AND YOU A VERY PLEASANT HOLIDAY.[26]

It is unlikely MacMillan's telegram contributed to Clyne's enjoyment of his vacation. Having succeeded in removing from the executive committee the major obstacle to his corporate agenda, Clyne was not about to reinstate him. However, he had no choice but to bow to MacMillan's wish to remain on the board.

MacMillan made another attempt to get reinstated on the executive committee in February 1970, apparently without success. A few weeks later,

MacMillan and John Hemmingsen at Nanaimo River logging site, July 1968.

after he and VanDusen had talked it over, they agreed to submit their resigna-
tions from the board of directors. VanDusen sent Clyne his notice, but
MacMillan managed to delay his final departure for an additional month
before submitting his resignation to Clyne. Nor could he resist offering a final
piece of advice.

> I have your memo of March 6th enclosing a copy of Van's resignation. I wish you
> to accept this letter as my formal resignation from the Board of MacMillan
> Bloedel Limited.
>
> I would like to inform you of an opinion which I hold: I think we should not
> proceed beyond a liquidation programme in Alabama. This territory is far from
> our experience and from our Head Office, these facts handicap us in the supervi-
> sion of important expenditures and operations. I wish you and the company
> every success. P.S. I would like my resignation publicized.[27]

The postscript may have been MacMillan's way of assuring Clyne he would
not change his mind again. His resignation was accepted with alacrity, and he
was given the purely symbolic position of honorary director.

Even now, MacMillan found it difficult to let go. For some time the company had been making a rocky transition from a proprietorship to professional management. By this time the largest single block of MB shares was held by Canadian Pacific Investments, which obtained them from MacMillan and the Bloedel family for stands of E&N timber on Vancouver Island. Bloedel, and to a lesser extent MacMillan and VanDusen, were still the largest individual shareholders, although since the timber deal, CP had steadily been buying shares on the open market.

Specht had resigned as president in 1968 upon learning that Clyne had no intention of making him chief executive officer. For the next three years there was no president, with Clyne and two vice-presidents attempting—none too successfully—to work as a joint executive team. Later that year the decision was made to expand from forest products into related industries. The 1969 annual report showed profits had gone up to $42.5 million, a record, but well below projections. Significantly, dividends were way down, with Clyne insisting on more expansion investment. The company was already established at several locations outside BC, and when the provincial opposition party, the New Democratic Party (formerly the CCF), called for increased forest industry taxes, Clyne threatened to curtail further investments in the province.

MacMillan, no longer on the company's internal information circuit, learned about most of these events from the newspapers. Concerned about the way he, as a shareholder, was being treated, he wrote to Bloedel:

> I never hear anything about what MB Ltd is doing or what is planned—excepting what I see in the newspapers. I am still greatly interested in the Company and would value any opportunity to discuss things with you. Having taken the initiative in starting the Company, I feel that what has happened to me is very different from what appears to be the practice in other successful companies.[28]

That summer, MB began dabbling in Australian real estate with its Asian partner, Jardine Matheson. Although it was not immediately obvious, a long downhill slide had begun. The Alabama operations were in trouble, and United Fruit pulled out of the partnership. Profits plunged to $17 million for the year and for the first time in history, the company paid no dividends. Clyne's response was to initiate even more expansion outside BC, including an attempt to exchange control of the company for acquisition of a large German paper company and a cash infusion to finance further acquisitions.

In late November, MacMillan wrote to VanDusen.

I understand that Prentice is coming up next week. I wish that you, he and I could sit down and do some thinking about our venture. I still feel a sense of responsibility to the shareholders, but don't know what to do about it—other than what we are doing. Some new ideas might come out of a discussion.[29]

He did not find out about the Australian real estate venture until the fall of 1971, when he read an item about it in the newspaper. At about the same time Bloedel, who was also being shunted to one side by Clyne, wrote asking for information. MacMillan responded:

I am very much out of touch with everything so there is very little you can learn from me, excepting perhaps that I think, considering our very large interest in MB Ltd, we have not been in close enough touch with it. I think it is important that when you are next up here we have a chat—to which I hope (if my brain is sufficiently alive) I can make a contribution.

I enclose a clipping which is the first news I have had of our financing hotel projects in Australia. It occurs to me that perhaps you may not have heard of this. It could be possible that our Company is not authorized to undertake this kind of business, and also that we do not have experienced staff to guide us. I feel that the publication of this information will not do us any good with those persons who may have a considerable investment in our Company. This item has caused me to do some thinking which I would like to discuss with you.[30]

MacMillan also wrote to VanDusen, expressing his concerns.

We have quite a few shareholders who probably put their money in MB Ltd. because they were favourably impressed by what we have accomplished in the forest business—to which business we should probably confine ourselves. What do you think?[31]

VanDusen shrugged it off, pointing out he had long thought the company should move out of BC and into non-forest industries.[32] Having spent fifty years deferring to MacMillan, he was not willing to support him now in a confrontation with Clyne.

final years

After MacMillan's eighty-fifth birthday, he left the second floor of his house less and less frequently. Although by now he was more or less confined to his home, he had not become a recluse, bemoaning his fate and lamenting his lack of authority. On the contrary, he was busily occupied with his prodigious correspondence, and he was reading prolifically, talking with visitors and going for occasional drives with his daughters.

It was probably the only period since he had lain recuperating in the sanatorium when he could reflect quietly about events in his life and think about new ideas. After each of her daily visits, Miss Dee departed with dictation for a sheaf of letters, such as one he sent to A.Y. Jackson, a Group of Seven painter and an old acquaintance.

I spend all my time in my home in Vancouver. I am now eighty-six. I am in good health but am not likely to go to eastern Canada again. . . . I would like to see you again, but I do not suppose that you are likely to come to Vancouver. Therefore, as we are not likely to meet again, I take this opportunity to send you my best wishes.

As I sit here dictating this letter, I am looking at the four small paintings which I acquired from you some years ago, and which face me on my livingroom wall. I am proud to own them.[1]

On another occasion, he described his state of mind to Don Bates, who came up from Portland on regular visits and with whom he still shared books and ideas.

Seeing you and a few other friends occasionally sustains me in my "job", which is to carry on in my endeavour without getting discouraged. If that happened, it would be dreadful. Of course, I should never get discouraged, as long as I have

MacMillan at his eighty-fifth birthday party at the Timber Club, Vancouver, September 11, 1970.

the ready opportunity of looking out of my window at the garden and trees, and have lots to read, and the pleasure of exchanging thoughts with friends.[2]

Books also sustained him, and still could fire his mind with new ideas. He was quick to read Rachel Carson's *Silent Spring*, one of the first eloquent warnings of environmental danger, and had Miss Dee scour the city for copies to send to friends, including one to Johnny Lecky. "Man is showing a tremendous capacity to invent ways of doing harm," he wrote to Lecky. "It would be a great joke on him if some of his inventions worked back through various channels, killing him off—which possibly is going to happen."[3]

And he tried to keep an eye on "my outfit," as he still called MB. This was not an easy task. In 1972 the much-dreaded socialists came to power in BC, but apart from threatening to confiscate MB's assets, they introduced few changes in the forest sector before they were voted out of office three years later.

In the spring of 1972, Clyne gave up the chief executive position at MB, promoting Robert Bonner from vice-president to president. Bonner, for eighteen years attorney general in the Bennett government, was cut in the Clyne mould—a man with a long legal career and no business experience. MacMillan learned of these changes from the newspapers, including the appointment of a new president, and commented to Bloedel:

> I am somewhat astonished that no one has ever spoken to me about Bonner's appointment, which sounds all right to me. I scarcely know him. Because, as you know, I am not going down town these days, I am thinking about asking him to come in for a chat.[4]

Two weeks later he wrote again to Bloedel, who had just resigned from the board and declined an honorary directorship.

> I feel that I am out of touch with the realities of MB Ltd and am very anxious to have a chat with you. During a chat we could probably develop a better view of what happens next than I can do by myself. I feel that I cannot "see" what is in the chairman's mind—partly because I have almost no communication with him; nor do I know what he is planning next for our business, of which I hear nothing either directly or indirectly.[5]

VanDusen, by this time, had withdrawn from the picture, with no desire for any involvement in MB affairs, even though he was still a substantial shareholder. Bloedel, after retiring from the board, avoided Vancouver and the possibility of having to confront Clyne. But he continued to be MacMillan's chief sounding board. In a November 1972 letter, MacMillan wrote to him:

> I think there might be a little planning we could do. Clyne is getting on in years and perhaps we should be thinking about his successor. I assume that his health is good, but I scarcely ever see him, or hear of or from him. He undoubtedly has it all planned out, but I have no idea who might be the "crown prince."
>
> My interest in MB Ltd is very important to my family, and I have to consider also (as is the case with you) that my name forms part of the Company name, and the Company has a large responsibility to the investing public. I no longer see anyone from whom I might get information respecting the direction in which the Company is headed. I am no longer invited to any meetings, therefore, it occurs to me that perhaps I should work toward getting my name deleted from the Company name.[6]

The following spring, Clyne retired as chairman. He persuaded the board to elect Bonner in his place, but board members balked at appointing someone with no business experience as president and chief executive officer as well. Denis Timmis, an executive with wide and varied experience in the forest products industry, was appointed president and CEO. Within a year he and Bonner were at each other's throats; Bonner quit the company and was replaced by George Currie, who had come to MB a few years earlier from the Canadian Imperial Bank of Commerce.

By this time rumours were circulating in the company and around the city that MacMillan, now in his eighty-eighth year, was becoming senile. It was, for those blindly pushing MB over the brink of financial disaster, a convenient way to disarm a still-influential voice of constraint on their actions. Judging by his correspondence, most of which was concerned with matters other than company affairs, MacMillan was as sharp and perspicacious as ever. And while his letters may have been refined by Dorothy Dee when she typed them, MacMillan also sent numerous handwritten notes and letters which were as lucid as the ones he dictated. He was, however, increasingly isolated, as he pointed out to Don Bates in early 1973.

> I am still in good health, but am beginning to find the days tiresome. I spend almost all my time reading, I have remained in this condition for so long now that very few people call on me. I suppose it is because I no longer have any news. I hope you will be up this way again before too long.[7]

A few weeks later, when Timmis was appointed president, MacMillan was not informed. He wrote to Bloedel:

I don't quite understand why I have to read in the newspaper who has been named President of MB Ltd. I am not happy about this. (The telephone is easy and convenient and is available to everyone.) It does not seem to be the right way to do things, and my experience tells me that one who does not do things the right way will bring upon himself some unnecessary trouble. . . . I cannot rid my mind of the thought that the chief owners of the business, no matter how inactive, should be in close enough touch with affairs to be aware, before it happens, of who is going to be the top executive officer.[8]

While this game of executive-suite musical chairs was going on, Clyne's two-pronged expansion plan—diversification into non-forestry ventures and investments outside BC—continued apace. With profits back up in 1973 and 1974, a major investment in a Brazilian forest operation was initiated. On the other side of the world, the company was involved in an all-out assault on Asian tropical forests, with most of the logs being shipped to Japan for processing. It was the kind of investment MacMillan had deplored when it occurred in BC twenty years previously—a big multinational corporation moving in to abuse the forests for a quick, easy profit elsewhere.

In 1974, the diversification plan was extended to businesses unrelated to the forest sector. The Ventures Group was set up to find investments in a wide array of fields, including aircraft manufacturing and pharmaceuticals. The biggest move, however, was the expansion of the wholly-owned subsidiary, Canadian Transport Company, into general shipping. Since its formation in 1924, CTC's primary function had been to carry MB products. Its secondary purpose was to haul additional cargoes, and over the years it had shown a small but steady profit by chartering ships on a short-term basis. Now a decision had been made to enter into the general shipping business, which involved signing long-term charters for ships. By mid-1974, the company had concluded numerous lengthy time charters, just as the world shipping business collapsed. CTC began leaking money like a rusty old Liberty ship, and dragging its parent, MacMillan Bloedel, down with it.

When news reached him in early 1974 of the Canadian Transport plan to enter the general shipping business, MacMillan saw clearly the dangers involved in the decision, and he hastened to warn Bloedel.

I judge from the enclosed clipping that MB Ltd has adopted a new policy respecting ship-owning. I wish to assure you that I had not heard of this until I saw it in

MacMillan at age ninety with his daughter Jean (right), his granddaughter Stephanie (centre), and two of his great-granddaughters, Joanna and Vanessa.

the newspaper. I have not heard any of the arguments pro or con. . . . We have done very well over the years in our business, namely, the lumber and forestry business. I hope our comparative success does not lead a new generation of ambitious and rising managerial talent to carry our capital and our name into a completely new field of ventures without at least giving such persons as you and me a little time to talk with them about it.[9]

Although MB's directors and executives were still assuring their shareholders, the public and each other that the company was on course, it was clear to MacMillan that it was headed for the rocks, and that his interests were not being well represented. "I doubt if we can rely upon the Board of Directors to protect us against anything," he wrote Bloedel. "I think if we don't give direction we will see a period of drift, therefore I think we should have more to say on our own behalf and not rely upon anyone else."[10]

Bloedel was not to be stirred. He responded to MacMillan's warnings only in the blandest of generalities. In July, having become aware of the move to Brazil, MacMillan wrote to Clyne, who was still serving as a director.

> I have been sitting around these past few days in very good health and doing a lot of thinking about my affairs. I have not arrived at any conclusion but I think that my work is about over and that therefore I should withdraw my name from any responsibility with MB Ltd.
>
> Beyond what I read in the newspapers, I do not have any information about what the Company is doing or planning to do. I have read in the press that we are planning an expansion in South America, which does not impress me—although I must confess that I am ignorant of the possibilities in that region. . . .
>
> I have no idea what MB Ltd may be undertaking in the next few years. I think that under the circumstances, it is only fair to say that I may wish to withdraw my name from the enterprise.[11]

In 1975, Clyne's chickens came home to roost. Under MacMillan, the company had been the largest in the province but was rarely viewed negatively because of its size; now it was a lumbering, global giant, portrayed as and often behaving like a stereotypical corporate ox. It not only symbolized corporate insatiability, its appetite was boundless. That summer, the confrontational labour climate Clyne had helped create over the previous few years degenerated into a three-month strike. Now heavily dependent on the US market, the company was shaken by the collapse of American lumber demand. Then the ticking bomb at Canadian Transport went off. In late July, MacMillan sent the new chairman, George Currie, a handwritten note. "What I read in the press about the CTCo disturbs me," it read. "I would appreciate it if you would please come up and explain it to me."[12]

In November, MacMillan sent Bloedel a note attached to a clipping from the Vancouver *Province*.

> Re the attached tearsheets which I think will interest you: I knew that business this year was not as good as in previous years, but I did not expect such a very large loss. I hope you come up soon, so that I can have a chat with you.[13]

The newspaper story described the greatest financial reversal not only in the company's history, but in all of Canadian corporate history since World War Two. MB had lost more than $32 million in the third quarter, down from a $72 million profit the previous year. The company was in dire straits; it was the first time since it had opened for business in 1919 that it had lost money. Even during the worst years of the depression it had been managed in such a way as to show a small profit.

The recriminations, retrenching and rebuilding which followed were no concern to MacMillan after the company hit bottom. The winter of 1975, like most before it, took its toll on him. He was ninety now, and withdrew into the privacy of his home and family. Over the spring and summer, a few old friends and colleagues came around for one last visit. Another damp Vancouver winter settled in, and by the new year MacMillan could speak only with difficulty. Still, he listened intently to his visitors, his eyes bright, grasping their hands, tears sometimes flowing down his cheeks when they left. At 8:30 on the morning of February 9, 1976, he quietly died.

"God is very wise," MacMillan had once said. "When a man has lived a long while and has learned a great deal about how to do things, God arranges to take him away. It is a good rule."

Epilogue

When he was a young man first setting out in business, MacMillan used an expression which became a guiding principle for him: "Making a name, and making it good." The phrase served almost as a personal admonition, a rule by which he lived and did business. By the time he died, he had fulfilled it in ways he could never have imagined; and his name, almost twenty years after his death, is still one of the most readily recognized in Canada.

MacMillan's legacy is extensive and varied. Most of the corporations he built, particularly his forest company, and the public institutions and facilities he established, are in British Columbia, especially Vancouver. The less tangible elements of his legacy, the ideas he adopted and fought to realize, have a much wider reach.

MacMillan Bloedel Limited, MacMillan's "outfit" and the company with which he is most closely identified, is still Canada's largest and best-known forest company. Since MacMillan's time it has gone through a series of changes in its organization, operation and corporate philosophy—not all of which would have had his support.

MacMillan built his forest company, by and large, with public approval. Although he was occasionally criticized for the size of his operation, most people believed his business endeavours benefitted the public. He provided jobs and public revenues, and did so in a manner which set standards for the entire forest industry. The company's attitudes toward labour relations, employee benefits, community responsibility and the forests it owned or controlled were, with rare exceptions, applauded. This has not always been the case since MacMillan left the helm.

During the 1950s and early 1960s, public sentiment favoured the growth of large corporations. They could raise the investment funds needed for rapid industrial growth, and they could provide jobs. Governments and bureaucrats—who assumed after their wartime experience with the Munitions and Supply model of public–private co-operation that government was responsible for directing the private sector—liked big companies because they were easier to deal with than numerous smaller firms. It was in this climate that Jack Clyne's vision of MacMillan Bloedel Ltd. as a rootless, multinational corporate predator was able to prevail for almost thirty years.

Ironically, one of the first to warn British Columbians of the dangers posed by the growth of large corporations was MacMillan himself. A few other prominent business leaders, such as Prentice Bloedel, also thought the growth of corporate oligopolies was undesirable, but they did not voice their opinions in public. MacMillan did, to little avail at the time.

A few decades' experience with large corporations operating in resource industries, such as BC forestry, altered public perceptions again, and the criticisms and warnings MacMillan once voiced became common. In recent years, there has been a widespread public perception that large corporations, owned by far-flung faceless institutions and run by professional managers, do not necessarily act in the best interests of the people who live and work near their operations.

As MacMillan warned during the second Sloan commission, a few companies did come to dominate the forest industry. They did form monopolies and all but eliminated free-enterprise, citizen businesses. They did not care for the forests of BC as well as they could have—although policy, practice and law hardly encouraged them to do so.

By about the mid-1970s, the people of BC had begun to question the principles and activities of the big forest corporations. This was because of what they perceived these companies to be doing to the province's forests, and also because they no longer saw such corporations as benevolent providers of economic development and security. Through the 1980s, thousands of forest and mill workers were laid off throughout the province, mostly because of new technology introduced into the woods and processing plants. The big companies continued to consume each other in their rampant growth. Mills were closed; communities were disrupted; people suffered. MacMillan Bloedel Ltd., the largest forest company in the country, had become as voracious as any. Because of the ambitions of Clyne and others, it became a symbol of everything the public disliked about large forest corporations.

Perhaps the worst blow to the company's reputation came from the enormous hostility it engendered among the public and its own employees on Vancouver Island, particularly in Port Alberni, the heart of its forestry operations. The company was shutting down mills like the Alberni plywood plant, while opening others outside BC. In the face of growing public hostility in BC, its executive officers were threatening to close more mills in the province and build new ones elsewhere. Control and direction of the company had shifted from Vancouver into the hands of the eastern Canadian Edper group, run by lawyers and accountants—the kind of people MacMillan had described as "never having had rain in their lunch buckets." By the late 1980s, forest workers, civic bodies and citizen organizations were suggesting the company pack up and leave its operations to others with a longer-term interest in the local economy. MacMillan's direst warnings were being realized, and even exceeded.

If the story had ended there, MacMillan's life, or at least an important part of it, could have been viewed as a tragedy. During his last years, he watched his "outfit," the company he had created in his own intellectual and ethical image, be subverted into the kind of corporate entity he abhorred. He went to his grave with Clyne's vision of MB intact.

But since that time, another change of direction has taken place at

MB, and a corporate philosophy much more in keeping with MacMillan's appears to have resurfaced. Control of the company has shifted to a wider assortment of shareholders, and direction over the company's operations has shifted back to BC. Executives have expressed a renewed commitment to forest management, product development and plant modernization. The company is once again making major investments in the province. MacMillan's name, his ideas and his standards are being invoked by senior officers. The company's responsibilities to the forests it controls and the communities in which it operates are, as they were in MacMillan's day, publicly acknowledged. Today, MacMillan would be much more likely to identify with the company that bears his name than he was during the last years of his life, when he talked of dissociating himself from it.

At least two generations of British Columbians have grown up with the institutional bequests that have made MacMillan's name part of the provincial landscape. The H.R. MacMillan Planetarium in Vancouver, an elegant civic luxury when it was first built, has performed a role that never could have been predicted at the time—certainly not by MacMillan. It became a focal point of imagination during the first, heady years of space exploration in the 1970s, when the public, particularly young people, engaged in a re-examination of man's place in the universe. The planetarium developed an international reputation for multi-media presentations providing mind-expanding experiences of outer space (enhanced for many young patrons by the psychedelic substances widely used at the time). It was one of the few public institutions to span the cultural chasm that split North American society during the 1960s and 1970s.

At the University of British Columbia, numerous facilities, libraries, scholarship funds, academic programs and collections were financed by MacMillan. Ironically, the only edifice bearing his name—the MacMillan Building, which houses the faculties of forestry and agriculture—is one he adamantly refused to pay for. His most enduring legacy to the university may turn out to be one the university did not receive until after his death—his large collection of personal and business papers, which will exert a profound influence on public policy in BC long after his other benefactions have been forgotten.

Inasmuch as ideas outlast physical facilities, some of MacMillan's most important contributions to UBC were barely noticed at the time. His early recognition of the importance of aboriginal culture, and the artifacts representing it, have already had an immeasurable influence on the political and social life of the province. His diligent acquisition of Native artifacts, at a time when few people in Canada cared about them and they were rapidly disappearing into private collections outside the country, were invaluable in the formative stages of the Museum of Anthropology, now one of the country's major cultural institutions.

If he could look back, MacMillan might find his hopes best realized in the stand of old-growth Douglas fir trees at Cathedral Grove, in the park bearing his name on Vancouver Island. He always had a profound faith that given the opportunity, the people of BC would choose to put the welfare of their forests ahead of short-sighted financial gains, and he thought of Cathedral Grove as a lasting reminder of the province's magnificent forest resources. The awe and wonder of the many people who walk through this forest each year has in no small way fuelled public concern for the forests over the last twenty years. That this public interest often has caused MB executives and managers considerable difficulty in their pursuit of short-sighted goals would no doubt amuse MacMillan a great deal.

MacMillan's philanthropic activities grew out of his basic belief in democratic society and his ideas about his responsibilities as a person of wealth and influence. They were not just the whims of a rich man—an approach to charity he thought was irresponsible. Nor were they inconsistent with his support of the Kidd Report's call for a drastic reduction in public spending, construed by some as a callous capitalistic lumber baron's indifference to the sufferings of the less fortunate or able.

When MacMillan began to make regular charitable contributions, and urged others to do the same, there was no philanthropic tradition in BC. The province's wealth was controlled from distant centres, and local managers tended to ship profits back to their head offices rather than contribute to a civil society in BC. The depression and the war convinced MacMillan that BC needed more social and cultural institutions and services. He was one of the first in his position to understand that if people were not assisted in times of distress, the society would degenerate into a totalitarian condition like those that had appeared in Europe and Asia. He felt that if he and others like him did not provide the social safety nets, cultural facilities and educational institutions necessary to civilized society, these needs would eventually be met by government. He did not dispute the validity of the needs or the importance of meeting them, but he felt strongly they should not be the responsibility of the state.

MacMillan thought that if government grew by providing these services, along with other interventions, the survival of democratic society was at risk. He opposed the welfare state not because he would have to pay higher taxes to support it: his voluntary charities cost him far more than any taxes he might have paid. Furthermore, he made his largest contributions during a time when most of his business associates and fellow citizens had given in to the notion that such services were the responsibility of government. To prevent a socialist state taking hold in BC or in Canada, he was prepared to meet what he saw as his personal obligations in a free, democratic society.

A measure of the depth of MacMillan's commitment to this principle is his philanthropic activities in the 1960s, when he began to dispose of a large portion of his wealth. To fund the many programs and projects he initiated or supported, he sold his shares in the business he had created. In doing so he not only gave away accumulated wealth, he also gave up power and authority in the corporate world. Many of his friends and colleagues used their wealth mainly to preserve and protect their power, some going as far as leaving the country to avoid paying taxes. MacMillan was very much opposed to this course of action. Instead, he decided to reinvest more than $30 million of his own money in the community where he had earned it.

The final disposition of the $23 million estate he left when he died, as detailed in his will, reveals much about how he viewed his civic responsibilities. First, he provided for his daughters and their families with substantial trust funds. Then he took care of various employees, distant relatives and old friends. Dorothy Dee received $750 per month for the rest of her life. He left the remainder to charitable endeavours, including a substantial amount for the advancement and education of northern Canadian Natives, contributions to the Vancouver Foundation, and a large bequest to UBC. Smaller endowments went to a wide array of organizations in Canada and the United Kingdom, including almost every small-town library on Vancouver Island.

MacMillan's most valuable charitable contribution to his country may have been his philanthropic principle, the idea that if one does well in this society, it is incumbent on him to support and sustain the social and cultural institutions which maintain it.

MacMillan's role in World War Two has been largely forgotten, although at the time it was widely recognized and applauded. It is difficult, in our cynical era, to appreciate how so many of the country's most successful and powerful business leaders—the people most able to protect themselves from the consequences of a distant war—could set aside their own interests and take on the exhausting task of mobilization.

MacMillan was one of the first public figures in Canada to recognize and speak out against fascism in Europe and Japan. One of the most difficult tasks in the early days of the war was to alert the country and rouse it from its indifference. At the beginning of the war, most Canadians were comfortably ensconced in their customary mid-Atlantic complacency, lulled by European appeasement on the one hand and North American isolationism on the other. They did not want to engage in a repeat of the previous war, and were slow to realize that totalitarian forces were on the march and must be fought. From the middle of 1939 on, MacMillan was one of the country's most passionate and effective advocates of all-out mobilization. That he did so without descending into the racism that was rampant in the country, and on the West Coast in particular, is to his credit.

His role in the events leading up to the expulsion from the coastal zone of Canadians of Japanese ancestry is more problematic. His correspondence from that period reveals that he felt strongly that the West Coast Japanese community posed a problem, if not a threat, to Canadian security. As a prominent public figure with substantial influence, his ideas and advice to government on this matter carried considerable weight, and there can be little doubt his opinions contributed to the ill-advised decision to expel the entire Japanese Canadian population after the attack on Pearl Harbor. But to pass judgement on MacMillan for his role in these events, from a morally comfortable vantage point a half-century later, is a difficult if not fruitless task.

MacMillan was criticized for playing a disruptive role when he was chairman of the Wartime Requirements Board. It was suggested he was too much the headstrong, individualistic businessman, creating dissension and failing to be the team player circumstances demanded. In some accounts, this criticism overshadowed his entire wartime experience. But the full story dispels the notion of a rift between MacMillan and C.D. Howe. The political crisis in which MacMillan played a central role was largely a product of an overly enthusiastic press corps, encouraged by opportunistic opposition politicians. Questions remain concerning MacMillan's willingness and ability to work in tandem with others, as a member of a group instead of its leader. Perhaps his dominating and independent personality, well-suited to the world of business, was a liability in the wartime world of public affairs, where diplomacy, tact and the ability to pull together were required.

MacMillan's most important role in the war was at Wartime Merchant Shipping. It was a critical task of enormous complexity and difficulty to provide a fleet of transport ships to the Allied effort. He not only performed the job capably, and with little fuss and fanfare, but did so with much broader interests in mind. He was not concerned merely with building ships, but with overall Allied strategy and, looking ahead, to the entire post-war period. In this he shone as a team player. Winning the war was a group effort, and MacMillan was an outstanding member of the team.

If he failed at anything in Ottawa, it was in his inability to master the byzantine craft of politics. If he did not know before the war he was not suited for the political life, he certainly learned it during the first few months of his wartime service. Curiously, he came away from the controversy and his time in the east with a profound faith in democratic government. This faith in his fellow citizens was central to his thinking in the post-war period.

Many of MacMillan's apparently contrary ideas and actions—his criticisms of the domestic war effort as head of the Wartime Requirements Board, his reservations and eventual dislike of post-war Liberalism, his distrust of large corporations, and his hesitations while at the NATO Defence Production Board—came from his opposition to the Munitions and Supply model being extended into post-war society. MacMillan had only minor concerns with this model under wartime conditions. But he felt the integration of the public and private sectors—government, industry and, to a lesser extent, labour—was a dangerous arrangement for peacetime. It was the type of social and economic restructuring that underlay European fascism, and he felt it could be entertained in Canada only as a temporary measure.

After his early experience with government and his wartime work, MacMillan could not accept the prevalent post-war notion that the government could direct the economy, through the active participation of its mandarins, in co-operation with industry and labour. He saw this kind of arrangement as inefficient—but he was even more worried that it might be efficient, in which case it would be a threat to individual freedom. If MacMillan erred in his thinking, it was not in his awareness of a threat to Canada's social and economic fabric, but in his perception of the characteristics of the threat. He thought the danger lay in socialism, a reasonable assumption in the context of his time, and believed the election of a socialist political party like the CCF would spell the end of the kind of society he valued. He did not know that the New Democrats, when they were elected in BC a few months before he died, would hold office only three years, and would have little effect on the influence of business interests.

What MacMillan was observing and warning against was the evolution in Canada of a corporatist state. Whereas socialism would create a monolithic state by replacing private ownership of industry with state ownership, corporatism creates a monolithic state—the combined and co-ordinated efforts of government, large corporations and labour unions. Howe's department during the war was a model of corporatism; the extent to which it prevailed after the war is a measure of the corporatist evolution over the past fifty years.

MacMillan held to a more pluralistic version of political economy, in which a diverse mix of small and large firms, supported by an array of autonomous public institutions and a state apparatus of modest scale, reinforced each other's strengths. Such a society, he believed, would allow its individual members to live free and productive lives. But MacMillan was no mere unreconstructed nineteenth-century liberal capitalist. His experience in the world, including two wars, had convinced him of the need for collective strength in a country. He believed that a truly diverse, liberal capitalist system would allow the growth of the most flexible, adaptable and capable society possible, providing the country with the strength to survive and prosper in the modern world.

After the war, MacMillan's most visible and important public work involved his participation in the decade-long forest policy debate at the centre of BC's political life. To assess events on the basis of public policies that prevailed until his death and beyond, is to consider his life a tragedy. His ideas were derided, his advice was ignored, and a set of forest policies to which he was adamantly opposed was adopted.

It has been forty years since MacMillan stood before Chief Justice Gordon Sloan and predicted what would happen if forest policies built around the Forest Management Licence system were implemented. To the great detriment of the people of BC, he turned out to be right. By 1975, when Peter Pearse, the first recipient of the MacMillan prize in forestry, conducted another royal commission on forestry, many of MacMillan's predictions had come true. The extent of corporate concentration had become, as Pearse stated, a matter of "urgent public concern." Pearse also warned that existing policies were leading to future timber shortages, which would result in mills being closed and workers losing their jobs.

By the early 1990s, MacMillan's predictions and Pearse's added warnings—both of them ignored by successive governments—could no longer be disregarded. As MacMillan and his colleagues had anticipated, under a tenure system characterized by public ownership of the land, no adequate forest management capability could develop. Private investment in publicly owned forests had not occurred, and successive governments had spent their forest revenues elsewhere. Although new forests had been started on logged-off lands, they were producing nowhere near their potential volumes or values. Annual timber harvests had to be drastically reduced, mills were closed, workers were laid off, and communities faced economic hardship, if not outright dissolution.

In the face of these economic and social calamities, MacMillan's vision for BC of a diversified economy based on stewardship of the province's renewable natural resources continues to be relevant. His alternative to the system which has prevailed, and failed, for the past forty years is still a valid choice. MacMillan's ideas—the potential he saw in British Columbia, the policies he proposed, and the society he envisioned—may well turn out to be his most enduring legacy.

The following abbrevations have been used:
P (Box Number)-(File Number) — H.R. MacMillan Personal Papers at University of
BC Library, Special Collections;
MB (Box Number)-(File Number) — MacMillan Bloedel corporate collection at
University of BC Library, Special Collections;
HS — Harvey Southam collection;
JL — John Lecky private collection;
CVA — City of Vancouver Archives;
BCARS — British Columbia Archives and Records Service.

Chapter 1: Ontario Boyhood

1. P 84-2, HRM to Mrs. D.J. Sinclair, June 24, 1968.
2. P 84-2, HRM to Mrs. D.J. Sinclair, June 18, 1968.
3. There is some confusion about the origin of this family. In later years
MacMillan would claim the name was spelled with one "l" and that the family was
not related to the David Willson clan. At other times he spelled the name with a double "l" and was less certain about its origins.
4. P 83-1, HRM to Religious Society of Friends, Oct. 23, 1970.
5. HS collection, G.R. Stevens MS, Chapter 1, p. 3.
6. P 84-5, HRM to Mrs. D.J. Sinclair, Aug. 25, 1972 and May 31, 1972.
7. P 81-7, HRM to D. Druery, Apr. 16, 1957.
8. P 85-3, HRM to A. Finlayson, Apr. 25, 1961.
9. P 84-2, HRM to Mrs. D.J. Sinclair, Mar. 4, 1968.
10. P 34-21, HRM to J.O. Wilson, June 22, 1964.
11. HS collection, Joanna to HRM, June 30, 1895.
12. HS collection, Joanna to HRM, Sept. 2, 1895.
13. P 84-2, HRM to Mrs. D.J. Sinclair, June 18, 1968.
14. *Ibid.*
15. HS collection, Joanna to HRM, Sept. 15, 1895.
16. P 81-7, notes from Mrs. J. Strachan, Sept. 1966.
17. P 29-18, HRM to J.V. Clyne, July 22, 1974.

Chapter 2: Birth of a Forester

1. MB 450-3, "My Week," by R.J. Deachman, Feb. 9, 1948.
2. HS collection, HRM diary, Sept. 23, 1903.
3. P 35-9, HRM to E.J. Zavitz, Apr. 10, 1968.
4. Bernhard Fernow, "Fleeting Thoughts on the Need of Forestry," *OAC Review*
(Dec. 1902), p. 8.
5. For a full discussion of the evolution of this policy and its implications, see H.V.
Nelles, *The Politics of Development* (Toronto: Macmillan, 1974). Also see Ken
Drushka, "Forest Ownership and the Case for Diversification," in Bob Nixon *et al*,
eds., *Touch Wood* (Madeira Park, BC: Harbour Publishing, 1993).
6. *OAC Review* (Jan. 1906), p. 180.
7. Quoted in Andrew Denny Rodgers III, *Bernhard Eduard Fernow: A Story of North
American Forestry* (Princeton, NJ: Princeton University Press, 1951), p. 396. Much of
this chapter is drawn from material in this book and from K.G. Fensom, *Expanding
Forestry Horizons* (Montreal: Canadian Institute of Forestry, 1972).

8. P 83-8, HRM to E.G. Shorter, Mar. 1969.

9. P 28-24, HRM to D. Bates, Feb. 14, 1973.

10. HRM, "Forest Conditions in Turtle Mountains," University of Guelph Archives, Mar. 21, 1906, p. 46.

Chapter 3: Graduate School

1. P 85-3, HRM to Mrs. M. Sagi, Dec. 8, 1969.

2. MB 176-18, Report of R. Craig.

3. P 21-1, HRM to Dr. F. Dockstader, Aug. 27, 1962.

4. *Ibid.*

5. *Marijean* log, Vol. 2, p. 59.

6. MB 418-43, HRM to Editor, MacMillan Bloedel & Powell River Co. newsletter, Feb. 15, 1965.

7. P 24-1, HRM to Mazo de la Roche, July 9, 1937.

8. *Ibid.*

9. D.K. Coupar interview with W.J. VanDusen, 1969, CVA, Add MSS 710.

10. *Ibid.*

11. HS collection, HRM to E. Mulloy, Sept. 13, 1908.

12. HS collection, HRM to Mazo de la Roche, Nov. 11, 1937.

Chapter 4: Between Life and Death

1. HS collection, HRM to Mazo de la Roche, Nov. 11, 1937.

2. HS collection, "Fighting the White Plague," *Canadian Magazine*, p. 74.

3. *Ibid.*

4. HS collection, HRM to Mazo de la Roche, Nov. 11, 1937.

5. *Ibid.*

6. *Ibid.*

7. Quoted in Andrew Denny Rodgers III, *Bernhard Eduard Fernow: A Story of North American Forestry* (Princeton, NJ: Princeton University Press, 1951), p. 459.

8. *Ibid,* P. 462.

9. HS collection, "Canada's Fire Brigade," *Collier's* (July 1910).

10. HS collection, "Further Reciprocity in Timber Products Unnecessary," n.a., n.d.

11. HS collection, HRM to E. Mulloy, Dec. 24, 1910.

12. HS collection, HRM to Mazo de la Roche, Nov. 11, 1937.

13. *Ibid.*

14. *Ibid.*

15. P 36-10, O. Price letters to HRM and R.H. Campbell, Feb. 14, 1912.

16. HS collection, HRM to M.A. Grainger, Feb. 16, 1912.

17. P 36-19, M.A. Grainger to HRM, Feb. 22, 1912.

18. HS collection, HRM to M.A. Grainger, Feb. 29, 1912.

19. P 36-19, R.H. Campbell to HRM, Apr. 24, 1912.

20. P 36-19, W.R. Ross telegram to HRM, May 22, 1912.

21. P 36-19, M.A. Grainger telegram to HRM, May 22, 1912.

22. *Ibid.*

23. P 36-19, M.A. Grainger telegram to HRM, May 27, 1912.

Chapter 5: Chief Forester

1 P 39-19, M.A. Grainger to HRM, June 2, 1912.
2. Report of the Chief Forester, B.C. Sessional Papers, 1913, p. 65.
3. P 36-17, B.E. Fernow to HRM, June 8, 1912.
4. P 36-17, HRM to B.E. Fernow, June 15, 1912.
5. P 36-19, R. Craig to HRM, June 5, 1912.
6. P 36-19, M.A. Grainger to HRM, June 25, 1912.
7. P 36-19, HRM to M.A. Grainger, June 12, 1912.
8. MB 116-8, D.K. Coupar interview with W.J. VanDusen, 1969.
9. Victoria *Colonist* (July 18, 1912), p. 1.
10. HS collection, unpublished MS by Harvey Southam, p. 29.
11. Andrew Denny Rodgers III, *Bernhard Eduard Fernow: A Story of North American Forestry* (Princeton, NJ: Princeton University Press, 1951), p. 506.
12. Sessional Papers, 1913, p. 65.
13. *Yale Forest School News* (Apr. 1914), p. 6.
14. *Marijean* log, Vol. 2, May 13, 1955, p. 139; and P 22-8, HRM to J.A. Lundel, *The Digester*, Sept. 8, 1960.
15. B.C. Sessional Papers, 1914, p. 47.
16. D.K. Coupar interview with W.J. VanDusen, 1969, CVA, Add MSS 710.
17. W.C. Gibson, *Wesbrook and his University* (Vancouver: UBC Press, 1973), p. 107.
18. HS collection, "Memo for Dr. Wesbrook," Dec. 3, 1914.
19. P 31-7, reprint of proceedings of Commission of Conservation fifth annual meeting, 1914.

Chapter 6: Around the World

1. BCARS, GR948, Box 1, Files 6 & 7.
2. HS collection, HRM to E. MacMillan, July 26, 1915.
3. MB 116-8, D.K. Coupar interview with W.J. VanDusen, 1969.
4. HS collection, HRM to E. MacMillan, July 30, 1915.
5. HS collection, HRM to E. MacMillan, Aug. 2, 1915.
6. HRM, "Conversion Methods—A visit to the Forests of Chaux and Faye de la Montrond, France," *Forestry Quarterly* (Dec. 1916), p. 599.
7. HS collection, HRM to E. MacMillan, Oct. 27, 1915.
8. Marion Hawley collection, E. MacMillan diary, 1915–16.
9. HS collection, HRM to E. MacMillan, Nov. 20, 1915.
10. HS collection, HRM to E. MacMillan, Nov. 21, 1915.
11. HS collection, HRM to E. MacMillan, Dec. 25, 1915.
12. P 36-20, HRM to F. Wesbrook, Jan. 15, 1916.
13. HS collection, HRM to E. MacMillan, Jan. 15, 1916.
14. HS collection, HRM diary, Feb. 9, 1916.
15. HS collection, HRM diary, Feb. 22, 1916.
16. P 36-20, HRM to F. Wesbrook, Mar. 15, 1916.
17. *Forestry Quarterly* (Dec. 1916), p. 624.
18. HS collection, HRM to E. MacMillan, Mar. 26, 1916.
19. HS collection, HRM diary, Apr. 12, 1916.
20. HS collection,, HRM diary, Apr. 19, 1916.

21. MB 440-9, HRM, "Developing a World Market for Timber," *Yale Forest School News* (July 1916).

22. HS collection, HRM diary, Apr. 22, 1916.

23. HS collection, HRM diary, Apr. 23, 1916.

24. P 36-20, F. Wesbrook to HRM, July 4, 1916.

25. Victoria *Colonist* (July 29, 1916).

26. HS collection, HRM speech to BC Lumber and Shingle Manufacturers Association, Aug. 1916.

27. P 36-20, HRM to F. Wesbrook, June 7, 1916.

28. P 16-17, HRM to A.E. Bryan, Oct. 31, 1968.

29. MB 440-9, HRM, "Developing a World Market for Timber," *Yale Forest School News* (July 1916).

30. P 36-19, HRM cable to W.R. Ross, Sept. 12, 1916.

31. HS collection, proceedings of Royal Commission, Sept. 22, 1916.

Chapter 7: In the Private Sector

1. W.H. Olsen, *Water over the Wheel* (Chemainus Valley Historical Society, 1963), p. 162.

2. *Ibid.*

3. *Ibid.*

4. *Ibid.*

5. *Ibid.*

6. *Ibid.*

7. *Ibid.*

8. *Forestry Quarterly* (Dec. 1916), p. 785.

9. P 36-20, HRM to F. Wesbrook, Nov. 7, 1916.

10. P 36-20, F. Wesbrook to HRM, Mar. 26, 1917.

11. HS collection, HRM to E.J. Palmer, Aug. 1917.

12. *Ibid.*

13. HS collection, unpublished MS by Harvey Southam, p. 41.

14. P 36-14, Bird & Co. to HRM, Jan. 8, 1918.

15. P 36-6, B.E. Fernow to Sir Robert Falconer, Sept. 27, 1917.

16. P 36-20, F. Wesbrook to HRM, Dec. 4, 1917.

17. *Marijean* log, Vol. 2, p. 161.

18. HS collection, unpublished MS by Harvey Southam, p. 42.

19. P 36-6, B.E. Fernow to HRM, Mar. 2, 1918.

20. P 36-6, HRM to B.E. Fernow, Feb. 24, 1919.

21. P 36-6, HRM to B.E. Fernow, Apr. 7, 1919.

22. W.C. Gibson to P. Pearse, personal communication, Jan. 18, 1994.

Chapter 8: Founding the Export Company

1. MB 176-8.

2. MB 176-7, M. MacMillan to HRM, Oct. 12, 1920.

3. MB 176-9, W.J. VanDusen to HRM, July 5, 1920.

4. MB 176-8, HRM to M. MacMillan, Sept. 21, 1920.

5. MB 176-7, M. MacMillan to HRM, Oct. 12, 1920.

6. MB 176-9, W.J. VanDusen to HRM, Oct. 29, 1920.

7. MB 176-9, M. MacMillan to HRM, Nov. 3, 1920.
8. MB 176-8, HRM to M. MacMillan, Mar. 10, 1921.

Chapter 9: Uneasy Partnership

1. *Marijean* log, Feb. 7, 1955.
2. *Ibid.*
3. Ken Drushka interview with Hon. H.P. Bell-Irving, Mar. 1995.
4. From an account by Ray Smith to Ken Drushka, Feb. 1994.
5. D.K. Coupar interview with W.J. VanDusen, 1969, CVA, Add MSS 710, p. 13.
6. MB 411-13, HRM to M. MacMillan, July 16, 1924.
7. MB 406-50, HRM cable to P. Sills, Aug. 12, 1994.
8. MB 408-13, P. Sills to HRM, Oct. 15, 1924.
9. MB 176-7, M. MacMillan to HRM, Feb. 25, 1925.
10. MB 176-8, HRM to M. MacMillan, Mar. 17, 1925.
11. MB 407-7, HRM to H.H. Stevens, Jan. 29, 1925.
12. MB 402-43, HRM to R.B. Teakle, Mar. 25, 1925.
13. MB 411-14, M. MacMillan to HRM, Jan. 10, 1928.
14. P 1-4, HRM to J.A. Ferguson, Feb. 2, 1927.
15. P 1-4, HRM to W.N. Millar, Mar. 4, 1927.
16. MB 411-13, HRM to M. MacMillan, Mar. 12, 1924.
17. MB 440-12, Jan. 1929.
18. MB 401-3, A. Flavelle to HRM, Apr. 26, 1929.
19. MB 405-26, HRM to J.D. McCormack, Mar. 30, 1935.
20. MB 407-46, HRM to President, Vancouver Board of Trade, July 8, 1929.
21. MB 440-12, *Vancouver Sun* (July 4, 1929).

Chapter 10: Taking on the Depression

1. MB 411-15, HRM to M. MacMillan, Oct. 21, 1929.
2. MB 116-8, D.K. Cupar interview with W.J. VanDusen, 1969.
3. MB 411-14 and 411-15, correspondence between HRM and M. MacMillan, 1930.
4. MB 441-1, HRM memo to several business and political figures, Apr. 15, 1932.
5. Grainger's role in this campaign and various articles and letters he wrote about the area are to be found in M.A. Grainger, *Riding the Skyline* (Victoria: Horsdal & Schubart, 1994).
6. MB 407-37, HRM to F.P. Burden, June 14, 1930.
7. MB 407-37, HRM to R.H. Pooley, Sept. 10, 1931.
8. MB 440-15, Montreal *Gazette* (Jan., 1931).
9. MB 405-18, letter to Premier S.F. Tolmie from HRM, W.J. Blake Wilson, Chris Spencer, Austin Taylor and George Kidd, Nov. 23, 1931.
10. MB 409-9, HRM to L.R. Scott, Jan. 7, 1932.
11. MB 404-14, HRM to T.D. Pattullo, Mar. 21, 1932.
12. MB 405-18 & 19.
13. MB 405-19, HRM to A.H. Williamson, Oct. 11, 1932.
14. *Vancouver Sun* (May 28, 1932).
15. J.H. Thompson and Allan Seager, *Canada 1922–1939: Decades of Discord* (Toronto: McClelland and Stewart, 1985), pp. 219–21.
16. MB 441-1, HRM memo to CMA, Apr. 15, 1932.
17. *Vancouver Sun* (Oct. 22, 1932).

Chapter 11: Lumber Wars

1. P 55-16, HRM to E. Bell, Mar. 30, 1933.
2. P 61-7, A. Jarvis to HRM, May 2, 1933.
3. P 61-7, HRM to A. Jarvis, May 9, 1933.
4. Ken Drushka interview with R. Shaw, Apr. 21, 1993.
5. P 52-8, HRM to C.D. Howe, July 26, 1933.
6. *Vancouver Sun* (Oct. 28, 1933).
7. P 3-1, correspondence between HRM and J. Stewart, Oct. 1933.
8. P 3-2, HRM to G.R. Stevens, Dec. 2, 1933.
9. P 3-2, HRM to Sir Joseph Flavelle, Dec. 6, 1933.
10. P 3-2, HRM to G.R. Stevens, Dec. 2, 1933.
11. P 3-2, HRM to Sir Joseph Flavelle, Jan. 15, 1934.
12. P 3-2, HRM to Sir Joseph Flavelle, Jan. 25, 1934.
13. Ken Drushka interview with Bert Hoffmeister, Mar. 22, 1993.
14. MB 402-28.
15. MB 116-8, D.K. Coupar interview with W.J. VanDusen, 1969.
16. MB 405-26, HRM to J.D. McCormack, Mar. 30, 1935.
17. Quoted in E.G. Perrault, *Wood & Water: The Story of Seaboard Lumber & Shipping* (Vancouver: Douglas & McIntyre, 1985), p. 96.
18. See Donald MacKay, *Empire of Wood* (Vancouver: Douglas & McIntyre, 1982), p. 124.
19. P 30-7, HRM to Sir Joseph Flavelle, Oct. 18, 1935.

Chapter 12: Life Becomes Art

1. See Donald MacKay, *Empire of Wood* (Vancouver: Douglas & McIntyre, 1982), pp. 132–33, for account of purchase.
2. HS collection, "Sequence in Comedy of Errors," Aug. 10, 1936.
3. P 24-4 & 24-5.
4. P 24-1, HRM to Mazo de la Roche, Nov. 11, 1937.
5. *Ibid.*
6. *Ibid.*
7. Mazo de la Roche, *Growth of a Man* (Boston: Little, Brown & Co., 1938), p. 380.
8. P 58-ll, HRM to J.M. Buchanan, Aug. 22, 1939.
9. From notes in HS collection and MB 441-8, Feb. 26, 1937.
10. London *Daily Telegraph* & *Morning Post* (May 22, 1939).
11. Vancouver *Province* (Mar. 26, 1938).
12. P 24-7, HRM to W.L. Runciman, July 4, 1939.
13. *Vancouver Sun* (July 5, 1939).
14. P 24-9, HRM to B.K. Sandwell, Aug. 7, 1939.
15. HRM to S. Lett, Aug. 31, 1939.
16. HS collection, Harvey Southam interview with R. Shaw, May 2, 1982, and unpublished MS by Harvey Southam, Chr. 8.
17. HS collection, "Thoughts After a Month in England."
18. P 100-6, HRM diary.

Chapter 13: Dollar-A-Year-Man

1. P 100-8, "Report of the Timber Controller, July–Aug.–Sept., 1940," p. 1.

2. P 24-13, HRM to S. McLean, May 28, 1940.
3. P 100-6, HRM diary, June 14, 1940.
4. P 100-6, HRM diary, June 1, 1940.
5. P 100-6, HRM diary, June 18, 1940.
6. MB 404-16, HRM to C.G. Power, June 20, 1940.
7. P 99-3, HRM to C.D. Howe, June 21, 1940.
8. P 7-1, HRM to S. McLean, June 24, 1940.
9. P 99-4, HRM to C.H. Grinnell, June 28, 1940.
10. P 99-4, HRM to G. Farrell, July 4, 1940.
11. MB 404-6, HRM to W.J. Asselstine, July 4, 1940.
12. P 75-7, E.C. Manning to HRM, Aug. 10, 1940.
13. P 75-7, HRM cable to E.C. Manning, Aug. 12, 1940.
14. P 75-7, HRM to E.C. Manning, Aug. 13, 1940.
15. P 75-7, HRM to E.C. Manning, Oct. 26, 1940.
16. P-15, HRM to C.D. Howe, Sept. 1, 1940.
17. P 24-9, HRM to G. Towers, Oct. 16, 1940.
18. P 99-12, HRM to W.A. Mackintosh, Oct. 26, 1940.

Chapter 14: Power Struggles in Ottawa

1. HS collection, HRM to W.J. VanDusen, Oct. 1940.
2. *Saturday Night* (Oct. 12, 1940).
3. National Archives, Minutes of War Committee of Cabinet, Nov. 7, 1940.
4. National Archives, Privy Council order 6601.
5. Robert Bothwell and William Kilbourn, *C.D. Howe: a Biography* (Toronto: McClelland and Stewart, 1979).
6. *Vancouver Sun* (Nov. 25, 1940).
7. MB 444-3, "Men, Materials, Money, Management and Imagination."
8. *Maclean's* (Jan. 1, 1941).
9. P 9-16, E.C. Manning to HRM, Jan. 2, 1941.
10. P 100-5, HRM to E.C. Manning, Jan. 7, 1941.
11. P 34-10, HRM to C.G. Swartz, May 5, 1967.
12. HRM, "Weighing Our War Capacity," *Saturday Night* (Jan. 2, 1941).
13. *Financial Post* (Feb. 8, 1941).
14. *Vancouver Sun* (Feb. 13, 1976).
15. *Saturday Night* (Feb. 22, 1941).
16. Bothwell and Kilbourn, *C.D. Howe*, pp. 146–47.
17. National Archives, Minutes of Cabinet War Committee, Feb. 18, 1941.
18. *Hansard* (Feb. 26, 1941), p. 1056–57.
19. National Archives RG19, Vol. 3984, File N-2-5-4, HRM to C.D. Howe, Mar. 10, 1941.
20. P 100-115, HRM to C.D. Howe (draft), Mar. 19, 1941.
21. P 7-5, HRM to C.D. Howe, Apr. 3, 1941.
22. Cf. Bothwell and Kilbourn, *C.D. Howe*, Chr. 10, in which various sources are quoted repeating statements allegedly made by MacMillan.

Chapter 15: Canada's "Buzz Saw"

1. *Time* (Apr. 14, 1941).
2. P 100-16, HRM memo to W.J. Bennett, May 27, 1943.
3. P 100-16, HRM to C.D. Howe, Apr. 21, 1941.
4. Vancouver *Province* (Apr. 25, 1941).
5. MB 407-44, HRM to W.J. VanDusen, Oct. 18, 1943.
6. P 7-9, HRM to W.J. VanDusen, Mar. 10, 1942.
7. P 75-9, HRM to W.J. VanDusen, May 7, 1942.
8. P 75-9, HRM to W.J. VanDusen, May 12, 1942.
9. *Ibid.*
10. P 75-9, HRM to W.J. VanDusen, May 13, 1942.
11. P 75-9, HRM to W.J. VanDusen, Aug. 27, 1942.
12. P 7-10, G.K. Shiels to HRM, May 28, 1942.
13. P 7-9, E.P. Taylor to HRM, Sept. 21, 1942.
14. P 100-113, HRM letter to "Dear Friend," Nov. 6, 1942.
15. P 90-13, HRM speech, Dec. 21, 1942.
16. MB 407-44, HRM to W.J. VanDusen, Jan. 30, 1943.
17. MB 407-44, W.J. VanDusen to HRM, May 8, 1943.
18. MB 407-44, HRM to W.J. VanDusen, Oct. 24, 1943.

Chapter 16: Debating Forest Policy

1. HS collection, "Memoirs of MacMillan Bloedel Ltd.," by C.G.
Chambers, unpublished MS, Jan. 9, 1975.
2. Orchard Collection, University of BC, Special Collections, Box 8-15.
3. P 100-5, HRM to R. Cameron, Dominion Forester, Jan. 14, 1941.
4. P 7-4, HRM to F.D. Mulholland, Jan. 14, 1941.
5. MB 413-33, HRM to J. Gilmour, Aug. 29, 1945.
6. *West Coast Lumberman* (Apr. 1944).
7. MB 419-32, HRM to T.T. Munger, July 3, 1944.
8. HS collection, Lawson, Clark & Lundell to E.P. Taylor, June 15, 1944.
9. Proceedings of B.C. Royal Commission on Forestry, Vol. 4, p. 1279.
10. MB 416-42, R.O. Sweezey to HRM, Dec. 5, 1944.
11. *Vancouver Sun* (June 21, 1944).
12. *Hansard* (June 30, 1944), p. 4472.
13. Vancouver *Province* (July 4, 1944).
14. MB 415-5, HRM to J.C. Ramsay, May 16, 1944.
15. Ken Drushka interview with V. Williams, Sept. 1991.
16. *Vancouver Sun* (Mar. 23, 1949).
17. Author's collection, Record of addresses, 6th Annual Truck Loggers'
Convention, Jan. 12–14, 1949.
18. MB 417-24, HRM to E.P. Taylor, Oct. 3, 1944.
19. HS collection, HRM notes for 25th anniversary speech.

Chapter 17: The Good Life

1. MB 416-37, HRM to J. Hart, Nov. 20, 1944.
2. P 81-6, Narrative by D.H. Bates, May 3, 1978; HRM to S. McLean, Oct. 25, 1945.

3. HS collection, HRM to W.J. VanDusen, L.R. Scott, E.B. Ballentine and H.H. Wallace, Oct. 15, 1945.

4. MB 412-40, HRM to Acting President, Vancouver Board of Trade, Nov. 13, 1945.

5. MB 413-31, Agenda for management meeting, Aug. 27, 1945.

6. HS collection, unpublished MS by Harvey Southam, Chr. 9.

7. P 57-2, HRM to J.M. Buchanan, Aug. 19, 1946.

8. P 24-2, Mazo de la Roche to HRM, Feb. 28, 1945.

9. MB 416-25, HRM to W.J. VanDusen, L.R. Scott, E.B. Ballentine and H.H. Wallace, Apr. 17, 1946.

Chapter 18: The Elder Statesman

1. *Maclean's* (Mar. 25, 1948).

2. MB 414-2, J.E. Liersch, "Report on Preliminary Meeting to Discuss Forest Legislation in BC," Sept. 13, 1946.

3. MB 414-2, F.D. Mulholland to J.E. Liersch, Oct. 17, 1946.

4. HS collection, HRM to E.T. Kenney, Nov. 18, 1946.

5. Orchard collection, University of BC Library, Special Collections, "Some more detail from diaries," Jan. 22, 1947.

6. MB 449-7, HRM address to Canadian Society of Forest Engineers, Feb. 27, 1947.

7. MB 412-74, HRM to C.D. Orchard, Sept. 13, 1947.

8. MB 413-27, E.J. Kenney to HRM, Dec. 24, 1947.

9. MB 413-36, J.D. Gilmour to HRM, July 27, 1948.

10. MB 412-75, correspondence between HRM and E.T. Kenney, May 3, 5 & 8, 1948.

11. MB 413-37, HRM to J.D. Gilmour, Feb. 4, 1949.

12. P 75-5, HRM to W.J. VanDusen, Feb. 15, 1949.

13. MB 413-37, HRM to J.D. Gilmour, Aug. 31, 1949.

14. MB 413-40, HRM to J.D. Gilmour, Feb. 4, 1950.

15. MB 413-40, HRM to J.D. Gilmour, Aug. 22, 1950.

16. MB 413-40, HRM to J.D. Gilmour, Aug. 30, 1950.

17. Victoria *Colonist* (Sept. 24, 1950).

18. *Vancouver Sun* (Aug. 30, 1949).

19. London *Evening Standard* (May 24, 1949).

20. University of BC, Spring Convocation Addresses, 1950, pp. 9–17.

21. P 102-11, C.D. Howe to HRM, Jan. 25, 1951.

Chapter 19: Cold Warrior

1. P 102-15, HRM cable to C.D. Howe, Feb. 1, 1951.

2. P 102-15, B.M. Hoffmeister to HRM, Feb. 5, 1951.

3. P 102-11, C.D. Howe cable to HRM, Feb. 9, 1951.

4. MB 417-78, HRM to L.R. Scott, Mar. 21, 1951.

5. HS collection, HRM to R.G.C. Smith, Mar. 15, 1951.

6. P 102-11, HRM to C.D. Howe, Mar. 19, 1951.

7. Vancouver *Province* (Apr. 14, 1951).

8. P 102-14, HRM to E. Gill, July 11, 1951.

9. Information on HRM's relationship with Dorothy Dee is taken from interviews with John Lecky, Marion Hawley, Jean Southam and Bert Hoffmeister.

10. P 102-15, HRM to M.W. MacKenzie, Deputy Minister, Department of Defence Production, June 27, 1951.

11. P 102-14, HRM to E. Gill, July 18, 1951.

12. *Ibid.*

13. MB 452-1, Announcement, Apr. 10, 1951.

14. Harvey Southam taped interview with P. Bloedel, 1981.

15. *Ibid.*

16. *Vancouver Sun* (Aug. 6, 1951).

17. P 102-10, HRM to W.J. VanDusen, Aug. 13, 1951.

18. P 102-15, HRM to M.W. Mackenzie, Sept. 26, 1951.

19. HS collection, HRM to R.G.C. Smith, Oct. 30, 1951.

20. HS collection, R.G.C. Smith to HRM, Feb. 26, 1952.

21. P 103-2, HRM to C.D. Howe, Mar. 24, 1952.

22. HS collection, HRM to C.D. Howe, Apr. 21, 1952.

23. MB 417-19, HRM to G. Murray, July 14, 1952.

24. David Mitchell, *W.A.C. Bennett and the Rise of British Columbia* (Vancouver: Douglas & McIntyre, 1983), p. 166.

25. MB 417-19, HRM to G. Murray, Sept. 16, 1952.

26. P 49-14, HRM to R. McMaster, Chairman, Steel Company of Canada, Apr. 8, 1953.

27. P 49-14, HRM to W.C. Mainwaring, Sept. 10, 1954.

28. P 57-4, G.M. Shrum to H. Doyle, May 9, 1956.

29. P 41-16, P.A. Larkin to HRM, June 18, 1956.

30. P 31-9, HRM to E.H. Williams, Aug. 27, 1952.

Chapter 20: The Second Sloan Commission

1. Cf. Ian Mahood and Ken Drushka, *Three Men and a Forester* (Madeira Park, BC: Harbour Publishing, 1990), Chr. 11.

2. MacMillan & Bloedel brief to Royal Commission on Forestry, Nov. 3, 1955, p. 11.

3. P 29-4, HRM to W.A.C. Bennett, Jan. 22, 1955.

4. P 14-4, HRM to R.E. Sommers, Jan. 24, 1955.

5. P 75-10, HRM to B.M. Hoffmeister, Jan. 24, 1955./

6. *Marijean* log, Vol. 2, p. 89.

7. MacMillan & Bloedel brief to Royal Commission on Forestry, Nov. 3, 1955, p. 32.

8. *Ibid.,* p. 52.

9. *Ibid.,* p. 47.

10. Victoria *Colonist* (Nov. 5, 1955).

11. Vancouver *Province* (Dec. 15, 1955).

12. *Marijean* log, Vol. 2, p. 209.

13. Proceedings of Royal Commission on Forestry, pp. 18,994–998.

14. Quoted in Ian Mahood and Ken Drushka, *Three Men and a Forester* (Madeira Park, BC: Harbour Publishing, 1990), p. 170.

15. See Ray Travers, "Forest Policy, Rhetoric and Reality," in Bob Nixon *et al*, eds., *Touch Wood* (Madeira Park, BC: Harbour Publishing, 1993).

16. By far the best account of the Sommers affair is found in David Mitchell, *W.A.C. Bennett and the Rise of British Columbia* (Vancouver: Douglas & McIntyre, 1983), p. 166.

17. *Marijean* log, Vol. 2, p. 284.

18. *The Forest Resources of British Columbia*, report of the commissioner, 1957, pp. 45–114.

19. MB 418-42, HRM notes, Dec. 13, 1957.

20. Ken Drushka interview with B.M. Hoffmeister, Mar. 1993; and Harvey Southam interview with Hoffmeister, 1980.

21. Ken Drushka interviews with Ian Mahood, 1992—94.

22. Ken Drushka interview with D. Cooke, Feb. 1995.

23. Ken Drushka interview with R. Shaw, Apr. 1993.

Chapter 21: Succession

1. See Ken Drushka and Harvey Southam interviews with R. Shaw and B.M. Hoffmeister.

2. P 82-1, "Looking Back On This Business," Dec. 1958.

3. Ken Drushka interview with Ian Mahood, Apr. 1, 1993.

4. See Donald MacKay, *Empire of Wood* (Vancouver: Douglas & McIntyre, 1982), pp. 221–22.

5. J.V. Clyne, *Jack of All Trades* (Toronto: McClelland and Stewart, 1985), p. 211.

6. See MacKay, *Empire of Wood*, p. 224.

7. CVA, H. Foley collection, Memorandum of agreement.

8. CVA, H. Foley collection, Untitled notes, Aug. 19, 1960.

9. CVA, H. Foley collection, John Liersch to Harold Foley, "Saturday."

10. CVA, H. Foley collection, J. Foley to H. Foley and C. Brooks, May 30, 1961.

11. CVA, H. Foley Collection, J. Foley to J.V. Clyne, May 30, 1961.

12. P 41-8, HRM to W.J. VanDusen, Jan. 24, 1958.

13. *Ibid.*

14. HS collection, HRM to J.V. Clyne and R. Shaw, July 17, 1958.

15. See Ian Mahood and Ken Drushka, *Three Men and a Forester* (Madeira Park, BC: Harbour Publishing, 1990), Chr. 13.

16. HS collection, "Thoughts," Apr. 20, 1959.

17. MacKay, *Empire of Wood*, p. 186.

18. *Marijean* log, Vol. 3, p. 179.

19. Ken Drushka interview with P. Larkin, Apr. 28, 1994. Also see HS notes.

Chapter 22: The Philanthropist

1. JL, HRM to J. Lecky, Apr. 30, 1962. In the first paragraph quoted, HRM states Brooks had asked for 2.5 million of his shares, but in the second one says he wanted 2.5 million shares from HRM, Bloedel and VanDusen. D. Cooke believes HRM meant that the request was for shares from all three owners.

2. Ken Drushka interview with J. Lecky, Aug. 1994.

3. Ken Drushka interview with J. Southam, Dec. 1, 1993.

4. See HS collection and HRM personal collection.

5. P 41-5, HRM to R.G. McKee, Aug. 25, 1965.

6. P 42-7, HRM to T.G. Wright, Aug. 8, 1963.

7. P 42-1, HRM to Dr. J.B. Macdonald, Aug. 9, 1965.

8. HS collection, correspondence between HRM and O. Roberts, Sept. 28, 1963 and on.

9. HS collection, HRM to University Resources Council, Oct. 6, 1969.

10. Harvey Southam interview with L. Stevenson, 1981.

11. JL, HRM to J.M.S. Lecky, Jan. 11, 1966.

12. JL, HRM to J.M.S. Lecky, Apr. 28, 1965.

13. Quoted in Donald MacKay, *Empire of Wood* (Vancouver: Douglas & McIntyre, 1982), p. 237.

14. P 32-17, HRM to R.S. McLaughlin, Oct. 4, 1965.

15. P 53-4, C.W. Brazier to W.G. Rathie.

16. HS collection, HRM to D.A. Rodger, June 30, 1969.

17. P 32-11, HRM to M.W. McCutcheon, Oct. 29, 1966.

Chapter 23: Losing Control

1. P 13-12, HRM to F. Howard, Apr. 19, 1967.

2. P 29-18, JVC to HRM, Dec. 28, 1973.

3. HRM to M.W. McCutcheon, Mar. 27, 1967.

4. P 16-8, HRM to J.S. Bates, July 28, 1967.

5. P 14-9, HRM to H.D. Scully, Mar. 23, 1967.

6. P 29-20, HRM to T.A. Crerar, Nov. 21, 1968.

7. P 32-17, HRM to R.S. McLaughlin, Mar. 19, 1968.

8. P 21-3, HRM to J. Tomb, Nov. 8, 1963.

9. Harvey Southam interview with J.V. Clyne, Aug. 16, 1982.

10. Harvey Southam interview with R. Shaw, May 11, 1982.

11. P 85-7, "Clyne's poem at my party," Aug. 30, 1965.

12. P 29-18, HRM to the Chairman, Mar. 6, 1966.

13. See Donald MacKay, *Empire of Wood* (Vancouver: Douglas & McIntyre, 1982), p. 267.

14. HRM to Dr. J.F. McCreary, Apr. 29, 1967.

15. JL, HRM to J.M.S. Lecky, Mar. 20, 1967.

16. JL, HRM to J.M.S. Lecky, Apr. 6, 1967.

17. P 32-17, HRM to R.S. McLaughlin, Oct. 4, 1967.

18. JL, HRM to J.M.S. Lecky, Jan. 8, 1969.

19. P 29-17, HRM to C. Clement, Sept. 10, 1966.

20. P 29-18, HRM to J.V. Clyne, Jan. 1, 1968.

21. P 29-18, "Rough Draft," Jan. 12, 1968.

22. P 29-18, HRM to J.V. Clyne, July 4, 1968.

23. P 29-18, HRM to J.V. Clyne, Mar. 20, 1969.

24. P 29-18, J.V. Clyne to HRM, Mar. 24, 1969.

25. P 29-18, HRM to J.V. Clyne, Apr. 25, 1969.

26. P 29-18, HRM to J.V. Clyne, May 9, 1969.

27. P 29-18, HRM to J.V. Clyne, Mar. 21, 1970.

28. P 34-14, HRM to P. Bloedel, Aug. 21, 1970.

29. MB 419-6, HRM to W.J. VanDusen, Nov. 20, 1970.

30. P 29-8, HRM to P. Bloedel, Oct. 12, 1971.

31. MB 419-6, HRM to W.J. VanDusen, Oct. 12, 1971.

32. P 34-14, W.J. VanDusen to HRM, Oct. 12, 1971.

Chapter 24: Final Years

1. P 15-6, HRM to A.Y. Jackson, Oct. 20, 1971.
2. P 28-23, HRM to D. Bates, Oct. 27, 1970.
3. JL, HRM to J.M.S. Lecky, Feb. 6, 1970.
4. P 29-6, HRM to P. Bloedel, May 9, 1972.
5. P 29-6, HRM to P. Bloedel, May 23, 1972.
6. P 29-6, HRM to P. Bloedel, Nov. 9, 1972.
7. P 28-24, HRM to D. Bates, Feb. 9, 1973.
8. P 29-6, HRM to P. Bloedel, Feb. 28, 1973.
9. P 29-8, HRM to P. Bloedel, Feb. 25, 1974.
10. P 29-8, HRM to P. Bloedel, Mar. 25, 1974.
11. P 29-18, HRM to J.V. Clyne, July 22, 1974.
12. MB 418-44, HRM to G.B. Currie, July 28, 1975.
13. P 29-8, HRM to P. Bloedel, Nov. 3, 1975.

The primary sources of material for this book were the collections of H.R. MacMillan's papers at the University of British Columbia Library, Special Collections. Additional material was obtained from a collection of papers, audiotapes and photographs compiled by Harvey Southam.

The following abbrevations have been used: P (Box Number)-(File Number) — H.R. MacMillan Personal Papers at University of BC Library, Special Collections; MB (Box Number)-(File Number) — MacMillan Bloedel corporate collection at University of BC Library, Special Collections; HS — Harvey Southam collection; JL — John Lecky private collection; CVA — City of Vancouver Archives; BCARS — British Columbia Archives and Records Service.

Published sources include numerous magazines, newspapers, journals and books. Most of these are cited in the Notes. The books most frequently used were:

Bothwell, Robert and William Kilbourn. *C.D. Howe: A Biography.* Toronto: McClelland and Stewart, 1979.

de la Roche, Mazo. *Growth of a Man.* Boston: Little, Brown & Co, 1938.

Gibson, William C. *Wesbrook and his University.* Vancouver: Library of the University of British Columbia, 1973.

Lyons, Cicely. *Salmon: Our Heritage.* Vancouver: BC Packers, 1969.

MacKay, Donald. *Empire of Wood.* Vancouver: Douglas & McIntyre, 1982.

Mahood, Ian and Ken Drushka. *Three Men and a Forester.* Madeira Park, BC: Harbour Publishing, 1990.

Mitchell, David. *W.A.C. Bennett and the Rise of British Columbia.* Vancouver: Douglas & McIntyre, 1983.

Perrault, E.G. *Wood & Water: The Story of Seaboard Lumber & Shipping.* Vancouver: Douglas & McIntyre, 1985.

Rodgers, A.D. *Bernhard Eduard Fernow: A Story of North American Forestry.* Princeton, NJ: Princeton University Press, 1951.

Taylor, G.W. *Timber: A History of the Forest Industry in BC.* Vancouver: J.J. Douglas, 1975.

Whitchurch History Book Committee. *Whitchurch Township.* Toronto: Stoddart, 1993.

A

Abbott, D.C., 291
Adams River, 49
Adirondack Forest Reserve, 35–36
Ailsa Craig, 47
Alaska pine, 245
Alberni Pacific Lumber Co., 142, 151, 171, 176–77, 226, 248–49, 279, 283
Algonquin Park, 37
Alpine Club, 281
Ambridge, Douglas, 280
American Can Co., 160–61
American Cellanese, 277–78
American Forest Congress, 35
Anderson, Ossian, 227–28
Andrews, L.R., 75
Anson aircraft, 214
Argus Corporation, 230, 263, 273
Arnold, T., 201
Ash River, 176–78
Asiatic Exclusion League, 47
Associated Timber Exporters of BC (Astexo), 115, 117, 139–43, 165–72
Auger, Fred, 358
Aurora, Ont., 19, 23, 27–28, 32, 45–46, 62–63, 83
Australia, 91–93, 119

B

Bailey, C.F., 33
Ball, Sir James, 115
Ballentine, E.B. (Blake), 175, 196, 226, 228, 296
Banff Springs Hotel, 53
Barclay, S.W., 75
Bateman, George, 200
Bates, Don, 125–26, 151, 257–58, 260–62, 267, 285, 288, 316, 379, 381
Beard, F.W., 71
Beaver Hills, Sask., 37, 53
Beaverbrook, Lord, 214
Bell, Ed, 161
Bell, Ralph, 196, 214
Bella Coola, 273
Bell-Irving, Henry, 130
Benedict, R.E., 75
Bennett, R.B., 153, 157–59, 164, 173
Bennett, W.A.C., 302–06, 308, 315–16, 320, 325, 339, 365
Bentley, Poldi, 245, 247, 314, 319
Berkinshaw, R.C., 216
Berryman, Harry, 315, 327
Big Bay Lumber Co., 116
Biltmore forest school, 51
bird collecting, 178
Blanchard Lumber, 368
Bloedel, Prentice, 241, 296–97, 299, 307, 308, 322, 327, 329, 335, 346, 368, 376–77, 380, 384
Bloedel, Stewart & Welch, 142–43, 176, 241, 245, 296–300
Bluebird Lumber Co., 127, 137
Bogarttown, 23
Bolshevik Revolution, 111
Bonner, Robert, 314, 320, 323, 338, 380–81

Bonney, P.S., 75
Borden, Henry, 201
Borden, Robert, 42, 64
Bow River, Alta., 37
Bowser, William, 89, 92
Bracken, John, 31
Brazier, Charles W., 359
Brentwood Bay, 140
British Admiralty, 84, 132, 186, 190
British Columbia; anti-Asian agitation, 188–89; economic conditions, 68–69, 78, 114, 144, 152, 305–06; forest industry in, 49, 69, 74, 77–78, 80–81, 90, 95, 115, 227–28, 244–45, 317–19; forest policies, 46, 60, 64, 69, 90, 183, 238–44, 246–47, 259–60, 275–83, 309–20, 323–24, 338–39, 374–75; politics, 78, 89, 153–56, 164, 249, 301–04, 314–16, 319–20, 323–35
BC Forest Club, 76
BC Forest Products, 264–65, 309, 311, 315, 323, 325
BC Forest Service, 60, 64, 67, 69–78, 84–85, 90, 92, 183, 276
BC Hydro, 304–05
BC Logging Association, 277
BC Lumber and Shingle Manufacturers' Association, 93, 185, 256, 284
BC Packers, 149–50, 160–62, 181–83, 206, 208, 226, 233–35, 250–51, 256, 265–66, 268, 270, 310
BC Plywoods Ltd., 175
BC Power Corporation, 237
British Commonwealth Air Training Plan, 194, 197, 214
Brooks, Anson, 347
Brooks, Conley, 335, 345–46
Brooks, Scanlon & O'Brien, 76
Brown, Loren, 135, 151, 168, 185, 210
Brown, Nelson, 47, 351
Bryan, Arthur, 127–28, 130
Bryant, Ralph, 46
Buchanan, John, 181–82, 206, 208, 235, 262, 265, 267–68, 288, 305, 310, 316, 353, 359, 370
Burden, F.P., 152
Burke, Stanley, 149, 160
Burma, 89–90
Burrard Drydock, 223, 225
Byers, Dr. Rodderick, 59, 61–62

C

Cameron, Colin, 249
Cameron Lake, 255
Campbell, R.H., 64–67, 71
Campbell, R.L., 75
Campbell River, 280
Camsell, Charles, 355
Canada Packers, 181
Canadian Chamber of Commerce, 191
Canadian Exporter, 124
Canadian Fisheries Council, 304
Canadian Forest Products, 314, 316
Canadian Forestry Association, 37, 41–42, 51, 162, 183

Canadian Foundation for Economic Education, 302–03
Canadian Imperial Bank of Commerce, 20, 153, 246, 267
Canadian Magazine, 57
Canadian National Drydock, 223
Canadian National Exhibition, 25
Canadian National Railway Co., 157
Canadian Pacific Investments, 376
Canadian Society of Forest Engineers, 51, 76, 240, 278
Canadian Transport Co., 131–32, 139, 156–57, 223, 328, 333, 382–84
Canadian Western Lumber, 167, 175, 241, 245, 278, 282
Canadian White Pine Co., 137, 151, 169, 172, 175–76, 178, 244, 246, 254, 279
Canusa Trading Co., 119, 132–34, 137–38, 168
Cardston, Alta., 53
Carney, Pat, 359
Carr, Emily, 274
Carson, Rachel, 380
Cascade Mountains, 125–26, 151–52
Cassiar Mountains, 285
Cathedral Grove, 255–56, 390
Caverhill, P.Z., 71, 75
Chalmers United Church, 349
Chambers, Charles, 238
Chapman, Herman, 46
Chehalis Logging Co., 137
Chemainus, 89, 95, 98–101, 247, 341–42, 359
Children of Peace, 18, 21
Chinese millhands, 101
Christie, H.R., 75
Clark, Clifford, 210
Clark, Judson A., 35–36, 40–42, 46, 60
Claxton, Brooke, 291
clear-cutting, 35–36, 241
Clement, Caroline, 25–26, 178, 181, 225, 263, 372
Clyne, J.V., 327–41, 347–48, 364, 366, 368, 370, 373–77, 380–81, 384, 387
commercial fishing industry, 149–50, 161
Commodore, 109
Community Chest, 163
Comox Logging Co., 241
Conservation Commission, 75, 79, 106–07
conservationism, 10–11, 34–35, 37, 59–61, 74, 184
Consolidated Paper Corp., 347
Continental Timber Co., 46
Co-operative Commonwealth Federation (CCF), 164, 173, 249–51, 301–03, 309
Cornell University, 35–36
corporatism, 394
Cottrelle, George, 200
Cowichan Lake, 101, 106, 171
Coyle, James, 107, 109
Craig, Roland, 36, 40, 46–49, 70, 75–76, 107
Crammond, Mike, 350
Creelman, George, 36
Cromie, Robert, 322
Crown Willamette, 278
Crown Zellerbach, 339

Crow's Nest Pass, 55
Currie, George, 381, 384

D
Daniels, Trevor, 265, 325
Davis, G.G., 107
Davis raft, 107, 109
Deachman, R.J., 31, 39, 42
Dee, Dorothy, 151, 187, 199–200, 221, 225, 267, 288, 292–94, 299, 323, 355, 369–70, 373, 381, 391
De la Roche, Mazo, 23–26, 39, 57, 178–81, 225, 263, 343
Denny, Mott & Dickson, 142, 176–77
Department of Aeronautical Supplies, 105
Department of Defence Production, 291
Department of Munitions and Supply, 192, 194–95, 200, 209–12, 214–16
Diefenbaker, John, 363, 371
Dingwall Cotts & Co., 131
Doane, Austin, 23
Dominion Mills, 176, 178
Dominion Trust Co., 78
Douglas fir, 123–24, 132, 139, 176, 183, 227, 239, 255
Doyle, Henry, 305

E
Eccott, Geoff D., 238, 296
Edmonton, 53
Edwards, William, 80
Eisenhower, Dwight, 291
Empress of Australia, 129
Empress of Canada, 186
Empress of France, 133
Empress of Russia, 128
Empress of Scotland, 133
Esquimalt & Nanaimo Railway, 99, 140, 233, 238–39, 245
Europa, 149

F
Falconer, Robert, 104, 112–13
Family Compact, 18
Fanny Bay, 233
Farrell, Gordon, 202
Farris, Bruce, 186
Farris, J.W. deB., 185
Federal Aircraft, 214–15
Fernie, 55
Fernow, Bernhard, 34–38, 42, 51, 59–60, 64, 69–70, 73–74, 89, 96, 102, 104–05, 112–13, 126, 138
Filberg, Bob, 241, 278–80
Financial Post, 211, 215–17
Fitch, Cyril, 357
Flavelle, Aird, 47–49, 117, 137, 142, 176
Flavelle, Joseph, 144, 153, 164–66, 173, 211
Foley, Harold, 227–28, 234, 280, 303, 333–38, 346–48, 357
Foley, Joe, 334–36
Forest Commission, 276–77
forest fires, 61, 64

Forest Management Licences (FMLs), 241, 259, 275–283, 307–20, 323–24, 338–39, 395
forest ownership, 38, 42, 60, 64, 163, 238–43, 259, 276–83, 308–09, 312–13, 318–19
forestry, 34–38, 40–42, 44, 50–55, 59–60, 64, 76, 79–80, 91, 138, 162, 183–84, 240, 278–79
Forestry Branch, federal, 37, 39–40, 42, 44, 49–50, 55, 61, 63–64, 66
Foster, George, 94–96
Fraser River Hydro and Fisheries Research Project, 304
free trade, 61–62
Fulton, Fred, 60, 64
Furness Withy & Co., 131, 147

G
Galapagos Islands, 316, 342
Gang Ranch, 79
Garibaldi Park, 281
Gattie, Brian, 197
George, Henry, 283
Gibson, Gordon, 310–11, 363
Gibson, Gordon, Jr., 363
Gibson, William, 359
Gill, Evan, 295
Gilmour, J.D., 71, 75, 241, 260, 276, 280–83, 314, 317
Gladwin, W.C., 75
Glencoe Lodge, 106
Gold, A.M.O., 75
Goldenberg, Carl, 210
Goodwin, Ginger, 111
Grainger, Martin Allerdale, 48, 60, 65–67, 69–72, 75, 90, 96, 107, 137, 142, 151–52, 162, 171, 176–77, 213, 225
Grauer, Dal, 302
Graves, Henry, 46
Gray, A. Wells, 197
Gray, Wilson, 325
Greer and Coyle Towing Co., 107
Grey, Gov. Gen. Earl, 41
Grinnell, Charles, 139–40, 142, 168, 190, 202
Growth of a Man, 23–24, 178–81
Guaymas, Mexico, 260–62
Guelph, Ont., 29–30
Gundy, J.H., 149

H
hand logging, 48, 107–08
Hanson, R.B., 211, 216
Harmac, 274, 286–87, 289
Harris, Arthur "Bomber," 342
Harris, Lawren, 273
Harrison, Bill, 212
Hart, John, 238, 255–56, 259, 276
Hartman, Clare, 27
Hastings Mill, 100, 342
Hecate Strait, 107, 110
Hell's Gate slide, 114
hemlock, 227, 245
Hemmingsen, John, 102, 373, 375
Hemmingsen, Matt, 101–02, 140
Henry, R.A.C., 210, 219

Herod, William, 291, 300
Herrmann, George, 124
Higgs, Leonard, 48
Hillcrest Lumber Co., 171
Hoffmeister, Bert, 151, 172, 254, 274, 285, 288–89, 296, 299, 315, 323, 325–27, 331, 366–67
Holland, 85, 154
Holland Landing, 17
Hong Kong, 150
Horton, Bill, 100–01, 341–42
Howard, Frank, 363
Howe, Alice, 225
Howe, C.D., 190, 192, 194, 196, 198, 201, 210–20, 222, 236, 263, 286–94, 299–301, 393
Howe, Clifton Durand, 50–51
Humbird, John, 167–68, 171, 176, 190–91, 245–47, 255
Humbird, John A., 99
Humbird, Tom, 103–04
Hutchison, Bruce, 211, 215, 220
Hyde Park Agreement, 226
Hyland, Norman, 359

I
Ilsley, J.L., 194, 210
Imperial Cannery, 368–69
Imperial Conference 1932, 157–58, 164
Imperial Munitions Board, 105, 107, 110, 113, 123, 158
India, 88–90, 104, 119, 122–23, 187
Indian Forest Service, 88
Indian Head, Sask., 36, 38–39
Indian stevedores, 100–01
Industrial Workers of the World (IWW), 110–111
International Pacific Salmon Fisheries Commission, 304, 353
International Woodworkers of America, 235
Iron River, 245
Irwin, H.S., 75

J
Jackson, A.Y., 378–79
Jamaica, 310
Japan, 127–29, 186–88, 206–07
Japanese relocation, 190, 206–07, 233–34, 392
Jardine Matheson, 150, 354, 376
Jarvis, Aemilius, 161
Jervis Inlet, 47–48

K
Kaniksu National Forest, Idaho, 54
Kenney, E.T., 275–77, 280–82
Kenny Lake, 350
Kew Gardens, 76
Kidd, George, 149, 153, 155–56, 159–60
King, William Lyon Mackenzie, 194, 211, 220, 255, 290
Kinghorn, Dr. Hugh, 57–58
Kinghorn, H.G., 75
Kirkland Island, 230, 355–56, 372
Klinck, Leonard, 29, 76, 78, 257
Koerner family, 245

Korean War, 287–92

L
Lafon, John, 70–71, 137
Larkin, Peter, 268–69, 305, 316, 335, 352
Larson, Ray, 214
Laurier, Wilfrid, 41–42, 59, 64
Lauzon, Que., 223
Leavitt, Clyde, 60, 106
Lecky, John, 186, 252, 266, 268
Lecky, Johnny, 205, 320, 346, 354, 356, 369, 380
Le Mare, Percy, 75
Leopold, Aldo, 11, 74
Lett, Sherwood, 190
Liersch, John, 276, 278, 317, 336
Lindquist Creek, 355
Lindsay, Casimir, 269
Lindsay, Ont., 46
Lodge, Henry Cabot, 364
Logsdail, A.J., 33
London, Eng., 83–84, 86, 120, 154, 292–93
Lord, Arthur, 311
Loveland, Walter L., 75
lumber exports, 75, 77–78, 80–81, 84–85, 93–97, 115, 117–18, 123, 169, 188, 283–85
Lundy, Daniel, 25
Lundy, Louisa, 25
Lusitania, 83
Lyons, Cicely, 359

M
Mabel Brown, 101
MacArthur, Douglas, 290
MacBean, Angus, 307
McBride, Richard, 46, 60, 63, 69, 78, 83–84, 89
McCormack, J.D., 165, 170
McCreary, J.F., 369
McCutcheon, Wallace, 267, 269, 316, 323, 363
Macdonald, J.B., 353
MacFayden, C., 76
MacKay, Alice, 349
MacKay, Donald, 341
McKelvie, Bruce, 358
Mackenzie, Norman, 256–57, 267, 269, 316
Mackenzie, Max, 299
Mackenzie, William Lyon, 18
Mackin, Henry, 167–68, 171–72, 175–76
McKinsey and Co., 366, 373
McLaughlin, R.S., 359
McLean, Stanley, 181, 192, 195, 198, 257, 274
Maclean's, 212–13, 275
McMaster, Ross, 302
MacMillan, Edna (Mulloy), 27–28, 31–32, 45, 53, 62–63, 67, 70–71, 73, 82–88, 90–91, 106, 113, 119, 129, 181, 188, 200, 267, 294, 321, 342–43
MacMillan, Harvey Reginald; and J.V. Clyne, 327–331, 339–41, 347–48, 367–68, 370, 373–75; and NATO, 287–96, 299–301, art collector, 273; assassination attempt, 129–30; at agricultural college, 29–44; at Yale, 45–47, 49–50; BC Packers, 149–50, 160–62, 181–83; character, 11–12, 25, 67, 125, 130, 200, 225, 366–67; Chief Forester in BC, 64–81, 89,
95–96; childhood, 22–27, 180; death, 385; family background, 18–22, 372; gentleman farmer, 270; head of Wartime Merchant Shipping, 218–19, 221–26; ideas of forest policy, 10–11, 85–86, 90, 151–52, 183–84, 240–41, 250–51, 260, 275–83, 309–20, 394–95; management style, 12, 111–12, 130–31, 147, 162, 183, 248–49, 326–27, 331–33; marriage, 63, 84, 199–200, 267, 293–94, 343; Mexican trips, 260–69, 267–69, 287–88, 301, 316, 320, 322, 343–45; parenthood, 67, 86, 91, 125–26, 171; philanthropy, 10, 144, 163, 255–56, 272–74, 349–53, 358–61, 372, 390–91; political views, 11, 136, 152–59, 163–64, 235, 249–51, 302–03, 319–20, 360–62, 393–94; relations with E.P. Taylor, 245–47, 263–65; religious views, 11, 31–32; round-the-world trip, 80–93, 95; Timber Controller, 192–210; tuberculosis, 49, 55–63, 79, 86–87; views on Japanese, 188–90, 234; Wartime Requirements Board, 210–20; wartime spruce procurement, 105–12; working for federal Forest Branch, 35, 37, 39–44, 46, 50–51, 61–65; working for Victoria Lumber & Manufacturing, 98–104
MacMillan, H.R., Export Co.; buying out Montague Meyer, 137–38; buying Victoria Lumber & Manufacturing, 245–47; competition with Soviets, 147–49, 153–54, 157–59, 283–84; expansion to Hong Kong, 150; exporting spruce, 132–34; financing, 150–51; formation, 115–17; lobbying efforts, 135–36, 143–44; mergers, 142–43, 234–35, 296–300, 308; operations, 117–24, 144, 233, 270, 285; post-WWII reorganization, 238, 244; pulp mills, 274, 286, 289; relations with Canusa Trading, 132–34; relations with sawmills, 139–43, 165–72, 175–78, 185–86, 202
MacMillan, Jean (Southam), 86, 91, 106, 125–26, 171, 177, 186–88, 200, 205, 216, 225, 322, 326, 334, 347–48, 354–55, 372, 383
MacMillan, Marion (Lecky), 71, 73, 106, 125–26, 171, 186, 200, 205, 293, 320, 322, 354
MacMillan & Bloedel Ltd., 9, 298–300, 307–08, 325, 327–41
MacMillan, Bloedel & Powell River Ltd., 335–41, 346–47, 366–68, 376–77, 380–85, 388–89
MacMillan Industries Ltd., 233
MacMillan Planetarium, 359–60, 389
McMillan, Edith (Willson), 20, 23
McMillan, Ellsworth, 22, 28, 30, 46
McMillan, Henry, 22
McMillan, Joanna (Willson), 21–25, 27, 30, 32, 44, 46, 58, 63, 73, 87, 98, 106, 124–25, 173–74
McMillan, John, 20–22
McMillan, John Alfred, 20–23
McRae, A.D., 270, 322
McTaggart-Cowan, Ian, 268, 273
Mahood, Ian, 307, 310, 333, 340
Mainwaring, Bill, 303
Makura, 93
Malkin, Phil, 149
Manning, E.C., 197, 203–04, 213
Manning Park, 213, 225, 242

Marijean, 266–70, 283, 288, 301, 310, 316, 320, 330, 335, 341–44, 347–48, 352, 354, 364, 369, 372
Martin, J.R., 75
Martin Paper Products, 324
Marvin, H.G., 75
Masset Timber Co., 110
Massey Harris, 265
Mayers, E.C., 129
Meighen, Arthur, 136
Merchant Marine, 123, 131, 136
Metropolitan Building, 117, 130, 168, 176
Mexico, 260–63, 267–69, 283, 287–88, 301, 216, 320, 322, 330, 343–44, 371–72
Meyer, John, 367
Meyer, Montague, 113, 115–23, 132–34, 137–40, 146–47, 150, 158, 172, 191, 200, 284–85, 343
Meyer, Tom, 231
Milford, Penn., 46
Millar, Willis "Pud," 46–47, 49, 55, 138
Mill Bay, 140
Minstrel Island, 48
Mitchell, David, 302
Mitchell, Howard, 223, 275, 298, 302, 358
Mitchell Press, 358
Monkman, R.K., 33
Moose Head Lake, 46
Moose Mountain, 37, 40
Moran Dam, 305
Mouat, Thomas, 75
Mulholland, Fred, 240, 276, 278, 317
Mulloy, Allan, 27, 34
Mulloy, Charles, 27
Munger, Thornton, 47, 243
Munro, H.G. (Hec), 265, 309, 315, 323, 325
Murray, Gladstone, 301–02
Murray, H.B., 75
Murray, Margaret "Ma," 359

N
Nanaimo, 286–87
New Democratic Party, 376, 380, 394
Newman, Murray, 269, 344–45
Newmarket, Ont., 17, 19, 22
Nimpkish River, 245
Norman, 86
North Atlantic Treaty Organization, 287–96, 299–301, 364
North Vancouver Ship Repairs, 223, 232
Northwest Bay, 245
Norway, 132, 154

O
O'Brien, George, 335
O'Brien, John, 76–77
Ocean Falls, 264
Oliver, John, 115, 322
Olsen, H.W., 359
Onassis, Aristotle, 355
Ontario Agricultural College, 27–45, 368
Ontario, Simcoe and Huron Railway, 19

Orchard, C.D., 238–39, 241–44, 251, 275–77, 279–80, 309, 311, 317, 319
Orduna, 83–84

P
Pacific Cedar Co., 137
Pacific Great Eastern Railway, 155
Pacific Mills, 264
Palmer, E.J., 89, 95, 99–104
Panama Canal, 78, 115
Papineau, Louis-Joseph, 18
Paris, 85
Pattullo, Duff, 154–56, 164, 185
Pearl Harbor, 226
Pearse, Peter, 352, 395
Pearson, Lester, 290–91, 364–65
Pemberton and Son, 149
Pendleton, Frank, 107, 137, 142
Pendleton, Ross, 142
Penn, William, 17
Pictou, NS, 223
Pigott, J.M., 229
Pinchot, Gifford, 10, 35, 41–42, 46, 64, 74
pine, 137
Pine Orchard, Ont., 20
Pioneer Envelopes, 252
Pooley, R.H., 152
Port Clements, 107, 113
Port Harvey, 48
Powell Lake, 47
Powell River, 48, 115, 296
Powell River Company, 227, 234, 280, 303, 326, 333–35
Powell River Sales Co., 336
Power, C.G. "Chubby," 197
Premier Lumber Co., 116
Prentice, John, 245
Presbyterianism, 21, 32
preservationism, 74
Price, Overton, 64–67, 70–71
Priest Lake, Idaho, 52, 55
Prince, G.H., 75
Prince Albert National Park, 53
Prince Rupert, 109, 111, 223
Prinses Juliana, 85
Pulp and Paper Industrial Relations Bureau, 330

Q
Quakerism, 11, 17–18, 20–21
Qualicum, 12, 127, 178–79, 235, 270, 340, 372
Queen Charlotte Islands, 107–09, 197, 205
Queen Elizabeth, 341–42
Queen Mary, 177

R
Ralston, J.L., 194, 213, 216
Rathie, W.G., 359
Rayonier Canada, 352
Reconstruction Party, 136, 173
reforestation, 239–41, 243–44
Reid, Escott, 290
Reid, Tom, 233–34, 247–48
Responsible Enterprise, 301

Riding Mountain, Man., 37, 43–44
Riley, Fenwick, 107
Rivers Inlet, 267
Robinson, H.K., 75
Robson, Bert, 273
Rockefeller, John D., Jr., 176–77
Rondeau Park, 37
Roosevelt, Theodore, 11
Ross, Norman, 36, 41
Ross, W.R., 63–64, 67, 76, 80, 89, 91, 96
Rowell-Sirois Commission, 184
Royal, Loyd, 353
Rush, Frank, 350

S
Saanich Inlet, 140
St. Laurent, Louis, 290
St. Regis Paper Co., 346
Ste-Agathe-des-Monts, Que., 58–60, 62, 179
Salmon River, 48–49
Sandwell, B.K., 189–90
Saranac Lake, NY, 56–58, 179
Saturday Night, 189, 209, 214–15
sawmilling, 77–78, 85, 93, 96, 109, 142
Scanlon, Joseph, 334
Schultz, C.D., 185, 325
Scott, F.C., 33
Scott, H.W., 33
Scott, Leon R., 107, 117, 127, 150, 154, 196, 289, 296
Scully, Hugh, 200, 208, 365
Seaboard Lumber Sales, 140, 142–43, 165, 168–70, 172, 185–86, 190–91, 202, 247, 264, 284, 297
Seaforth Highlanders, 254
Shanghai, 186–87
Sharon, Ont., 18, 23, 25
Sharp, Mitchell, 295
Shaw, Keith, 307, 310, 367
Shaw, Ralph, 151, 190–91, 285, 296, 323, 326–27, 329, 335, 338, 366–67
Shawnigan Lake Lumber Co., 244, 246
Shiels, G.K., 229
shipbuilding, 218–19, 222–26, 229–30, 232, 236
Shives, A.K., 75
Shorter, Ernie, 338
Shrum, Gordon, 304–05
Sifton, Clifford, 26, 60
Sills, Percy, 116, 124, 127–28, 130, 132–34, 137, 225, 366
Simcoe, John Graves, 17–18
Simcoe, Lake, 16
Simpson, Robert, 19
Sinclair, Georgina, 359, 372
Sino-Japanese War, 186–87
Sloan, Gordon, 238, 309, 324–25
Sloan Commission; 1944/45, 240–41, 243–44, 246, 259–60, 276; 1955, 309–20, 323–24, 339, 395
Smith, A.P., 33
Smith, Eustace, 358
Smith, Guy, 300–01
Smith, Sid, 142, 176–77
Smith, Davidson and Wright, 252
Smithsonian Institution, 345

Smoot-Hawley Act, 157, 165
Social Credit Party, 301–03, 320, 365–66
Sommers, Robert, 309, 314–16, 320, 323–25
Sorel, Que., 223
South Africa, 86–87
Southam, Gordon, 187, 205, 252, 268
Southam, Gordon, Jr., 320
Southam, Harvey, 322
Soviet Union, 147–49, 153–54, 157–59, 283
Society of American Foresters, 251
Specht, Charles, 366–67, 376
spruce, 105–07, 109, 132–34
spruce procurement; WWI, 105–12; WWII, 197, 202, 205
Spruce Woods, Man., 37
Steen, C.S., 268
Steeves, Dorothy, 249
Stevens, G.R., 358–59
Stevens, Henry H., 135–36, 159, 173
Stevenson, H.S., 131–32, 150, 223
Stevenson, Leigh, 267–68, 355
Stewart, Elihu, 37
Stewart, James, 163, 267
Stilwell, Jack, 76
Stone, Carlton, 171
Stone, Hector, 171
strikes, 111, 330, 384
Stuart Island, 78
Sturdy, David, 314, 320, 325
Sutherland, J.G., 75
Symington, Herbert, 201

T
tariffs; American, 157, 165, 168, 170; UK, 157–58, 166
Taylor, Austin, 105–07, 109, 113, 116–17, 153, 223, 234, 264
Taylor, E.P., 201, 212, 230–31, 246, 250, 263–65, 270
Teakle, R.B., 136
Temagami Reserve, 37
Temple of Peace, 18
Terminal City Club, 117, 176
Thomas, Bert, 151
Thomsen and Clark Timber Co., 233
Timber Control Board, 190–91, 197, 211–12
Timber Distributors, 148, 158
timber leases, 46–47, 60, 67
Time, 221
Timmis, Denis, 381–82
Tofino, 309
Tokugawa, Ikemasa, 165
Tolmie, Simon Fraser, 153–56, 164
Tolmie, W.F., 358
Toumey, James, 46
Towers, Graham, 205–06, 210
Truck Loggers' Association, 259–60
Trudeau, Dr. E.L., 56
Trudeau, Pierre E., 365
tuberculosis, 20–22, 55–63, 86–87
Turner, John, 365
Turtle Mountains, Man., 37, 40, 43
Tweedsmuir Park, 355

U

Union Steamships Ltd., 202
unions, 110–11, 161, 229, 231, 235, 248
United Fruit Co., 367, 376
United States Forest Service, 69
University of British Columbia, 11, 76, 78–79, 126,
    144, 155, 256, 286, 351–53, 361, 369, 389–90;
    forestry school, 76, 79, 89–90, 92, 102–03, 112,
    272–73, 351–52, 370; Museum of Anthropology,
    273, 390
University of Guelph, 368
University of New Brunswick, 51
University of Toronto, 44, 46, 50–51, 104–05, 112–13,
    138, 162

V
Vancouver Aquarium, 11, 269, 344–45
Vancouver Art Gallery, 274
Vancouver Board of Trade, 160, 163, 184, 188, 192
Vancouver Club, 190, 275
Vancouver Creosoting Ltd., 124
Vancouver Foundation, 349, 351
Vancouver Welfare Federation, 349
Vandorf, Ont., 22
VanDusen, W.J., 51, 71, 75, 77, 84–85, 107–08, 117,
    119–23, 128, 147, 150, 168, 171, 185, 196, 203,
    208–09, 227, 234–35, 251, 285, 296, 299, 327,
    329–30, 335, 346, 349, 375, 377, 380
Vanier, Georges, 360
Victoria Lumber & Manufacturing Co., 89, 95, 97,
    99–104, 138, 167, 245–47, 255
Victoria Lumber Co., 246, 279
Victoria Manufacturing Depot, 223
Vietnam War, 364

W
Wallace, Harold, 117, 151, 176–77, 196, 238
Wartime Housing Ltd., 229
Wartime Industries Control Board, 200–01, 208–10
Wartime Merchant Shipping Co., 218–26, 229–30, 232,
    236, 393
Wartime Requirements Board, 210–20, 223, 392–93
Waterton Lake, 37, 53–55
Wesbrook, Frank, 76, 78–79, 88–89, 92, 94, 102–03,
    106, 112, 257, 359
Wesgufka, Ala., 49–50
Western Lumber Manufacturers' Association, 185,
    284
Western Prince, 212
Weyerhaeuser, Frederick, 99
Weyerhaeuser, George, 137
Whitchurch Township, Ont., 17, 359
Whitcroft, Boswell, 171
White, James H., 105
White River, 48
Whitesail Lake, 355
Whitford, H.N., 75
Whitman, Walt, 13
Wilgress, Dana, 291
Williams, Harewood, 306
Williamson, Alan, 149
Willson, David, 18–19, 21
Willson, Hugh L., 21

Willson, J. Wellington, 21
Wilson, Harold, 284–85, 342, 362, 367, 370
Winch, Harold, 249
Winnipeg General Strike, 114
Wismer, Gordon, 313
Woodsmen of the West, 48, 72
Woodsworth, J.S., 164
Woodward, W.C., 201, 212
working circles, 259, 309–10, 312–13, 324, 339

Y
Yale Forestry School, 35–36, 44–47, 49–50, 138
Yarrows Shipyard, 223
Yoho National Park, 37
Yokohama, 127–29, 132
Yonge Street, 17–19
Yorkton, Sask., 53
Yuculta Rapids, 48

Z
Zavitz, Charles, 27, 29, 31–32
Zavitz, Edmund J., 36, 40–41, 43, 5130, 232, 236